:he Contributors

10 Chen received his Ph.D. in Electrical Engineering at the University Knoxville in August 2009, and his B.S. and M.S. in Computer Science on Engineering from Fu-Jen Catholic University, Taiwan in 1997 and vely. In April 2001, he joined the Panasonic Taiwan Laboratory Com- research and development engineer where he remained until August st 2009, he joined in the department of Math and Computer Science ina Central University as an assistant professor. Since August 2011, department of Electrical and Computer Engineering at Old Dominion he is currently an associate professor. His research interests include nated surveillance systems, pattern recognition, image processing, igent systems, data analysis and mining, and network performance as published more than 70 peer reviewed articles.

Yang received the B.S. degree from Central Police University in 1997, from National Central University in 2002 and Ph.D. degree from Tung University in 2014. He worked for the Forensic Laboratory at University from 1997 through 2005 and became an information system mputer Center at Central Police University from 2005 through 2014. ned the Department of Forensic Science at Central Police University ant Professor. His major research interests include image processing, ding, and forensic sciences. Dr. Yang is a founder of the Multimedia boratory at Central Police University and a frequent forensic appraiser in the area of image, video, and digital evidence. He has published les related to steganography, multimedia forensic, and computer aided nation. Dr. Yang is a senior member of Taiwan Academy of Forensic he Chinese Cryptology and Information Security Association. He is y General of Taiwan Academy of Forensic Sciences.

is an Associate Professor of Communication Engineering in the sics and Electronics, Henan University, Henan Province, China. At

present, Dr. Chen is the leader of the first-class specialty and exce pilot specialty of communication engineering in Henan Province, th excellent experimental class of communication engineering, the m innovation team of Henan Province, and the deputy director of the teaching center of electronic and communication engineering in He Director of Kaifeng Intelligent Manufacturing Engineering Techno Center, member of Henan Communication Society and member of He manufacturing Committee. In addition, he is serving as Director of Gr in Detection Technology and Automation, head of Communication En member of the research group in the Laboratory of advanced comput and intelligent applications. His research interests are in Evolution tion, Communication Network, Digital Signal Processing and Evolut Methods on Electronic Circuits and Systems. He has published ov papers, over 10 conference papers and 7 granted patents.

* * *

Chiou-Shann Fuh received the BS degree in computer science ar engineering from National Taiwan University, Taipei, Taiwan, in 1983, in computer science from the Pennsylvania State University, Universi 1987, and the PhD degree in computer science from Harvard Universit MA, in 1992. He was with AT&T Bell Laboratories and engaged in monitoring of switching networks from 1992 to 1993. He was an ass sor in Department of Computer Science and Information Enginee Taiwan University, Taipei, Taiwan from 1993 to 2000 and then pror professor. His current research interests include digital image process vision, pattern recognition, mathematical morphology, artificial inte learning, chatbot, big data analysis, cloud computing, and their a defect inspection, automatic optical inspection, industrial automatic camera, digital video camcorder, surveillance camera, and camera m color interpolation, auto exposure, auto focus, auto white balance, co and color management.

Md Ariful Haque is a PhD candidate in the Department of C Modeling and Simulation Engineering at Old Dominion University. working as a graduate research assistant in the Virginia Modeling Simulation Center. He has earned Master of Science (M.S.) in Modeli and Visualization Engineering from Old Dominion University (ODU has received Master of Business Administration (M.B.A) degree fro of Business Administration (IBA), University of Dhaka in 2016. He

e degree in Electrical and Electronic Engineering from Bangladesh Engineering and Technology in 2006. Before joining Old Dominion a graduate student, he has worked in the Telecommunication indus- ears. His research interests include but not limited to cyber-physical y, power grid security analytics, cloud computing, machine learning,

u received his doctoral degree in Sociology at Florida State Uni- 4, and he is a tenure track faculty in the department of Sociology at State University since 2015. As an assistant profession of Sociology, ious courses at the undergraduate/graduate level, including Social oduction to Ethnic Studies, and Sociology of Aging. He is interested ion of mental health, aging, gender difference, and marital status in ife course perspective. He has advanced statistical training including ation modeling, latent growth curve analysis and managed multiple ojects throughout his time in graduate school. His scholarly work Sociological Spectrum and Asian Population Studies. His research association between marital status and mental health, with special riations across age groups and to dimensions of psychological well- he elderly, using data from the Midlife and Development in the United S). His research on these topics has won Sociological Spectrum 2013 aper Award.

Huang got his master degree from Graduate Institute of Networking a, National Taiwan University in 2020.

nar received his Ph.D. in Forensic Chemistry from the King Georges rsity, India in 2015. He received his Postgraduate degree in Forensic from Dr. B. R. Ambedkar University, Agra, India. Kumar holds a ree in Chemistry (Hons) from the University of Delhi, India, 2008. as Guest Lecturer at the Department of Bio-Technology, Babasaheb edkar University, India, from 2015 to 2016 and has taught many nsic Sciences to Masters students. Currently, he works as an Assistant e Department of Forensic Sciences, College of Criminal Justice, Naif y for Security Sciences, Riyadh, Saudi Arabia. His research interests of forensic chemistry, which focuses on the temperature-dependent gradation of cardiac Troponin-T and its association with postmortem s worked extensively with researchers in a variety of other forensic lines, including personal identification using advanced imaging ir analysis for time mapping of drug use, and evidentiary aspects

of bite-mark analysis. Dr. Kumar has been qualified the UGC-Natio Test for Lectureship, and UGC-National Eligibility Test for Junior lowship (J.R.F.) in the subject Forensic Science. He has been award grants from national funding bodies, including the International T Scheme, Science and Engineering Research Board (SERB), Departm and Technology (DST), Government of India, 2013. Kumar has p than 30 peer reviewed scholarly research articles, as well as authore chapters published by IGI Global, Springer and Walter De Gruyter. many posters and oral lectures at various national and internationa He is on numerous Editorial Boards and acts as a peer reviewer fo Forensic and Medicine journals.

Sachin Shetty is an Executive Director for Center of Secure a Critical Systems in the Virginia Modeling, Analysis and Simulation Dominion University. He holds a joint appointment as a Professor with t of Computational, Modeling and Simulation Engineering. Sachin S his PhD in Modeling and Simulation from the Old Dominion Univ under the supervision of Prof. Min Song. Prior to joining Old Dom sity, he was an Associate Professor with the Electrical and Compute Department at Tennessee State University. He was also the associate Tennessee Interdisciplinary Graduate Engineering Research Institut the Cyber Security laboratory at Tennessee State University. He als appointment as an Engineer at the Naval Surface Warfare Center, Cra research interests lie at the intersection of computer networking, ne and machine learning. His laboratory conducts cloud and mobile sec and has received over $12 million in funding from National Science F Office of Scientific Research, Air Force Research Lab, Office of N Department of Homeland Security, and Boeing. He is the site lead o ber Security Center of Excellence, the Department of Homeland Sec Center of Excellence, the Critical Infrastructure Resilience Institu Department of Energy, Cyber Resilient Energy Delivery Consortium has authored and coauthored over 140 research articles in journals a proceedings and two books. He is the recipient of Fulbright Speciali Cybersecurity Research Challenge award, CCI Fellow, DHS Scienti Award and has been inducted in Tennessee State University's million has served on the technical program committee for ACM CCS, IEE IEEE ICDCN, and IEEE ICCCN. He is a Senior Member of IEEE.

Anu Singla has 19 years of total work experience in teaching a the field of Forensic Science. She was also the founder faculty mem

ience & Criminology, Bundelkhand University, Jhansi and continu- in the development and growth of the department since its inception. is working as Associate Professor at Institute of Forensic Science & Bundelkhand University, Jhansi, India.

master student in Graduate Institute of Networking and Multimedia, an University.

Index

A

C

D

E

F

G

I

P

R

S

T

W

www.ingramcontent.com/pod-product-compliance
Lightning Source LLC
LaVergne TN
LVHW081330300926
849161LV00008B/1937

Nemátode–Plant Interactions and Controlling Infection

Waleed Fouad Abobatta
Horticulture Research Institute, Agriculture Research Center, Egypt

Rehab Yasin Ghareeb
Plant Protection and Biomolecular Diagnosis Department, Arid Lands Cultivation Research Institute, City of Scientific Research and Technological Applications, Egypt

A volume in the Advances in Environmental Engineering and Green Technologies (AEEGT) Book Series

IGI Global
PUBLISHER of TIMELY KNOWLEDGE

Published in the United States of America by
IGI Global
Engineering Science Reference (an imprint of IGI Global)
701 E. Chocolate Avenue
Hershey PA, USA 17033
Tel: 717-533-8845
Fax: 717-533-8661
E-mail: cust@igi-global.com
Web site: http://www.igi-global.com

Library of Congress Cataloging-in-Publication Data

Names: Abobatta, Waleed Fouad, 1968- editor. | Ghareeb, Rehab Yassin, 1981- editor.
Title: Nematode-plant interactions and controlling infection / edited by Waleed Fouad Abobatta, and Rehab Yasin Ghareeb.
Description: Hershey, PA : Engineering Science Reference, [2023] | Includes bibliographical references and index. | Summary: "Nematode-Plant Interactions and Controlling Infection illustrates the techniques and recent methodologies as well as the interaction between host and nematodes to achieve nematode invasion in plants. It further investigates the role of the plant in confronting nematodes upon penetration, the challenges that face infected plants to resist nematode invasion, and the risk of transmission of nematodes. Covering topics such as biological control, molecular plant pathology, and organic farming systems, this premier reference source is an essential resource for crop producers, agrochemical professionals, agricultural scientists, botanists, plant breeders, biologists, students and academicians of higher education, librarians, researchers, and academicians"-- Provided by publisher.
Identifiers: LCCN 2023008197 (print) | LCCN 2023008198 (ebook) | ISBN 9781668480830 (hardcover) | ISBN 9781668480847 (paperback) | ISBN 9781668480854 (ebook)
Subjects: LCSH: Nematode diseases of plants. | Plant nematodes--Control. | Nematode-plant relationships.
Classification: LCC SB998.N4 N376 2023 (print) | LCC SB998.N4 (ebook) | DDC 632/.6257--dc23/eng/20230509
LC record available at https://lccn.loc.gov/2023008197
LC ebook record available at https://lccn.loc.gov/2023008198

This book is published in the IGI Global book series Advances in Environmental Engineering and Green Technologies (AEEGT) (ISSN: 2326-9162; eISSN: 2326-9170)

British Cataloguing in Publication Data
A Cataloguing in Publication record for this book is available from the British Library.

All work contributed to this book is new, previously-unpublished material.
The views expressed in this book are those of the authors, but not necessarily of the publisher.

For electronic access to this publication, please contact: eresources@igi-global.com.

Advances in Environmental Engineering and Green Technologies (AEEGT) Book Series

ISSN:2326-9162
EISSN:2326-9170

Editor-in-Chief: Sang-Bing Tsai, Zhongshan Institute, University of Electronic Science and Technology of China, China & Wuyi University, China; Ming-Lang Tseng, Lunghwa University of Science and Technology, Taiwan; Yuchi Wang, University of Electronic Science and Technology of China Zhongshan Institute, China

MISSION

Growing awareness and an increased focus on environmental issues such as climate change, energy use, and loss of non-renewable resources have brought about a greater need for research that provides potential solutions to these problems. Research in environmental science and engineering continues to play a vital role in uncovering new opportunities for a "green" future.

The **Advances in Environmental Engineering and Green Technologies (AEEGT)** book series is a mouthpiece for research in all aspects of environmental science, earth science, and green initiatives. This series supports the ongoing research in this field through publishing books that discuss topics within environmental engineering or that deal with the interdisciplinary field of green technologies.

COVERAGE

- Water Supply and Treatment
- Radioactive Waste Treatment
- Electric Vehicles
- Air Quality
- Industrial Waste Management and Minimization
- Green Technology
- Waste Management
- Biofilters and Biofiltration
- Renewable Energy
- Green Transportation

IGI Global is currently accepting manuscripts for publication within this series. To submit a proposal for a volume in this series, please contact our Acquisition Editors at Acquisitions@igi-global.com or visit: http://www.igi-global.com/publish/.

The Advances in Environmental Engineering and Green Technologies (AEEGT) Book Series (ISSN 2326-9162) is published by IGI Global, 701 E. Chocolate Avenue, Hershey, PA 17033-1240, USA, www.igi-global.com. This series is composed of titles available for purchase individually; each title is edited to be contextually exclusive from any other title within the series. For pricing and ordering information please visit http://www.igi-global.com/book-series/advances-environmental-engineering-green-technologies/73679. Postmaster: Send all address changes to above address.

Titles in this Series

For a list of additional titles in this series, please visit:
http://www.igi-global.com/book-series/advances-environmental-engineering-green-technologies/73679

Perspectives on Global Biodiversity Scenarios and Environmental Services in the 21st Century
Naveen Kumar Chourasia (Government College, Bichhua, India) and Kavita Chahal (Government College, Bichhua, Chhindwara, India)
Engineering Science Reference • copyright 2023 • 300pp • H/C (ISBN: 9781668490341) • US $250.00 (our price)

Promoting Sustainable Management Through Technological Innovation
Supriya Lamba Sahdev (ISBR Business School, India) Chitra Krishnan (Amity University, Noida, India) and Ahdi Hassan (Global Institute for Research Education and Scholarship, India)
Engineering Science Reference • copyright 2023 • 320pp • H/C (ISBN: 9781668499795) • US $300.00 (our price)

Environmental and Technological Aspects of Redox Processes
Gheorghe Duca (Institute of Chemistry, Moldova State University, Moldova) and Ashok Vaseashta (International Clean Water Institute, USA)
Engineering Science Reference • copyright 2023 • 435pp • H/C (ISBN: 9798369305126) • US $295.00 (our price)

Climate Change Mitigation With Hydrogel in Agriculture
Elsayed Gamal Zaki (Egyptian Petroleum Research Institute, Egypt) and Shimaa Mohamed Elsaeed (Egyptian Petroleum Research Institute, Egypt)
Engineering Science Reference • copyright 2024 • 350pp • H/C (ISBN: 9781668498774) • US $250.00 (our price)

Machine Learning and Deep Learning for Smart Agriculture and Applications
Mohamamd Farukh Hashmi (National Institute of Technology, Warangal, India) and Avinash G. Kesakr (Visvesvaraya National Institute of Technology, India)
Engineering Science Reference • copyright 2023 • 257pp • H/C (ISBN: 9781668499757) • US $265.00 (our price)

For an entire list of titles in this series, please visit:
http://www.igi-global.com/book-series/advances-environmental-engineering-green-technologies/73679

IGI Global
PUBLISHER of TIMELY KNOWLEDGE

701 East Chocolate Avenue, Hershey, PA 17033, USA
Tel: 717-533-8845 x100 • Fax: 717-533-8661
E-Mail: cust@igi-global.com • www.igi-global.com

Table of Contents

Detailed Table of Contents

Nematodes are multicellular organisms that may shed their cuticle. The Ecdysozoa category of organisms usually are tiny and subterranean. Nowadays, plant parasitic nematodes are known as the main agricultural pathogens and are recognized to attack plants and propose crop yield damage over the world. Some assessments suggest they cause over 70 billion dollars of losses worldwide each year. Nematodes are soil-borne (with exceptions like Aphelenchoides, a foliar nematode). Even though many saprophytic nematodes are embedded in soils, they vary from plant pathogenic nematodes in that the last have stylets. Soil nematodes, exceptionally the nematodes feeding on bacteria and fungi, can help to maintain sufficient levels of plant-available nitrogen in farming. Plant-parasitic nematodes are costly losses in crop yield production. Ubiquitous in nature, phyto-parasitic nematodes are specific to every important agricultural crop and reveal a significant restriction on global food security.

It is closely related to zoology and botany, which are the origin of all biological sciences. The study of plant-pathogenic nematodes requires a great understanding and knowledge of general zoology, invertebrate zoology, parasitology, and animal behavior science, which helps understand the nutritional behavior of nematodes on their plant hosts and thus the effects of different control methods on nematodes, as well as being familiar with botany and its various branches, especially the functions of organs, appearance, and internal anatomy, especially of the root system, and these are requirements for understanding the different and mutual influences between nematodes and plants.

Plant pathogenic nematode science is a relatively recent science that is related to many applied and research sciences. It is closely related to zoology and botany, which are the origin of all biological sciences. The study of plant-pathogenic nematodes requires a great understanding and knowledge of general zoology, invertebrate zoology, parasitology, and animal behavior science, which helps understand the nutritional behavior of nematodes on their plant hosts and thus the effects of different control methods on nematodes, as well as being familiar with botany and its various branches, especially the functions of organs, appearance, and internal anatomy, especially of the root system, and these are requirements for understanding the different and mutual influences between nematodes and plants.

Plant-parasitic nematodes are a high-priced burden in agricultural crop manufacturing. Over 4100 species of plant-parasitic nematodes have been recognized. Together, they represent an envisioned $80–$118 billion dollars annually in damage to crops. Plant-parasitic nematodes are identified as one of the finest threats to vegetation globally. The most damaging plant-parasitic nematodes (PPNs) are taken into consideration to be the root knot nematodes (RKNs), Meloidogyne sp., which are accountable for losses in vegetable plants throughout the world and determine the common use of chemical pesticides. The major genera of phytoparasitic nematodes discovered to cause crop losses were Heterodera, Hoplolaimus, Meloidogyne, Pratylenchus, Rotylenchulus, and Xiphinema. In different reports, M. incognita suppression turned into observed in field soil treated with P. penetrans in contrast to untreated soil. Opportunity means of pest control consisting of the use of biological controls are great interest for crop producers.

Infection by entomopathogen nematodes (EPN) occurs by an infectious juvenile that establishes itself in the body cavity of an insect and suppresses the immune response of the insect. Temperature variations and desiccation are the main abiotic factors affecting EPN survival, and UV light destroys EPNs more quickly than chemical insecticides. Spraying directly on the soil surface is the most powerful strategy for applying EPNs. IJs enter the insect body through natural pores and produce hydrolytic enzymes. Entomopathogenic nematodes (EPNs) have been shown to have a variety of impacts on insect immunology, including the destruction of hemocytes, the destruction of proteins that recognize nematodes, and the disruption of prophenoloxidase activation. The mode of action of bacteria inside the nematodes has been discovered through extensive Bt strain screening and cry gene sequencing. Cry proteins can enter nematodes through the stylet and create lytic holes in their cell membranes. LPS has an antagonistic action against nematodes.

The variation in different fruit plant groups morphologically, anatomically, and physiologically beside its type in addition to the preferred place of cultivation all make a difference in the response to the infection by nematodes. In addition, the conditions in the orchard around the fruit plants affect the infection by nematodes.

Some plant fruits are big trees (pears), but others are small (grapes). Some of them have a woody stem and others don't (banana). Deciduous fruits can grow in a cold climate, but evergreen fruits grow in a worm places. Orchards of pome and stone fruits are found in cold areas, but bananas grow in hot places and need a lot of water while olives prefer sandy soil with small amounts of water. The nematodes could be affected, and the amount could be decreased in sunny places with the decrease of water and by using mulch. Starting with clean and healthy seedlings or by grafting on tolerated rootstocks could minimize the nematode infection in orchards.

Plant-parasitic nematodes are one of the biggest threats to the production of various fruit crops worldwide. Nematodes cause adverse impacts on various crops, such as declining plant growth, and may lead to the destruction of large areas, like in banana, peach, and citrus orchards. Yellowing, stunting, and wilting are common symptoms of nematode infection in fruit orchards, and they are all associated with tree decline and decreasing yield. The efficiency of controlling nematodes varies due to many biotic and abiotic factors, such as soil type, nematode species, host plant physiology, and climate conditions, which affect nematode dynamics and infestation in the soil. Therefore, using integrated strategies with new management practices for controlling nematode infection is required to reduce economic loss for fruit producers. Hence, using biological control field management practices such as fallow, cover crops, and mulching, could be an effective way to inhibit nematodes' growth. Thus, they sustain the production of various crops, protect the environment, and preserve human health.

Citrus slow decline nematode, Tylenchulus semipenetrans, causes significant damage to all citrus-producing regions of the world and limits the production of citrus fruit under a wide range of environmental and edaphic conditions. T. semipenetrans is a semiendoparasite with a fairly narrow host rang that includes olive, grape, and citrus. No species in the genus citrus has been found to have resistance to T. semipenetrans but the only source of resistance incorporated into citrus is a dominate trait derived from Poncirus trifoliate. This chapter covers host preference among the citrus species and other hosts, biotypes, or races of T. semipenetrans and the relationship between different initial densities of T. semipenetrans and nematode infestation. Consequently, it is necessary to understand the environmental factors affecting parasitism to T. semipenetrans and management practices.

There are various challenges facing citrus orchards, including abiotic and biotic threats. In citrus orchards, nematodes are one of the significant biotic stressors that cause severe threats on growth and fruiting of citrus trees. Managing nematodes in citrus orchards reduces tree decline, decreasing various hazards of chemical nemacides on environment and humanity. Using biological control enhances tree growth, improves productivity, and protects the environment. There are different methods to manage the nematode problem in different crops, including physical techniques, crop rotation, resistant varieties, in addition to biological control. Using multiple strategies to control nematodes is beneficial and safer through reducing the use of nematicides and utilizing environmentally friendly substances, which are safer for human health and protect the environment. Presently, there is more interest in using organic nematicides to control plant-parasitic nematodes because of their positive impacts on crop productivity, soil, and the environment.

Sustainable agriculture relies on meticulous soil health assessment, including nematode indicators. They serve as vital bio-indicators, yet the study of beneficial nematodes remains unexplored. The goal of this chapter is to explore nematodes' significance in agriculture, covering their role in soil health, responses to climate change affecting crop production, and contributions to SDGs. Nematodes are valuable indicators due to their roles in nutrient cycling. Nematodes enhance soil health through various strategies, contributing to plant growth. Resistant pests are driving a shift towards eco-friendly pesticides, promoting biological control for plant protection and environmental preservation. EPNs rely on understanding adaptation and establishment in agricultural ecosystems. EPNs, known for their virulence and adaptable cultivation, combat insect pests effectively across various agricultural systems. The connections between nematodes, soil health, and pest management promise to advance sustainable agriculture, bolster environmental stability, and ensure food security.

Preface

Plant-parasitic nematodes are one of the dangerous threats to the agricultural sector worldwide, often invisible to the naked eye, wielding a substantial impact on agriculture, disrupting nutrient availability, reducing crop yields, and influencing soil health. This book, *Nematode-Plant Interactions and Controlling Infection*. endeavors to unravel the complex web of interactions between these microscopic organisms and the plants they affect.

This book focuses on nematodes, both plant-parasitic and free-living, through careful exploration of in- vitro and in vivo techniques. Furthermore, we delve into the mechanisms of nematode invasion in plants, shedding light on the strategies employed by nematode their host plants. The challenges faced by infected plants in resisting nematode invasion are dissected, and the ominous risk of nematode transmission are carefully examined.

Throughout the pages of this book, readers will embark on a journey through the ecology and distribution of nematodes, gaining insights into the impact of phytoparasitic nematodes on plants and the intricate dance that ensues once infection takes hold. Innovative applications aimed at bolstering plant tolerance to nematode invasion are discussed, alongside strategies for biological control in real-world agricultural settings. We also uncover the often-overlooked beneficial role of free-living nematodes, emphasizing their potential in soil health and plant productivity.

In a rapidly evolving field, we cover a wide array of topics that reflect the latest advancements. From diagnostic techniques, this book leaves no stone unturned. Practical management strategies for tackling nematode infections in economically significant crops are detailed, providing valuable insights for crop producers, agrochemical companies, agricultural international organizations, plant breeders, academic researchers, and students in the agricultural field. In addition to the exploration of various methodologies and applications that are used to fight nematode infection across various crops and soil types.

This book provides a comprehensive overview of the intricate world of plant nematodes and their interactions with crops. It includes a fundamental understanding of plant nematodes and their ecological distribution. The diverse division of plant parasitic nematodes and the often-overlooked beneficial free-living nematodes are explored, shedding light on their roles in soil health. The nefarious plant pathogenic nematodes and the dynamic between plants and nematodes upon penetration are analyzed, highlighting the challenges faced by infected plants. Readers gain insights into the risk of nematode transmission and strategies for biological control, particularly in citrus orchards, with a focus on the pathogenicity of citrus nematodes and effective management in fruit orchards. The threat posed by specialized nematodes to their hosts is discussed, along with the molecular plant pathology of plant-parasitic nematodes. The book delves into the impact of phytoparasitic nematodes on crops and the fiercest nematode species that parasitize them, respectively. The utilization of modern methods to control nematodes and the economic losses they inflict complete this comprehensive exploration of nematode-plant interactions.

We hope that this book serves as a valuable resource for all those engaged in the fight against nematode-induced challenges in agriculture. By combining fundamental knowledge with cutting-edge research and practical insights, we aspire to empower our readers with the tools they need to address this critical issue effectively.

Waleed Fouad Abobatta
Horticulture Research Institute, Agriculture Research Center, Egypt

Rehab Yasin Ghareeb
Plant Protection and Biomolecular Diagnosis Department, Arid Lands Cultivation Research Institute, City of Scientific Research and Technological Applications, Egypt

Acknowledgment

People who want to develop and encourage others' educational skills contribute to making the world a better place. Therefore, people who share their ideas and time to guide the brains of the future, help others to improve their situation for the better, and encourage development and creativity are the ones who deserve thanks and appreciation worldwide.

The book's editors, Prof. Dr. Waleed Fouad Abobatta and Prof. Dr. Rehab Yassin Ghareeb, thank and praise God for the completion of this book.

We could thanks and gratitude to all colleagues who contributed their knowledge to the publication of this book. We also thank all the reviewers who spent their time and expertise in producing this book in a way that befits the reader and provides useful and advanced scientific information to contribute to the advancement of humanity.

Chapter 1
Overview of Plant Nematodes:
Introduction to Nematodes Over-World

Rehab Yassin Ghareeb
https://orcid.org/0000-0002-3428-0536
Arid Lands Cultivation Research Institute, City of Scientific Research and Technological Applications, Egypt

Yara Yassin
Plant Protection and Biomolecular Diagnosis Department, Arid Lands Cultivation Research Institute, City of Scientific Research and Technological Applications, Egypt

Nader R. Abdelsalam
Faculty of Agriculture, Saba Basha, Alexandria, Egypt

ABSTRACT

Nematodes are multicellular organisms that may shed their cuticle. The Ecdysozoa category of organisms usually are tiny and subterranean. Nowadays, plant parasitic nematodes are known as the main agricultural pathogens and are recognized to attack plants and propose crop yield damage over the world. Some assessments suggest they cause over 70 billion dollars of losses worldwide each year. Nematodes are soil-borne (with exceptions like Aphelenchoides, a foliar nematode). Even though many saprophytic nematodes are embedded in soils, they vary from plant pathogenic nematodes in that the last have stylets. Soil nematodes, exceptionally the nematodes feeding on bacteria and fungi, can help to maintain sufficient levels of plant-available nitrogen in farming. Plant-parasitic nematodes are costly losses in crop yield production. Ubiquitous in nature, phyto-parasitic nematodes are specific to every important agricultural crop and reveal a significant restriction on global food security.

DOI: 10.4018/978-1-6684-8083-0.ch001

1. INTRODUCTION

Nematodes, also known as roundworms, are one of the most abundant and diverse groups of animals on Earth., members of the phylum Nematoda (round worms) have been in existence for an estimated one billion years, making them one of the most ancient and diverse types of animals on earth (Wang et al. 1999), Figure 1.

Figure 1. The classes and orders of plant parasitic nematodes

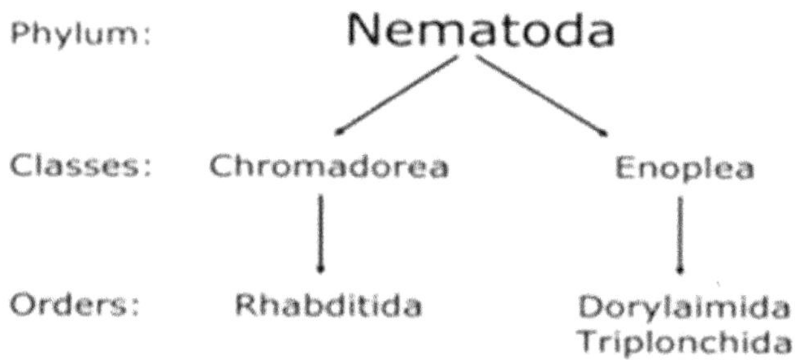

They are found in almost every environment, from the deep sea to the highest mountains. They have been around for hundreds of millions of years, and their fossil record dates back to the Cambrian period (over 500 million years ago). Nematodes are simple organisms with a cylindrical body and a single opening at one end for both eating and excreting waste (Subbotin et al., 2005) They have no respiratory or circulatory systems, but they do have a nervous system and muscles that allow them to move. Most nematodes feed on bacteria, fungi, and other small organisms in the soil or water. Some species are parasites that live inside other animals or plants. Nematodes were first described by Swedish botanist Carl Linnaeus in 1758. Since then, scientists have identified over 25,000 species of nematodes, Zhao et al. (2018). In recent years, they have become important model organisms in scientific research due to their simplicity and ease of study. Nematodes play an important role in many ecosystems by breaking down organic matter and cycling nutrients through the environment. In addition, some species are used as biological control agents to control pests in agriculture or forestry. Overall, nematodes are an incredibly diverse group of animals that have been around for hundreds of millions of years and play an important role in many ecosystems around the world. There are nematode species that feed on plants, bacteria, protozoa, fungi, and other nematodes. They have the ability to parasitize humans, animals, and insects. Plant parasitic nematodes (PPNs), also known as plant parasitic nematodes, are common in agricultural soils and feed on various plant parts (Subbotin, 2020, 2019).

Nematodes are the most predominant group of multicellular organisms on the earth but they are a less well-known group. They are a class of multicellular, translucent, pseudocoelomate organisms that resemble threads or worms and are either free-living or parasitic on plants or animals. They make up the meio- and mesofauna's most numerous phylum in terms of numbers. Nematodes are something unseen and unheard of for many of us, though. It is hypothesized that this is because of their diminutive size and propensity to hide in soil, water, plant, and animal tissues. Because they are so common, nematodes are connected to humans, domestic animals, plants, insects, and other invertebrate and vertebrate organisms. They exhibit a variety of life forms, including terrestrial, aquatic, entomopathogenic, free-living, predatory, parasitic (plant and animal), and free-living.

Plant nematodes are microscopic, non-segmented worms that live in the soil and feed on plant roots. They are one of the most important pests of agricultural crops, causing significant losses in crop yields. Nematodes can also cause diseases in plants, such as root knot nematode disease and cyst nematode disease. Plant nematodes are divided into two main groups: endoparasitic and ectoparasitic. Endoparasitic nematodes live inside the plant's root system, while ectoparasitic nematodes live outside the root system and feed on the surface of the roots. Both types of nematodes can cause damage to plants by feeding on their roots and reducing their ability to absorb nutrients from the soil. Nematodes have three life stages: eggs, juveniles, and adults. Nematodes can overwinter in any of these stages. The environment, initial nematode populations at planting, the pathogenicity of the nematode species, and the capacity of the plant to tolerate nematode feeding all interact in complex ways to cause crop damage.

The majority of PPNs feed by piercing and killing root cells using needle-like structures known as stylets. The root knot nematode (RKN) and soybean cyst nematode (SCN), two of the most economically harmful nematodes, enter roots and establish long-term feeding sites where they finish their life cycles without harming the nearby cells.

Nematode infections can mimic both biotic factors like stem and root rots as well as abiotic stresses like drought and nutrient deficiencies because the symptoms are similar to those of impaired root growth and function. Yellowing, stunting, and wilting are common nematode symptoms, and they are all accompanied by a decrease in yield. Although infections can go undetected until populations are well above economic thresholds, above-ground symptoms are not always obvious. Although the RKN causes root galling, the severity of the galling may vary depending on how the RKN species interact with the plant. Once a crop is planted and roots begin to grow, it will provide food for nematode reproduction and population growth.

The number of nematodes required to cause crop losses varies depending on the nematode and plant species, so the proper interpretation of the nematode assay results will be crucial for the implementation of management practises to minimise PPN impacts.

2. WHAT ARE NEMATODES?

Nematode is a compound word made up of the Greek words nema, which means "thread," and toid, which means "form." Unsegmented roundworms called nematodes are among the most ecologically diverse animal species on the planet. Most nematodes are vermiform, long, and slender, but some are swollen. Due to their rounded cross-section, the majority of people refer to them as roundworms. They range in size from 15–35 µm wide, 300–1000 µm long and some nearly 4 mm, they are so little that they cannot be seen with the naked eye. They causes crop losses amounting to 125 billion dollars each year (Chitwood 2003). PPNs are significant agricultural pests that worldwide. Phytonematodes have the ability to infect plants either directly or in conjunction with other pathogens like viruses, bacteria, and fungi (Elhady et al. 2017, Adam et al. 2014).

Figure 2. Morphology of male and female root-knot nematodes (Handoo et al., 2004)

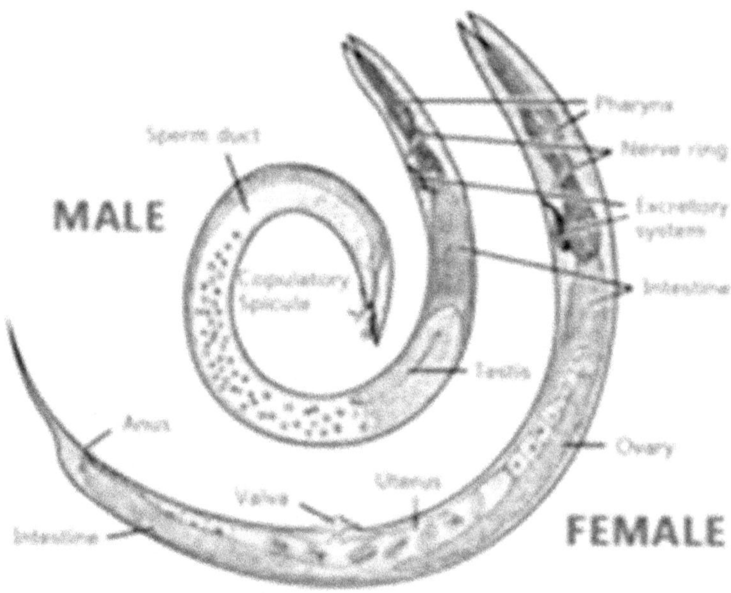

3. MORPHOLOGY OF NEMATODES

Nematodes are a group of unsegmented worms that are found in a variety of habitats. They have a cylindrical body shape and range in size from microscopic to several meters in length. The morphology of nematodes is highly adapted to their environment and lifestyle, and they can be divided into two main groups: free-living and parasitic.

Free-living nematodes have a simple body plan consisting of an elongated, cylindrical body with a cuticle, or outer covering that is composed of several layers. The cuticle is secreted by the epidermis and provides protection against desiccation and other environmental stresses. The head region contains the mouth opening, which leads to the pharynx, esophagus, intestine, and anus. Nematodes also possess a muscular system that allows them to move through their environment by peristalsis or ciliary action.

Parasitic nematodes have more complex morphologies than free-living species due to their adaptation for parasitism. These species typically possess specialized organs such as suckers or hooks for attachment to their hosts as well as specialized digestive systems for extracting nutrients from their hosts' tissues. In addition, some parasitic species possess reproductive organs that allow them to reproduce within their hosts' bodies, Siddiqi, (2000). Plant-parasitic nematodes use their stylet, which is made up of parts of the mouth and oesophagus, to pierce plant cell walls. The plant cell is then pumped directly into their digestive system. These nematodes only eat specific plant parts. Most consume roots. Some nematodes feed only on the root's outer tissue, while others penetrate deeper and completely enter the host. They harm the plants in a variety of ways. Some nematodes force their hosts to produce either increased amounts of the structures they live in or nutrients they need to survive. Some of them release metabolites that destroy host tissue. All of these actions cause plants to lose energy in order to support nematodes. Some plant-parasitic nematodes can harm plants so severely that crops may become unmarketable, have lower yields, or even fail entirely. Plant parasitic nematodes are a serious threat to a diversity of crop plants worldwide (Ali et al., 2014; Ghareeb et al., 2019, 2020).

Overall, the morphology of nematodes is highly adapted to their environment and lifestyle, allowing them to survive in a variety of habitats.

Figure 3. Characteristics of typical male and female plant parasitic nematodes in terms of morphology

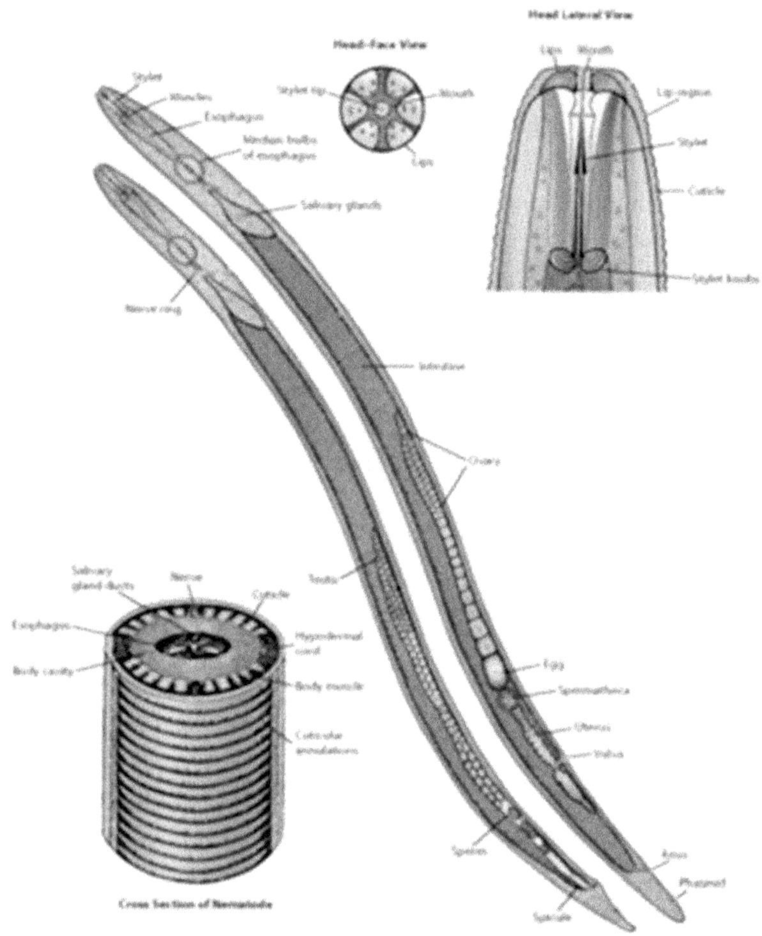

4. PLANT-NEMATODE INTERACTIONS

Plant nematodes can cause significant damage to crops and other plants, resulting in reduced yields and even plant death. Nematodes interact with plants in a variety of ways, including parasitism, predation, and mutualism. Parasitism is the most common interaction between nematodes and plants. Plant parasitic nematodes feed on plant tissues, often causing root damage or wilting of the plant. The most common type of parasitic nematode is the root-knot nematode (Meloidogyne spp.), which feeds on root cells and causes galls to form on the roots in Figure 4. Other types of parasitic nematodes include cyst nematodes (Heterodera spp.) Figure 5, which feed on root

Figure 4. Some of the most significant plant parasitic nematodes have interesting morphologies and sizes

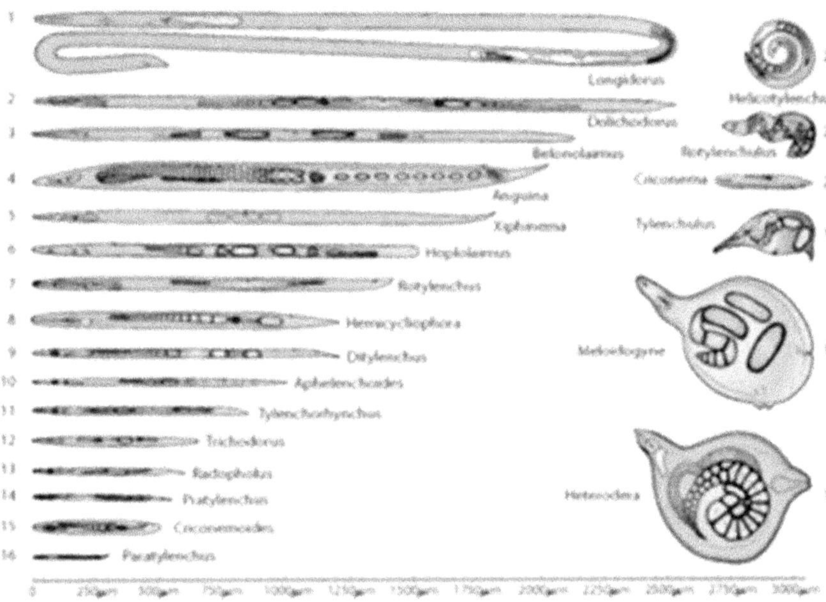

cells and form cysts on the roots; stem and leaf nematodes (Ditylenchus spp.) in Figure 6, which feed on stems and leaves; and stunt nematodes (Tylenchorhynchus spp.), which feed on root cells but do not form galls or cysts, Figure 7.

Predatory nematodes are another type of interaction between plants and nematodes. These predatory species feed on other organisms such as bacteria, fungi, protozoa, or other small animals found in soil. Predatory species can help control populations of plant-parasitic species by feeding on them directly or by competing for food resources. Mutualistic interactions between plants and nematodes also occur in nature (Smith et al., 2008;Jones JB et al., 2004). In this type of relationship, both organisms benefit from their association with one another. For example, some species of arbuscular mycorrhizal fungi form symbiotic relationships with plant roots that provide nutrients to the plant in exchange for carbohydrates produced by photosynthesis. These fungi also provide protection from certain types of parasitic nematode species by forming a physical barrier around the roots that prevents them from entering the plant tissues (Khan et al., 2017).

In conclusion, there are a variety of interactions between plants and nematodes that can have both positive and negative effects on crop yields or other plants in an ecosystem. Understanding these interactions is essential for developing effective

Figure 5. Root knot nematodes (Meloidogyne sp.) on garden tomato
Source: Yang and Eisenback (1982)

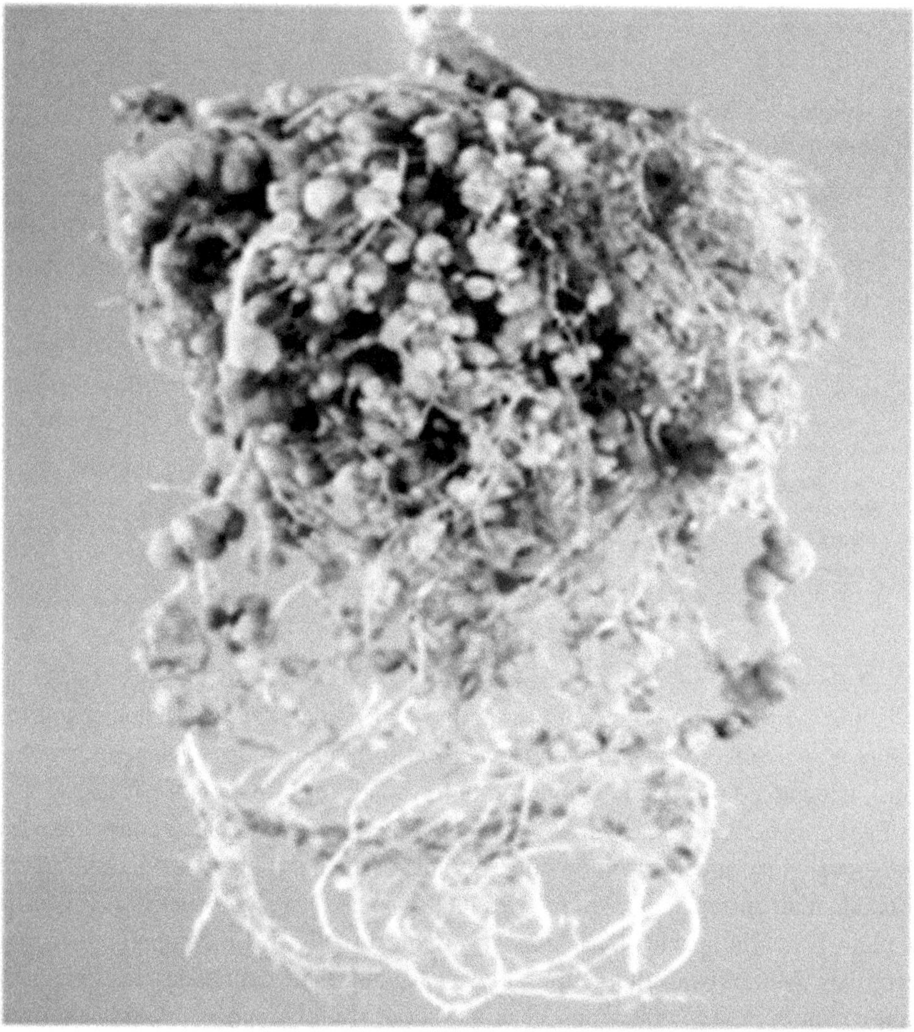

management strategies for controlling populations of damaging species while preserving beneficial ones.

Any agricultural crop, including staples like cereals, legumes, potatoes, sugar cane, sugar beets, bananas, coconuts, and sweet potatoes, can be impacted by the activity of some species of phytoparasitic nematodes, which lowers crop production and quality and results in significant economic losses. Production losses are typically estimated to be between 10% and 15%. The results are mostly influenced by the population

Figure 6. Potato infested by potato cyst nematodes (DuPont; picture updated 2017)

Figure 7. Stem and bulb nematode Ditylenchus dipsaci on garlic (Jones et al., 2017)

density in the soil and roots, the vulnerability of the cultivated plant species, and the environmental factors. The most significant are cyst and root-knot nematodes from an economic (Jones et al., 2013). The effect of nematodes on agriculture is often underestimated as they produce non-specific symptoms that are often confused with situations of water stress, nutritional disorders, soil fertility problems, or with other secondary fungal or bacterial infections that are usually favored by the action of these organisms (Mesa-Valle et al., 2020). Figure 6 revealed that the two most prominent species are Meloidogyne javanica and Meloidogyne incognita, which appear in more than 1900 articles each. Most of these articles deal with tomato crops. The nematodes of the family Meloidogynidae are the ones that mainly affect this crop, inducing galls in the roots, and affecting its functionality by altering water absorption and nutrient transport. A third species in this family is Meloidogyne hapla, and although it appears among the seven species with the greatest presence, its relative importance is not as high as that of the two previous species. The next species to appear is Heterodera glycines, a nematode that mainly affects soybeans. The next two species are Globodera rostochiensis and Globodera pallida, which are nematodes that mainly affect potato crops. The last species of the nematode to appear is Heterodera schachtii. It is mainly associated with sugarbeet and Arabidopsis thaliana. This is because this system Heterodera schachtii/Arabidopsis thaliana has become a powerful and versatile research model to study and characterize the possible interactions between the cyst nematodes and the plants they effect (Böckenhoff, 1994). In addition to these, there are two other species of nematodes that are not of such notable general importance but that affect specific crops, and many studies have been published on them.These are Bursaphelenchus xylophilus, which mainly affects pine plantations, and Radopholus similis, which affects banana plantations. There are 164 and 133 publications on these, respectively. Only the most frequently cited nematodes appear in this list; other nematodes do not appear, although they affect such important crops as orange trees (Marín-Buzón et al., 2020), which are of great economic and agronomic interest.

Finally, Plant nematodes are important economically because they can cause significant damage to crops and other plants. They feed on plant roots, stems, and leaves, causing stunted growth, wilting, and even death of the plant. Nematodes can also transmit viruses and other diseases to plants, leading to further losses in crop yield. In addition, some nematode species are vectors for plant-parasitic fungi that can cause serious damage to crops (Hussey et al., 2003). As a result of these damaging effects, nematode infestations can lead to significant economic losses for farmers and other agricultural producers.

CONCLUSION

Plant pathogenic nematodes are an important group of organisms that can cause significant damage to crops and other plants. They can cause stunted growth, wilting, and even death of the plant. Plant pathogenic nematodes can also spread diseases from one plant to another. The importance of plant pathogenic nematodes lies in their ability to cause significant damage to crops and other plants. They are a major pest problem in many parts of the world, causing billions of dollars in crop losses each year. Plant pathogenic nematodes can also spread diseases from one plant to another, making them a major threat to global food security.

Overall, plant pathogenic nematodes are an important group of organisms that can cause significant damage to crops and other plants if left unchecked. It is therefore essential for farmers and gardeners alike to be aware of these pests and take steps towards controlling them before they become a major problem in their fields or gardens. By understanding their biology and ecology as well as developing effective management strategies for controlling them, we can help protect our crops from the damaging effects of these pests while ensuring global food security for generations to come., In conclusion Plant nematodes are an important part of the natural environment, playing a vital role in the health and productivity of plants. They can be beneficial, providing essential nutrients to plants, or they can be harmful, causing disease and damage to crops. Understanding the importance of plant nematodes is essential for successful crop management and for maintaining healthy soils.

REFERENCES

Adam, M., Westphal, A., Hallmann, J., & Heuer, H. (2014). Specific microbial attachment to root knot nematodes in suppressive soil. *Applied and Environmental Microbiology*, *80*(9), 2679–2686. doi:10.1128/AEM.03905-13 PMID:24532076

Ali, A., Matthews, W., Cavagnaro, P., Iorizzo, M., Roberts, P., & Simon, P. (2014). Inheritance and mapping of Mj-2, a new source of root-knot nematode (Meloidogyne javanica) resistance in carrot. *Journal NNN/of Heredity, 105*, 288-291.

Böckenhoff, A., & Grundler, F. M. W. (1994). Studies on the Nutrient Uptake by the Beet Cyst Nematode Heterodera Schachtii by in Situ Microinjection of Fluorescent Probes into the Feeding Structures in Arabidopsis Thaliana. *Parasitology*, *109*(2), 249–255. doi:10.1017/S003118200007637X

Chitwood, D. J. (2003). Research on plant-parasitic nematode biology conducted by the United States Department of Agriculture–Agricultural Research Service. *Pest Management Science*, *59*(6-7), 748–753. doi:10.1002/ps.684 PMID:12846325

De Ley, P., Blaxter, M., Vierstraete, A., & Lockington, D. (2006). Nematode Morphology: An Overview with Emphasis on Free-Living Species. *International Journal for Parasitology*, *36*(13), 1327–1358. doi:10.1016/j.ijpara.2006.06.003

Elhady, A., Giné, A., Topalovic, O., Jacquiod, S., Sørensen, S. J., Sorribas, F. J., & Heuer, H. (2017). Microbiomes associated with infective stages of root-knot and lesion nematodes in soil. *PLoS One*, *12*(5), e0177145. doi:10.1371/journal.pone.0177145 PMID:28472099

Hussey, R. S., Barker, K. R., & Starr, J. L. (2003). Plant-parasitic nematodes. In J. J. Burdon & J. G. Ellis (Eds.), *Plant pathology* (pp. 551–569). Blackwell Science Ltd.

Jones, J. T., Haegeman, A., Danchin, E. G. J., Gaur, H. S., Helder, J., Jones, M. G. K., Kikuchi, T., Manzanilla-López, R., Palomares-Rius, J. E., Wesemael, W. M. L., & Perry, R. N. (2013). Top 10 plant-parasitic nematodes in molecular plant pathology. *Molecular Plant Pathology*, *14*(9), 946–961. doi:10.1111/mpp.12057 PMID:23809086

Jones, J. T., Haegeman, A., Danchin, E. G. J., Gaur, H. S., Helder, J., Jones, M. G. K., Kikuchi, T., Manzanilla-Lopez, R., Palomares-Ruis, J. E., Wesemael, W. M. L., & Perry, R. N. (2013). Top 10 plant-parasitic nematodes in molecular plant pathology. *Molecular Plant Pathology*, *14*(9), 946–961. doi:10.1111/mpp.12057 PMID:23809086

Khan, Z. R., & .... (2017). Predatory Nematodes As Biological Control Agents Of Plant Parasitic Nematodes. *Frontiers in Microbiology*, *8*, 1–14.

Marín-Buzón, C., Pérez-Romero, A., Tucci-Álvarez, F., & Manzano-Agugliaro, F. (2020). Assessing the Orange Tree Crown Volumes Using Google Maps as a Low-Cost Photogrammetric Alternative. *Agronomy (Basel)*, *10*(6), 893. doi:10.3390/agronomy10060893

Mesa-Valle, C. M., Garrido-Cardenas, J. A., Cebrian-Carmona, J., & Miguel, T. F. M.-A. (2020). Global Research on Plant Nematodes. *Agronomy (Basel)*, *10*(8), 1148. doi:10.3390/agronomy10081148

Rehab, Elnouby, Hasan, Ticu, Popa, Bungau, & Hafez. (2019). New Trend for Using the Reduced Graphene Oxide as Effective and Eco-friendly Nematicide. *Materiale Plastice*, *56*(1).

Siddiqi, M. (2000). *Topley & Wilson's Microbiology & Microbial Infections: Parasitology* (10th ed.). Arnold Publishers Ltd.

Subbotin, S.A. (2019). *Plant Pathogenic Nematode Species: Identification Keys to Genera from Russia and Adjacent Countries.* Springer International Publishing AG.

Subbotin, S.A. (2020). *Plant Pathogenic Nematode Species: Identification Keys to Genera from Russia and Adjacent Countries.* Springer International Publishing AG.

Subbotin, S. A., Madani, M., Krall, E., Sturhan, D., & Moens, M. (2005). Molecular diagnostics, taxonomy, and phylogeny of the stem nematode Ditylenchus dipsaci species complex based on the sequences of the internal transcribed spacer-rDNA. *Phytopathology*, *95*(11), 1308–1315. doi:10.1094/PHYTO-95-1308 PMID:18943362

Wang, X., Meyers, D., Yan, Y., Baum, T., Smant, G., Hussey, R., & Davis, E. (1999). In planta localization of a β-1,4-endoglucanase se-creted by Heterodera glycines. *Molecular Plant-Microbe Interactions*, *12*(1), 64–67. doi:10.1094/MPMI.1999.12.1.64 PMID:9885195

Zhao, J. H., Tu, G. J., Wu, X. B., & Li, C. P. (2018). *Characterization of the complete mitochondrial genome of Ortleppascaris sinensis (Nematoda: Heterocheilidae) and comparative mitogenomic analysis of eighteen Ascaridida nematodes.* Academic Press.

Chapter 2
Ecology, Distribution of Nematodes

Al Sadek M. Ghazala
Faculty of Agriculture, University of Tripoli, Libya

Abdullah Khder Mohammad
Mosul University, Iraq

Wasan S. Hussain
Department of Biology, Mosul University, Iraq

ABSTRACT

It is closely related to zoology and botany, which are the origin of all biological sciences. The study of plant-pathogenic nematodes requires a great understanding and knowledge of general zoology, invertebrate zoology, parasitology, and animal behavior science, which helps understand the nutritional behavior of nematodes on their plant hosts and thus the effects of different control methods on nematodes, as well as being familiar with botany and its various branches, especially the functions of organs, appearance, and internal anatomy, especially of the root system, and these are requirements for understanding the different and mutual influences between nematodes and plants.

ECOLOGY AND DISTRIBUTION OF NEMATODES

Nematodes are aquatic microscopic organisms that depend on thin layer of water to live and move within pathways of soil pores of 25–100 μm diameter. Soil nematodes can be a testing tool for and understanding of biological mechanisms in soil due to their central role in soil and linkage to ecological processes. Ecological succession

DOI: 10.4018/978-1-6684-8083-0.ch002

is one of the most tested community ecology concepts, and a variety of nematode community indices have been proposed for purposes of environmental monitoring.

Nematodes are microscopic worms that cause billions of dollars in crop losses annually, and all crops in the world are susceptible to at least one species of nematode parasites (Bozbuga *et al.*, 2018). Plant-parasitic nematodes are the major biotic stressor in crop cultivation (Treonis *et al.*, 2018). Plant-parasitic nematodes are called plant-parasitic because nutrients they receive come directly from plants. They have a needle-like structure called a stylet that helps them to pierce the plant cell walls and reach the juicy contents on which they feed (Bernard *et al.*, 2017). Correct identification of nematode species is essential for choosing the proper methods of control. The morphology-based taxonomy of nematodes is complex due to intraspecific variation, but the tools and techniques based on biochemical and molecular markers can successfully diagnose a wide range of nematode species (Carneiro *et al.*, 2017). The economic consequences of crop losses caused by nematodes come in many variations and are associated with a decrease in the crop quality and yield. The assessment of such losses and periodic updates of these estimates may be of great use to set research priorities. In addition to that, it can serve as a benchmark for policy planners/funding agencies to support research in this field, as well as for public/private sector to make appropriate investments in management (Kumar *et al.*, 2020).

Plant-parasitic nematodes are recognized as one of the greatest threats to crops worldwide. It has been found that nematodes, alone or in combination with other soil microorganisms, infect almost every part of the plant, including roots, stems, leaves, fruits and seeds. Although worldwide recognition of nematodes as important cause of plant diseases did not occur until the middle of the last century, nematodes were studied more than 150 years earlier. The most damaging plant-parasitic nematodes are considered to be the root-knot nematodes (RKNs), *Meloidogyne* sp. (Almohithef *et al.*, 2018), which are responsible for losses in vegetable crops throughout the world and determine the common use of chemical pesticides (Sikandar et al., 2020). In Spain, RKNs can be found in all horticulture areas (Archidona-Yuste *et al.*, 2018), causing yield losses in essential vegetable crops, such as cucumber (85%), tomato (59%), zucchini (40%), watermelon 36%, and lettuce (29%) (Gullino *et al.*, 2019). RKNs can also be seen posing a threat to vegetable production in other European and transcontinental countries such as Turkey, Slovenia, Brazil, and Greece. In Turkey, for example, they can attack numerous higher plants, including monocotyledonous, dicotyledonous, herbaceous, and woody plants, all thanks to it wide host range (Aydınlı *et al.*, 2013).

DISTRIBUTION AND ABUNDANCE OF NEMATODES

The disturbance amount in a host by parasites and the response of the host are influenced by the distribution of the parasites in the host. Their occurrence in a particular organ or tissue and their density at particular sites may profoundly affect the course of disease. Outside the host, the distribution of nematode is thus an important aspect of disease and even causal observation indicates that they are seldom distributed uniformly within or outside the host.

The distribution of eggs in the *Heterodera* cyst may approach uniformity whereas the distribution of cysts in on the host root may be non-uniform. Similarly, animals with strong territorial behavior may have fairly uniform distribution, although it may be non-uniform over a larger area. Non –uniform distribution may be on two general types, random and clumped, and of the two, clumped distribution is the most common in nature. (Southwood, 1986).

Some Aspects of Distribution

Little information is known on the micro-environment within the plant that influences the endoparasites. There is little information on the relations between biochemical and physical

gradients within roots and the distribution of endoparasites. We do know the plants are frequently infected by more than one parasite and that they often interact so that one parasite is suppressed Rishbeth (1963).

Within the host the picture of a parasite invading an animal host and then proceeding a long route through the body until it eventually reaches a place where it settle down, it is not applicable to most plant nematodes, even those that migrate within the host tissues. Thus distribution within the plant is often closely related to the invasion site. Kirkpatrick and his colleagues (1964) have explained that many plant nematodes show a preference of feeding or penetration at particular zones on the root or at specific stages in the roots ontogeny. Some species prefer the meristem, some the zone of elongation and others the region of maturation. Preference are also shown for tissues and cells within a particular region of the root e.g. the apical meristem, the epidermis, the cortex or the stele. The distribution of larvae of *Heterodera schactii* in the root of rape (*Brassica rapus*) and sugar beet (*Beta vulgaris*) was determined by Kampfe (1960)who examined the root tips, the points of origin of lateral roots and intervening regions separately. Although similar numbers occurred in these three regions, the juveniles were clumped in groups at the points of origin of lateral roots thus giving a patchy distribution within the whole root system. Presumably sites of lateral root information are particularly attractive to migrates of *Hereodera scahctii*.

Kinloch and Allen (1972) provide us with another example of differential attraction and invasion; *Meloidogyne hapla* is not attracted to and does no invade galled tissue of tomato roots to the same extent as *Meloidogyne javanica*. Consequently, there is a more rapid invasion of the host by *M. javanica* and, in a population containing both species, *M.javanica* is likely to predominate and its distribution in the host is likely to be more widespread. Although the distribution of nematodes within the plant is probably influenced to a large extent by preferences for particular types of tissues, it would be wrong to assume that other factors are of little importance. Thus *Heterodera trifolii* is usually found in the roots of its host but Ross (1960) found cyst of this species on both surfaces of white clover leaves. Giant cells were formed, so presumably development proceeded in a similar way to that in the root. *Meloidogyne incognita* can cause galls on the leaves of *Siderasis* (Tradescantia) *fuscata*. The galls contained adult females, viable eggs and juveniles; giant cells were formed; what is more; the roots were not infected. (Miller and Di Edwardo, 1962). Similarly, Wong and Willetts (1969) describe the information of galls in the serial parts of tomato, French bean and Capsicum seedlings inoculated with *Meloidogyne javnica*. In these species at least, distribution within the host is not restricted by tissue specificity; more likely it is influenced by the inability of the nematodes to migrate to and invade the leaves, except on those few occasions when the leaves touch the soil. Wallace and Doncaster (1964) have suggested that nematodes can be divided arbitrarily into three groups: those that are active enough to swim in deep water; those like *Ditylenchus* and *Aphelenchoides* that can swim in a thick water film, escape from the soil to ascend plants above ground level, finally those that are too inactive to swim, the crawlers, which are largely confined to the soil. Furthermore the nematodes that have propulsive power to escape from the soil are able to survive desiccation for long periods. They can therefore tolerate exposure to dry conditions when they are exposed to the air on the plant surface. Thus, locomotory activity and adaptations for survival can be added to the list of factors that influence the distribution of nematodes within the host plant. The presence of other organisms within the same host plant may also influence the distribution of a nematodes species. This is an obvious statement; however it is less obvious how frequently it happens. Plants infected with fungi were similar influences on nematodes. In some associations the numbers of nematode are reduced. Thus, in study of the association between the causal fungus of brown root rot of tomatoes and *Heterodera rostochiensis*.

. There is plenty of evidence that the numbers of nematodes in plant increase when another parasite is present. Higher population of developed on soybeans grown in soil infested with *Fusarium* than in *Fusarium* free–soil (Ross, 1965). Significantly higher numbers of *Pratylenchus pentrans* occurred in egg-plant roots infected with *Verticillium albo-atrum* than occurred in plants without the fungus (McKeen and

Mountain, 1960). Similarly *Verticillium dahlia* in egg plant and tomato was associated within increased rate of reproduction of *P. penetrans* (Mountain and McKeen, 1962). The authors suggest that the nematodes are influenced by changes in the physiology of the host plant induced by the parasitic activates of the fungus. In coincident infection with two or more nematode species most observations suggest that at least one species is suppressed. Freckman and Chapman (1972) however studied invasion by *Heterodera tifolii* and *Pratylenchus penetrans* in the same red clover plant and found no influence of either nematode on the other. Prior infection by *H. trifolii* did not affect invasion by *P. penetrans* and the distribution of *P. penetrans* was un affected by the presence of *H. trifolii*. In this example there is no indication of competition. when equal numbers of *P. penetrans* and *Tylenorhynchus martini* were introduced into red clover and Lucerne plants, *P. penetrans* increased as well in the presence of *T. martini* as it did when alone, but *T. martini* in the presence of *P. pentrans* increased only 10 to 25% as much as it did when alone (Chapman, 1959). There are similar kinds of observations with other nematodes: the reproduction of *Meloidogyne incognita* decreased in the presence of *Pratylenchus brachyrus* (Johnson and Nusbaum, 1970), *Heterodera tabacum* was suppressed by *Pratylenchus penetrants* when it occurred on 90% of a population of the two; however, *H. tabacum* at moderate to high levels, suppressed and eventually replaced a smaller number of *P. penetrants* in the same population (Miller, 1970). Experiments with three species of nematodes on the same host have been tried. Thus Johnson (1970) infected six Bermuda grasses with the ectoparasiic nematodes *Criconemoides ornatus*, *Tylenchorhynchus martini* and *Belonolmius longicaudatus*. The reproductive rate of all three was suppressed, *B. longicaudatus* least of all. Although the factors influencing the interactions between nematode species are unknown, it seem possible that if numbers and distribution are ever determined at the ecological level, it will be most apparent with ectoparasitic nematode which are likely to occur close together on the root feeding on the same cells, without that intimate physiological relationship between host and parasite as in the Hereteroderidae for example.

Outside the host information about the distribution and numbers of nematodes in the host helps us to understand the nature of disease by identifying the location of the initial disturbance in the individual plant. Outside the host, the distribution and abundance of nematodes is important in studies in the aetiology of disease in plant populations. We need to know whether the nematodes are distributed uniformly, randomly, or occur in patches and the use of the distribution provides us with a way of answering this question. Barker and Nusbaum (1971) provide excellent data to illustrate this aspect. To explain why the numbers of a nematode are higher in one place than another is a familiar but difficult ecological problem. There are so many environmental components that could influence the nematode, and in any situation it is likely that several do in fact have some influence on the distribution and abundance of a particular nematode. Satisfactory explanations of distribution must therefore take into account as many as possible of these components and furthermore, assess their relative contribution to the observed distribution. To make the problem more difficult the existing distribution of a nematode might have been decided many years ago by some factors that no longer exists and which cannot be measured at the time the nematodes samples are taken. However from previous records it may be possible reasons for distribution patterns. Thus, Brown 1958 showed that there was a correlation between the present distribution of *Heterodera goettingiana* in East Suffolk in England and those districts which grew the host beans during the latter half of the nineteenth and during the early twentieth century. Similarly in a survey in Bedfordshire, population of *Heterodera cruciferae* were greater and more fields were infested on farms of less than 20.234 hectare (fifty acres) than on larger farms (Winfield and Coppock 1970). Brassican crops were probably grown more frequently on the smaller farms. Such explanations of distribution are plausible but impossible to test experimentally because of lack of data on other factors that might have been important at that earlier time. If distribution is the result of previous introductions on planting stocks, then the main factor determining distribution might be missed altogether. Hannon (1962) suggest that *Tylenculus semipentrans* is distributed in citrus groves by infested nursery stock rather than because of its endemic expense Hannon (1962) Hannon argues, might account for the observation that infested groves often adjoin uninfected groves. Such an explanation would be valid if there was evidence of the level nematode infestation in the nursery stock when they were planted out. Without such data there are too many other possible explanations to allow us to accept the hypothesis of infested nursery material too readily. These examples serve to make the point in some situations the problem of accounting for distribution may be insoluble. We might measure and correlate all the environmental factors we can think of and find significant correlations and still miss the most important component because it no longer exists to be measured. Hopefully there are not too many situations like this and of course we can never know without historical data which are and which are not.. Let us now look at some of the environmental components that influences the distributions and abundance of nematodes outside the host. In making surveys of nematode distribution and in any field problem, the most obvious factor that asks to be measured is soil type. There are in fact a large number of references to correlations between the distributions and abundance of various plant nematodes and the sort of soil from which they were recovered. Rebois and Cairns (1968), for example took soil samples from soybean field in 28 major soybean producing counties in Alabama, Northern Florida and in Georgia. The number of genera and the total number of plant nematodes that were recovered per pint of soil were higher in the higher-textured soils than in the clay soils. Similarly there is a close relationship between the distributions of five species of *Xiphenema* and soil type in Israel, Cohn (1969) attributes this to the fact that the smaller species occur in the heavier soils whereas in the lighter soils, nematodes of greater body volume are more abundant. Cohn suggested that oxygen content and water holding capacity might determine the ability of species of *Xiphenema* to survive. Thus, larger nematodes might require a well aerated soil, he is presumably thinking of relative decrease in surface area to volume of the body as nematodes increase in size, and the consequent relative reduction in available area through which gases can diffuse through the cuticle. Thus, it is not soil type that determines distribution and abundance but some other factors associated with soil type. In Cohns studies, distribution of the *Xiphinema* species was associated with body size and soil aeration. This is plausible suggestion, although there is no evidence to test the hypothesis; we might have suggested that the larger size of pores in the soil might beter accurate the larger *Xiphinema* spp. By enabling them to move about in the soil from root to root more quickly. Study by, Chirchir *et al.*, 2008 showed that soil texture is a major factor that influences abundance and distribution of plant parasitic nematodes. The higher numbers of plant-parasitic nematodes associated with sandy soils could be attributed to high porosity and aeration that favor nematode mobility. The poor drainage associated with clay soils may explain the low numbers of nematodes recovered from these soils. Similar studies have revealed that the effect of soil texture on the distribution of some of the nematodes is more than that of climatic or topographic factors (Mateille *et al.*, 2002). According to Cadet and Spaull (2003), it is the quantity of sand and organic matter in soil that appears to affect the distribution of nematodes associated with sugarcane. It has been suggested that sandy soils enable easy movement of nematodes and provide optimum aeration which leads to increased pathogenicity. The negative effect of nematodes on plants is aggravated by the fact that sandy soils have a lower water and nutrient holding capacity (Avendaño *et al.*, 2004). This study showed that numbers of plant parasitic nematodes increased steadily in the crops and then declined after the third ratoon. Increase in nematode numbers could be due to multiplication of the nematodes as a result of continuous presence of susceptible crops (Kratochvil *et al.*, 2004). According to Quader *et al.* (2001) populations of plant parasitic nematodes were continuously building up in Australasian vineyard fields mainly due to the perennial nature of the crop. Loss of plant growth vigour coupled with competition among the nematodes, for limited reserves could explain the decline (Akhtar and Malik, 2000).

Understanding the interaction between soil physicochemical properties and plant-parasitic nematodes is critical for effective and environmentally-friendly management. The citrus nematode, *T. semipenetrans* was positively correlated with soil mineral nutrients (K, Ca, Na, and C) and organic matter content. A strong correlation was found between these parameters and the prevalence of *T. semipenetrans* in citrus growing areas in Spain (Sorribas *et al.*, 2008). In contrast, (Benjlil *et al.*, 2019) reported a negative correlation between the prevalence of palnt parasitic nematodes parasitizing saffron and soil organic matter. Soil organic matter accumulation could significantly reduce plant parasitic nematodes abundance in wheat via decreasing their vital proprieties (Hu *et al.*, 2013). Moreover, the prevalence of *Pratylenchus* spp., *Helicotylenchus* spp., and *X. americanum* was closely related to the mineral content including Fe, Ca, and Na (Karuri *et al.*, 2017). In addition, (Francl *et al.*, 1993) observed that the population density of *Heterodera glycines* was positively correlated with magnesium (Mg) content. In this study, most of the identified plant parasitic nematodes showed a negative correlation with nitrogen (N), except for *Criconemoides*, *Rotylenchus* spp., and *Tylenchorhynchus* spp. interestingly, the accumulation of nitrate via nitrification is considered destructive to plant parasitic nematodes (Rodriguez- Kabana *et al.*, 1986). Phosphorus content was positively correlated with the occurrence of *Meloidogyne* spp., *Xiphinema* spp., *P. thornei*, *P. vulnus*, and *P. neglectus*. (Nisa *et al.*, 2021) reported the same positive correlation between nematode abundance and soil P content. Soil pH also had a strong influence on the abundance of citrus nematodes. Low pH increased nematode abundance and diversity (Wang *et al.*, 2009). Soils with acidic pH increase the proliferation of the root-knot nematode *Meloidogyne* spp. (Korthals *et al.*, 1996). In contrast, (Salahi Ardakani *et al.*, 2014) found that the highest abundance of *T. semipenetrans* was observed in soils of pH 7. Moreover, Van Gundy and Martin (Van Gundy *et al.*, 1962) indicated that the density of *T. semipenetrans* in citrus was four times higher in neutral soils than in acidic soils.

Soil texture and structure significantly affect the movement of soil nematodes. (Salahi Ardakani *et al.,* 2014) found that the maximum abundance of the citrus nematode *T. semipenetrans* was found in clay soils. A recent study (Laasi et al., 2022) showed that Aphelenchoides spp., Merlinius spp., and Pratylenchus spp. were associated with sandy and silt soils in wheat fields. In contrast, *Longidorus* spp., and *Xiphinema* spp. were more common on soils with higher clay content. (Mokrini *et al.*, 2019) found that the abundance of several plant-parasitic nematodes affecting raspberries was strongly associated with soil granulometry. Kim *et al.* (2017) indicated that sandy soils favor the growth of nematodes such as *M. incognita* by promoting their motility and feeding activities.

Nematode distribution at different spatial scales is affected by several factors (Liu *et al.*, 2019). Temperature, moisture, land use, soil properties, availability of nutrients, root architecture, and biomass are some of the variables that affect vertical distribution of nematodes (Fan-xiang *et al.*, 2005; Liu *et al.*, 2019). The level of crop damage by plant parasitic nematodes is influenced by their distribution within the soil profile (Nguyen *et al.*, 2020).

Determination The effectiveness of control strategies determines by vertical distribution of plant parasitic nematode (Ingham *et al.*, 2000). *Helicotylenchus*,

Meloidogyne, and *Scutellonema* occurred in relatively high numbers at 0 cm to 30 cm and 30 cm to 60 cm depths in all the regions but the abundance was not significantly different. However, *Scutellonema* showed a specific association with 0 cm to 30 cm depth. Similarities in nematode genera at the two depths were also depicted in the method of analysis. These nematodes have been previously reported in association with sweet potato in Kenya (Njuguna and Bridge, 1998). Lack of significant differences at the two soil depths may be due to the influence of crop growth stage on plant parasitic nematodes population dynamics whereby similar numbers of nematodes may be present either in the soil or within the plant at a particular time (Sharma *et al.*, 1992). Root density (Robinson *et al.*, 2005) and architecture also influence the vertical distribution of plant parasitic nematodes (Pudasaini *et al.*, 2006, Wesemael and Moens, 2008). The population density of plant parasitic nematodes within plant roots at 0 cm to 30 cm and 30 cm to 60 cm depths was not tested in the current study, and this avenue of study warrants further investigation. Based on soil conditions, the sweet potato root system may occur up to 2-m deep (Woolfe, 1992), which may explain the uniform distribution of most of the plant parasitic nematodes genera within the soil profile (Fan-xiang *et al.*, 2005). At depths below 40 cm, Rodriguez-Kabana and Robertson (1987) also observed constant distribution of *Meloidogyne arenaria*, which they attributed to light soil texture.

Contrary to our observations, *Scutellonema* and *Helicotylenchus* have been reported to decrease with depth in most crops. In a groundnut field, >95% of *Scutellonema clathricaudatum* were found at 0 cm to 30 cm (Sharma *et al.*, 1992) while a higher population of *Helicotylenchus* was observed at 20 cm to 30 cm in a fallow field (Fan-xiang *et al.*, 2005). Similarly, Araya and De Waele (2011) observed a decline in the number of *Helicotylenchus* at higher soil depths in a banana field. In the current study, population densities of *Tylenchus* were variable at both depths in the three regions, which is similar to observations reported elsewhere in the literature (Fan-xiang *et al.*, 2005, Siddiqui, 2007). *Meloidogyne* species are considered the most economically damaging plant parasitic nematodes in sweet potato and they cause 10% to 20% of yield losses (Koenning *et al.*, 1999; Okechalu and Wonang, 2015). The losses resulting from Root Knot nematodes infestation may be variable depending on the nematode species (Overstreet *et al.*, 2009). Root-knot nematodes also cause necrosis and predispose the storage roots to cracking, which reduces their market value (Lawrence *et al.* 1986). High population density of *Helicotylenchus* reduces the weight of tubers, roots, and shoot biomass (Lopez *et al.*, 1981). Although *Scutellonema* (Njuguna and Bridge, 1998; Coyne *et al.*, 2003) and *Tylenchus* (Haougui *et al.*, 2011) are associated with sweet potato, there are limited studies on their effects on yield.

Effective control of plant parasitic nematodes that were observed in this study will require that information on their vertical distribution is considered. This was demonstrated by Ingham *et al.* (2000), where *Meloidogyne chitwoodi* was not controlled by metam sodium, ethoprop, and oxamyl at 0 cm to 30 cm due to migration of the nematode from 120-cm depth to the upper soil layers. In a different study, solarization at 40-cm soil depth was more effective in controlling *M. incognita* eggs (Nico *et al.*, 2003). Rodriguez-Kabana and Robertson (1987) also suggested that for control of *M. arenaria*, placement of fumigant nematicides that move upwards should be such that they cover the depth where the nematodes are most prevalent. Low-cost palnt parasitic nematodes management techniques in sweet potato would be preferred by smallholder farmers in Kenya. The optimal depth at which the treatments would have a higher efficacy against the plant parasitic nematodes should be tested on a case-by-case basis.

Shannon and Simpson diversities were higher at 0 cm to 30 cm across Central, Manyatta, and Nembure regions with the component having a higher contribution to γ diversity. Zheng *et al.* (2012) also observed a decrease in these indices at higher soil depths, which was corroborated by Zhong *et al.* (2015) in different cropping systems. Processes that shape plant parasitic nematodes populations such as niche differentiation, competition, reproduction, different strategies, and feeding methods affect their diversity and influence their response to control agents (Pérez *et al.*, 2000; Pudasaini *et al.*, 2006). Effect of occurrence of multiple plant parasitic nematodes on crop damage was demonstrated in sweet potato, whereby competition between *M. incognita* and *R. reniformis* during concomitant infection resulted in the dominance of one species (Thomas and Clark, 1983). Co-occurrence of multiple plant parasitic nematodes in soil influences the damage that they cause in crops (Thomas and Clark, 1983) and hence the need to consider these aspects when choosing nematode control strategies.

Regional differences that were observed in diversity indices and PPN abundance may be explained by several factors including soil properties (Hoogen *et al.*, 2019; Li *et al.*, 2020). Soil properties are an important factor in shaping distribution of nematode communities at different spatial levels (Nguyen *et al.*, 2020).

Biodiversity is an essential ecological phenomenon because it represents a complex set of interacting ecological, evolutionary, biogeographical and physical processes (Huston, 1997). Native biodiversity is being lost at a rapid rate owing to anthropogenic causes, including habitat destruction, pollution and the spread of non-native species (Barnosky *et al.*, 2011).

The collection of data on nematode distribution in space is much more useful if there are factors that we can correlate with such numbers because they suggest where we might look for causal factors. Analysis of the data by simple correlation, multiple and stepwise regressions is the next phase. Ferris *et al.* (1971), used two mathematical methods, a resemblance equation and community ordination, to study the community structure of plant nematode population in soybean soils in Indiana and Illinois. The methods enable the relative densities of different species to be measured at fourteen sites which were chacterized by latitude, soil type average frost free period, average annual temperature, average annual precipitation and average soybeans yield with good management. According to the authors the reasons for taxonomic resemblance or lack of it are probably determined by soil factors. Although the quantifiable characteristics of the environment were not correlated with the location of sites, some sites had similar communities of nematodes. In particular, community structure tended to be similar on the dark-coloured, highly productive soils; on the lighter soils, the communities were different from those in the darker soils and from each other, possibly reflecting the more diverse nature of the lighter soils. Here then is a method whereby populations in different areas can be measured and the possible influence of the environmental factors assessed through correlation.

We might ask however, whether the study of communities and the proportion of its members is more worth-while than the study of populations of different species. If the relative number of species within a population has any biological significance then we look for interactions between species in their effects on each other and on the plant. We have already mentioned interactions between different species in the host, and in the soil there are some indications of possible antagonism; for example Griffin and Darling (1964), suggest that *Xiphinema amrecanum* has a suppressing effect on *Criconemoides xenoplax* in ornamental spruce nurseries. However it seem to me that the task of elucidating, at this stage, community structure, a concept which some believe has contributed little to the understanding of the abundance and distribution of animals (Andrewwartha, 1970). The alternative approach is to consider one species at a time, to try to explain its distribution and to treat other species as further environmental components of the single species under study.

The Distribution and Abundance of Nematodes Time

The distribution and abundance of nematodes are never static and, just a numbers inside the host change as the host matures and senesces, so do numbers of nematodes in the soil change as the seasons progress. In studies on the spatial distribution of nematodes, it is therefore necessary to sample several times during the growing season of the host and several consecutive seasons. Account must also be taken of changes in distribution with time both horizontally and vertically. There are numerous references in the literature to seasonal fluctuations in nematode numbers with suggestions for the factors causing them. Griffin and Darling (1964) have shown that *Xiphinema americanum* has to potential peaks per season in ornamental spruce nurseries. One cycle extend from April to August, the other from September to January. The peaks become less obvious in the juveniles and non-gravid females as sampling depth increased, whereas the gravid adults showed two distinct cycle at all sampling depths. *Meliodgyne naasi* may also have two population peaks in the soil because there are few nematodes in winter, amarked increased in spring, a decrees when host roots are invaded in the summer and a moderate increase in late summer (Frankiln et al., 1971). Flegg (1968) found that eggs of *Longidorus* and *Xiphinema* spp. were produced only during a brief period in early summer but proportions of each juveniles and adult stage in the population showed no comparable annual cycle. In studies on the population of *Pratylenchus penetrans* in soil and roots in rye-tobacco rotations in Canda, Olthof (1971) showed that populations were low in the summer and high in the autumn, although seasonal changes were not consistent from year to year. There is a distinct annual periodicity of high populations of *Radopholus simils* in Citrus lemon rootlets with a peak in October to December and a trough in February to June. The particular month when peak populations occurred varied from grove to grove and from year to year (Ducharme, 1967). The seasonal emergence of *Heterodera avenae* juveniles was studied over three years in wheat in Victoria, Australia (Meagher, 1970). Juveniles numbers were highest during late autumn and winter. Meagher attributes the increase to changes in soil moisture and temperature that favour hatch. Banyer and Fisher (1971), on the other hand, attributes in the increase in such numbers to the onset of low soil temperatures in late autumn and they give experimental evidence that indicates that this change in temperature beaks dormancy in the eggs.

There are many other records in this vein and most authors suggest what factors might be responsible for the observed fluctuations. It is usual for example to include rainfall and temperature data with graphs showing the changes of nematode numbers with time. However there has been no attempt in nematology, as far as I know to account for population changes with time in a statistical way such as that done by Frris (1971) and co-workers for spatial distribution. Such analysis of data would indicate the relative contribution of the nematodes various environmental components to the observed fluctuations not only those periodic fluctuations associated with changes in the host but other less pronounced but important changes superimposed on the more uniform cycles. The statistical technique of periodic regression as described by Bliss (1958) would be a useful tool in such ecological studies. Although the task of elucidating the ecology of distribution of nematodes is a formidable one, it is not impossible if the is done by a team of workers. However, there is no one facet of the work that should be considered very seriously, the method used to extract the nematodes from the soil. The paper of Barker *et al* (1969) indicates that the method of extraction greatly influences the numbers recovered during certain periods of the year. In addition to extraction method, rainfall, season, crop and cultural practices that influence apparent shifts in population densities of nematodes the interaction of extraction method and time of sampling may often be important; in other words, different extraction procedures might give different pictures of seasonal fluctuations. The only answer to this difficulty is to use extraction techniques that give high recoveries without depending on the nematodes own movements as in the Baermann method.

Dispersal

Unlike insects nematodes do not move very far or very quickly by their own locomotory power, they have not evolved any characteristics that enable them to disperse quickly. In fact they appear to be more efficient at aggregating. Being obligate parasites, plant nematodes do not feed while in the soil in the absence of plants, hence, they have a limited time to actively move before their resources of chemical energy run out. Appear to be related to reducing the time the nematodes spends away from the host. Neither have they evolved any characteristics that enable them to escape from the soil nor for projecting themselves into the air like fungi. Some nematodes like *Ditylenchus*, *Anguina* and *Aphelenchoides* can escape from the soil and with their highly developed ability to resist desiccation, their chances of dispersal are correspondingly greater, but for the endoparasitic and ectoparasitic nematodes that attack roots such opportunities rarely occur. The conclusion can be drawn therefore, that the survival on nematodes in the evolutionary sense has been successful for seasons other than those related to rapid dispersal. As has been before, the usual arguments for the selective advantages of dispersal are that animals live in a changing environment and that if a population did not disperse it would stand chance of becoming extinct when the weather in any particular year was unusually severs or when ecological succession of plants in an area no longer provided sufficient amounts of the right food. We might therefore, ask whether nematodes have any particular adaptations that enable them to overcome these hazards. Most plant nematodes live in the soil where fluctuation in temperature and moisture decrease rapidly with depth, hence spectacular climatic changes above ground are likely to be less hazardous in the soil. Moreover, nematodes have the ability to migrate vertically in response to physical gradients and so escape the hazards, hence the seasonal changes in vertical distribution of nematodes that we have just discussed. When it comes to the hazard of diminishing food reserves the inability to disperse is a disadvantage in a natural ecosystem it is likely that isolated populations of a species do in fact become extinct as succession of plant species occurs. However many nematodes have a wide host range and can therefore occupy a particular area even though the plant population is changing. The efficiency of crop rotation as a nematode control measure is an indication of how changing plant populations can influence populations of nematodes, but it is also true that a significant proportion of the nematode population always survive to build up again when suitable hosts are eventually planted. So inability to disperse is offset to some extent by escape in the soil from detrimental climatic conditions and by the ability to persist, perhaps in low numbers, on alternative hosts during times when preferred hosts are few. Some dispersal does occur however, even over long distances but it is passive and by chance. Wind-borne dispersal has been reported by Orr and Newton(1971) who found *Meloidogyne* juveniles among 28 genera recovered from dust traps placed 2m above the ground in Western Texas. Numerous plant nematodes were found and the authors suggest that the deposition of nematode-laden dust might explain the extensive occurrence of root knot nematodes in the area. Dispersal by surface run-off of water and by irrigation is also possible (Faulkner and Bolander, 1966). In fact the most efficient ways of nematode dispersal appear to be a consequence of humans own activities, especially in his agricultural activities.. Erosion of the land by wind or its dust storms, the movement of water in irrigation channels, the transport of infected plant material between different countries and the spread of nematodes on farm implements and vehicles all contributes to nematode dispersal and produces disease problems at the same time . the in influence of dispersal on distribution of nematodes may be of little importance in areas of natural plant communities but human influence on the environment is now so extensive that, in agriculture at least, we cannot overlook the effects of our own activities on nematode distribution, Unfortunately, such activities are hard to measure and there unusually little quantitative evidence that enables us to assess this aspect of distribution.

However, these organismal groups do not exist in isolation, but from part of a complex network in the soil (Richter et al., 2019).The calculation of ecological indices. Consequently, neamtodes are considered valuable indicators of soil ecosystem health (Sanchez-Moreno and Ferris, 2018).

Nematode distribution at different spatial scales is affected by several factors (Liu *et al.*, 2019). Temperature, moisture, land use, soil properties, availability of nutrients, root architecture, and biomass are some of the variables that affect vertical distribution of nematodes (Liu *et al.*, 2019). The level of crop damage by PPN is influenced by their distribution within the soil profile (Nguyen *et al.*, 2020). Plant-parasitic nematodes damage crops through different modes of feeding within or outside the plant. They are broadly classified as ectoparasites, semi-endoparasites, migratory endoparasites, and sedentary endoparasites (Palomares-Rius *et al.*, 2017). There are several PPN that are associated with sweet potato, including the economically damaging *Rotylenchulus reniformis* and *Meloidogyne* species that cause 5% to 10% yield losses (Jatala, 2019).

In sweet potato, the use of resistant varieties (Kim and Yang, 2019), application of nematicides, plant extracts, green manure (Waisen *et al.*, 2020), and organic amendments (Stirling, 2020, 2021) are among the methods used to control PPN. As demonstrated in some studies, the depth at which the PPN occurs is important for the efficacy of some of the control strategies. For instance, in potato and cabbage, there was an interaction between the depth at which *Belonolaimus longicaudatus* occurred and the effect of nematicide on the PPN. Population of *B. longicaudatus* was higher at 20 cm to 40 cm compared to 0 cm to 20 cm after nematicide application (Pérez *et al.*, 2000).

The most popular methods to control nematodes include soil fumigants and nematicides (Giannakou and Panopoulou, 2019). However, in order to preserve the beneficial soil organisms, it is necessary to use alternative ways of neutralizing parasitic nematodes. A sustainable production system uses environmentally friendly alternatives to maintain and improve the health of antagonist organisms. Soils with high antagonistic potential lead to the suppression of soil-borne pathogens. In suppressive soil, pathogens do not form, persist or establish, but cause little or no damage at all (Topalovic *et al.*, 2020). Suppressive soils have already been described for many soil-borne pathogens, including plant-parasitic nematodes (Romanenko *et al.*, 2008). Soils that suppress the development of nematodes have been intensively studied over the past several years, but the mechanisms of suppression are not entirely understood. (Ghini and Morandi, 2006) summarized the biotic and abiotic factors associated with soil suppression. Among them; soil microbiota plays an important role (Topalovic *et al.*, 2020) in maintaining biological productivity (Dutta *et al.*, 2019).

REFERENCES

Aït Hamza, M., Moukhli, A., Ferji, Z., Fossati-Gaschignard, O., Tavoillot, J., Ali, N., Boubaker, H., El Mousadik, A., & Mateille, T. (2018). Diversity of plant-parasitic nematode communities associated with olive nurseries in Morocco: Origin and environmental impacts. *Applied Soil Ecology*, *124*, 7–16. doi:10.1016/j.apsoil.2017.10.019

Akhtar, M., & Malik, A. (2000). Roles of organic soil amendments and soil organisms in the biological control of plant-parasitic nematodes: A review. *Bioresource Technology*, *74*(1), 35–47. doi:10.1016/S0960-8524(99)00154-6

Almohithef, A. H., Al-Yahya, F. A., Al-Hazmi, A. S., Dawabah, A. A. M., & Lafi, H. A. (2018). Prevalence of plant-parasitic nematodes associated with certain greenhouse vegetable crops in Riyadh region, Saudi Arabia. *Journal of the Saudi Society of Agricultural Sciences*, *19*(1), 22–25. doi:10.1016/j.jssas.2018.05.001

Araya, M., & De Waele, D. (2011). Reaction of six Musa genotypes to root-parasitic nematodes. *International Journal of Pest Management*, *57*(3), 229–238. doi:10.1080/09670874.2011.581770

Archidona-Yuste, A., Cantalapiedra-Navarrete, C., Liebanas, G., Rapoport, H. F., Castillo, P., & Palomares-Rius, J. E. (2018). Diversity of root-knot nematodes of the genus *Meloidogyne* Göeldi, 1892 (Nematoda: Meloidogynidae) associated with olive plants and environmental cues regarding their distribution in southern Spain. *PLoS One*, *13*(6), e0198236. Advance online publication. doi:10.1371/journal.pone.0198236 PMID:29924813

Avendaño, F., Pierce, F. J., & Melakeberhan, H. (2004). The relationship between soybean cyst nematode seasonal population dynamics and soil texture. *Nematology*, *6*(4), 511–525. doi:10.1163/1568541042665214

Aydınlı, G., Mennan, S., Devran, Z., Sirca, S., & Urek, G. (2013). First report of the root-knot nematode *Meloidogyne ethiopica* on tomato and cucumber in Turkey. *Plant Disease*, *97*(9), 1262. doi:10.1094/PDIS-01-13-0019-PDN PMID:30722413

Banyer, R. J., & Fisher, J. M. (1971). Diagnostic and advisory programes. In Plant parasitic nematodes. Academic Press.

Barker, K. R., Nusbaum, C. J., & Nelson, L. A. (1969). Season population dynamics of selected plant-parasitic nematodes as measured by three extraction procedures. F. *Nematology*, *1*, 232.

Barnosky, A. D., Matzke, N., Tomiya, S., Wogan, G. O., Swartz, B., Quental, T. B., Marshall, C., McGuire, J. L., Lindsey, E. L., Maguire, K. C., Mersey, B., & Ferrer, E. A. (2011). Has the Earth's sixth mass extinction already arrived? *Nature*, *471*(7336), 51–57. doi:10.1038/nature09678 PMID:21368823

Benjlil, H., Elkassemi, K., Aït Hamza, M., Mateille, T., Furze, J. N., Cherifi, K., Mayad, E. H., & Ferji, Z. (2020). Plant-parasitic nematodes parasitizing saffron in Morocco: Structuring drivers and biological risk identification. *Applied Soil Ecology*, *147*, 103362. doi:10.1016/j.apsoil.2019.103362

Bliss, C.I. (1958). Periodic regression in biology and climatology. *Connecticut Agri. Exp. Stn. Bull., 615.*

Bozbuga, R., Lilley, C. J., Knox, J. P., & Urwin, P. E. (2018). Host-specific signatures of the cell wall changes induced by the plant parasitic nematode. *Meloidogyne incognita. Scientific Reports*, *8*(1), 17302. doi:10.103841598-018-35529-7 PMID:30470775

Brown, E. B. (1958). Pea root eelworm in the eastern countries of England. *Nematologica*, *3*(4), 257–268. doi:10.1163/187529258X00012

Carneiro, R. M. D. G., de Oliveira Lima, F. S., & Correia, V. R. (2017). *Methods and tools currently used for the identification of plant parasitic nematodes*. Nematology – Concepts, Diagnosis and Control. doi:10.5772/intechopen.69403

Chapman, R. A. (1959). Development of *Pratylenchus penetrans* and *Tylenchorhynchus martini* on red clover Alfa Alfa. *Phytopathology*, *49*, 357.

Chirchir, A. K., Kimenju, J. W., Olubayo, F., & Mutua, G. K. (2008). Abundance and Distribution of Plant Parasitic Nematodes Associated With Sugarcane in Western Kenya. *Asian Journal of Plant Pathology*, *2*(1), 48–53. doi:10.3923/ajppaj.2008.48.53

Clark, C., & Wright, V. (1983). Effect and reproduction of *Rotylenchulus reniformis* on sweet potato selections. *Journal of Nematology*, *15*, 198–203. PMID:19295791

Cohn, E. (1969). The occurrence and distribution of species of *Xiphinema* and *Longidorus* in Israel. *Nematologica*, *15*(2), 179–192. doi:10.1163/187529269X00614

Coyne, D. L., Talwana, H. A. L., & Maslen, N. R. (2003). Plant-parasitic nematodes associated with root and tuber crops in Uganda. *African Plant Protection*, *9*, 87–98.

Ducharme, E. P. (1967). Annual population periodicity of *Radopholus simils* in Florida citrus groves. *Dis. Reporter.*, *51*, 1031.

Dutta, T. K., Khan, M. R., & Phani, V. (2019). Plant-parasitic nematode management via biofumigation using brassica and non-brassica plants: Current status and future prospects. *Current Plant Biology*, *17*, 17–32. doi:10.1016/j.cpb.2019.02.001

Fan-xiang, M., Wen-ju, L., Wei, O., Yong, J., Qi, L., & Da-zhong, W. (2005). Vertical distribution of plant nematodes in an aquic brown soil under different land uses. *Journal of Forestry Research*, *16*(1), 39–42. doi:10.1007/BF02856852

Faulkner, L. R., & Bolander, W. J. (1966). Occurrence of large nematode populations in irrigation canals of South Central Washington. *Nematologica*, *12*(4), 591–600. doi:10.1163/187529266X00446

Ferris, V. R., & Bernard, R. L. (1971). Crop rotation effects on population densities of end parasitic nematodes. F. *Nematology*, *3*, 119. PMID:19322357

Flegg, J. J. M. (1968). Life cycle studies of some *Xiphinema* and *Lohgidorus* species in Southeastern England. *Nematologica*, *14*(2), 197–210. doi:10.1163/187529268X00426

Francl, L. J. (1993). Multivariate analysis of selected edaphic factors and their relationship to *Heterodera glycines* population density. *Journal of Nematology*, *25*, 270–276. PMID:19279768

Freckman, D. W., & Chapman, R. A. (1972). Infection of Red clover seedlings by *Heterodera trifolii* Goffart and *Pratylenchus penetrans* (Cobb). *Journal of Nematology*, *4*(1), 23–28. PMID:19319241

Ghini, R., & Morandi, M. A. B. (2006). Biotic and abiotic factors associated with soil suppressiveness to Rhizoctonia solani. *Scientia Agrícola*, *63*(2), 153–160. doi:10.1590/S0103-90162006000200007

Giannakou, I. O., & Panopoulou, S. (2019). The use of fluensulfone for the control of rootknot nematodes in greenhouse cultivated crops: Efficacy and phytotoxicity effects. *Cogent Food & Agriculture*, *5*(1), 1643819. doi:10.1080/23311932.2019.1643819

Griffin, G. D., & Darling, H. M. (1964). An ecological study of *Xiphinema americanum* Cobb in an ornamental spruce nursery. *Nematologica*, *10*(3), 471–479. doi:10.1163/187529264X00493

Gullino, M. L., Albajes, R., & Nicot, P. C. (2019). *Integrated pest and disease management in grreenhouse crops*. Springer Int Publishing., doi:10.1007/978-3-030-22304-5

Hannon, C. I. (1962). The occurrence and Distribution of the citrus-root nematode, *Tylenchulus semipentrans* Cobb., in Florida. *Pl. Dis. Reptr.*, *46*, 451–455.

Haougui, A., Doumma, A., Toufique, B. M., & Kollo, I. A. (2011). Survey of plant parasitic nematodes associated with sweet potato in Niger. *Asian Journal of Agricultural Sciences.*, *3*, 32–36.

Hu, C., & Qi, Y. (2013). Effective microorganisms and compost favor nematodes in wheat crops. *Agronomy for Sustainable Development*, *33*(3), 573–579. doi:10.100713593-012-0130-9

Huston, M. A. (1997). Hidden treatments in ecological experiments: Re-evaluating the ecosystem function of biodiversity. *Oecologia*, *110*(4), 449–460. doi:10.1007004420050180 PMID:28307235

Ingham, R., Hamm, P., Williams, R., & Swanson, W. (2000). Control of *Meloidogyne chitwoodi* in potato with fumigant and nonfumigant nematicides. *Journal of Nematology*, *32*, 556–565. PMID:19271010

Jatala, P., Jansson, R. K., & Raman, K. V. (2019). Sweet potato pest management. New York: CRC Press.

Kampfe, L. (1960). Die raumliche vertelung des primarbefaals von *Heterodera shachtii* Schmidt in den Wirtswurzeln. *Nematologica, 5*(1), 18–26. doi:10.1163/187529260X00208

Karuri, H. W., Olago, D., Neilson, R., Njeri, E., Opere, A., & Ndegwa, P. (2017). Plant Parasitic Nematode Assemblages Associated with Sweet Potato in Kenya and Their Relationship with Environmental Variables. *Tropical Plant Pathology, 42*(1), 1–12. doi:10.100740858-016-0114-4

Kim, E., Seo, Y., Kim, Y. S., Park, Y., & Kim, Y. H. (2017). Effects of soil textures on infectivity of root-knot nematodes on carrot. *The Plant Pathology Journal, 33*(1), 66–74. doi:10.5423/PPJ.OA.07.2016.0155 PMID:28167889

Kim, Y.-H., & Yang, J.-W. (2019). Recent research on enhanced resistance to parasitic nematodes in sweet potato. *Plant Biotechnology Reports, 13*(6), 559–566. doi:10.100711816-019-00557-w

Kinloch, R. A., & Allen, M. W. (1972). Interaction of *Meloidogyne hapla* and *M. javanica* infecting tomato.F. *Nematology, 4*, 7. PMID:19319239

Koenning, S. R., Overstreet, C., Noling, J. W., Donald, P. A., Becker, J. O., & Fortnum, B. A. (1999). Survey of crop losses in response to phytoparasitic nematodes in the United States for 1994. *Journal of Nematology, 31*, 587–618. PMID:19270925

Korthals, G. W., Alexiev, A. D., Lexmond, T. M., Kammenga, J. E., & Bongers, T. (1996). Long-term effects of copper and ph on the nematode community in an agroecosystem. *Environmental Toxicology and Chemistry, 15*(6), 979–985. doi:10.1002/etc.5620150621

Kumar, V., Khan, M. R., & Walia, R. K. (2020). Crop loss estimations due to plant-parasitic nematodes in major crops in India. *National Academy Science Letters, 43*(5), 409–412. doi:10.100740009-020-00895-2

Lawrence, G. W., Clark, C. A., & Wright, V. L. (1986). Influence of Meloidogyne incognita on resistant and susceptible sweet potato cultivars. *Journal of Nematology, 18*, 59–65. PMID:19294141

Li, X., Zhu, H., Geisen, S., Bellard, C., Hu, F., Li, H., Chen, X., & Liu, M. (2020). Agriculture erases climate constraints on soil nematode communities across large spatial scales. *Global Change Biology, 26*(2), 919–930. doi:10.1111/gcb.14821 PMID:31479174

Liu, T., Hu, F., & Li, H. (2019). Spatial ecology of soil nematodes: Perspectives from global to micro scales. *Soil Biology & Biochemistry*, *137*, 107565. doi:10.1016/j.soilbio.2019.107565

Lopez, E. A., Gapasin, R. M., & Palomar, M. K. (1981). Effect of different levels of *Helicotylenchus* nematode infestation on the growth and yield of sweet potato. *Annals of Tropical Research*, *3*, 275–280.

Maegher, J. W. (1970). Seasonal fluctuations in numbers of larvae of the cereal cyst nematode (*Heterodera avenae*) and of *Pratylenchus minyus* and *Tylechorhynchus brevidens* in soil. *Nematologica*, *16*(3), 333–347. doi:10.1163/187529270X00018

Mckeen, C. D., & Mountain, W. B. (1960). Synergism between *Pratylenchus pentrans* (Cobb) and *Verticillium albo-atrum* R and B in eggplant wilt. Canad.f. *Bot.*, *38*, 789.

McSorley, R., & Dickson, D. W. (1990). Vertical distribution of plant-parasitic nematodes in sandy soil under soybean. *Journal of Nematology*, *22*, 90–96. PMID:19287693

Miller, H. N., & Di Edwardo, A. A. (1962). Leaf galls on *Siderasis fuscata* caused by the root knot nematode, *Meloiodgyne incognita incognita. Phytopathology*, *52*, 1070.

Mokrini, F., Laasli, S. E., Iraqui, D., Wifaya, A., Mimouni, A., Erginbas-Orakci, G., Imren, M., & Dababat, A. A. (2019). Distribution and occurrence of plant-parasitic nematodes associated with raspberry (*Rubus Idaeus*) in Souss-Massa region of Morocco: Relationship with soil physico-chemical factors. *Russian Journal of Nematology*, *27*, 107–121. doi:10.24411/0869-6918-2019-10011

Mountain, W. B., & Mckeen, C. D. (1962). Effect of *Verticillium dahliae* on the population of *Pratylenchus penetrans*. *Nematologica*, *7*(4), 261–266. doi:10.1163/187529262X00549

National root and tuber crops development strategy 2019-2022. (2019). *Ministry of Agriculture, Livestock and Fisheries*. https://kilimo.go.ke/wp-content/uploads/2021/04/Roots-and-Tuber-Crops-Strategy-2019-2022.pdf

Nguyen, S. V., Nguyen, P. K. T., Araki, M., Perry, R. N., Tran, L. B., Chau, K. M., Min, Y. Y., & Toyota, K. (2020). Seasonal variation and vertical distribution of nematode communities and their relation to soil chemical property and rice productivity under triple rice cultivation in the Mekong Delta, Vietnam. *Nematology*, *23*, 33–46. doi:10.1163/15685411-bja10028

Nguyen, S. V., Nguyen, P. K. T., Araki, M., Perry, R. N., Tran, L. B., Chau, K. M., Min, Y. Y., & Toyota, K. (2020). Seasonal variation and vertical distribution of nematode communities and their relation to soil chemical property and rice productivity under triple rice cultivation in the Mekong Delta, Vietnam. *Nematology*, *23*, 33–46. doi:10.1163/15685411-bja10028

Nico, A. I., Jimenez-Diaz, R. M., & Castillo, P. (2003). Solarization of soil in piles for the control of Meloidogyne incognita in olive nurseries in southern Spain. *Plant Pathology*, *52*(6), 770–778. doi:10.1111/j.1365-3059.2003.00927.x

Nisa, R. U., Tantray, A. Y., Kouser, N., Allie, K. A., Wani, S. M., Alamri, S. A., Alyemeni, M. N., Wijaya, L., & Shah, A. A. (2021). Influence of ecological and edaphic factors on biodiversity of soil nematodes. *Saudi Journal of Biological Sciences*, *28*(5), 3049–3059. doi:10.1016/j.sjbs.2021.02.046 PMID:34025181

Njuguna, L., & Bridge, J. (1998). Plant parasitic nematodes of Irish potatoes (Solanum tuberosum) in Central Province and sweet potatoes (Ipomoea batatas) in Central, Nyanza and Coast Provinces of Kenya. *International Journal of Nematology*, *8*, 21–26.

Okechalu, O. B., & Wonang, D. L. (2015). The response of eleven sweet potato (*Ipomoea batatas* (L.) Lam) cultivars to infection by *Meloidogyne* spp. in Jos, Nigeria. *IOSR Journal of Pharmacy and Biological Sciences*, *10*, 42–49.

Olthof, T. H. A. (1971). Seasonal fluctuations in population densities of *Pratylechus pentrans* under a rye-tobacco rotation in Ontario. *Nematologica*, *17*(3), 453–459. doi:10.1163/187529271X00693

Overstreet, C., Loebenstein, G., & Thottappilly, G. (2009). The sweet potato. Dordrecht: Springer.

Palomares-Rius, J. E., Escobar, C., Cabrera, J., Vovlas, A., & Castillo, P. (2017). Anatomical alterations in plant tissues induced by plant-parasitic nematodes. *Frontiers in Plant Science*, *8*, 1987. doi:10.3389/fpls.2017.01987 PMID:29201038

Pérez, E., Weingartne, D., & McSorley, R. (2000). Niche distribution of *Paratrichodorus minor* and *Belonolaimus longicaudatus* following fumigation on potato and cabbage. *Journal of Nematology*, *32*, 343–348. PMID:19270987

Pérez,, E, Weingartner, D, & McSorley,, R. (2000). Distribution of *Paratrichodorus minor* and *Belonolaimus longicaudatus* following fumigation on potato and cabbage. *Journal of Nematology*, *32*, 343–348. PMID:19270987

Pudasaini, M. P., Schomaker, C. H., Been, T. H., & Moens, M. (2006). Vertical distribution of the plant-parasitic nematode, *Pratylenchus penetrans*, under four field crops. *Phytopathology*, *96*(3), 226–233. doi:10.1094/PHYTO-96-0226 PMID:18944436

Rebois, V. R., & Cairns, E. J. (1968). Nematodes associated with soybeans in Alabama, Florida and Georgia. *Pl.Dis.Rptr.*, *52*, 40.

Rishbeth, J. (1963). Stump protection against *Fomes annosus*.3.Inoculation with *Peniophora gigantean*. *Annals of Applied Biology*, *52*(1), 63–77. doi:10.1111/j.1744-7348.1963.tb03728.x

Robinson, A. F., Akridge, R., Bradford, J. M., Cook, C. G., Gazaway, W. S., Kirkpatrick, T. L., Lawrence, G. W., Lee, G., McGawley, E. C., Overstreet, C., & Padgett, B. (2005). Vertical distribution of *Rotylenchulus reniformis* in cotton fields. *Journal of Nematology*, *37*, 265–271. PMID:19262871

Rodriguez-Kabana, R., & Robertson, D. (1987). Vertical distribution of *Meloidogyne arenaria* juvenile populations in a peanut field. *Nematropica*, *17*, 199–208.

Romanenko, N. D., Zaets, V. G., Kozereva, N. I., Popov, I. O., & Tabolin, S. B. (2008). Biological agents for plant protection against plant parasitic nematodes and other pathogens, and perspectives of their use in the 21st century. *RUDN J Agronomy Animal Industries.*, *2*, 39–51.

Ross, J. P. (1960). *Heterodera trifolii* a foliage pathogen of white clover. *Phytopathogy*, *50*, 866.

Salahi Ardakani, A., Tanha Mafi, Z., Mokaram Hesar, A., & Mohammadi Goltappeh, E. (2014). Relationship between soil properties and abundance of *Tylenchulus semipenetrans* in citrus orchards, Kohgilouyeh va Boyerahmad Province. *J. Agric. Sci. Technol.*, *16*, 1699–1710.

Sharma, S., Waliyar, F., Subrahmanyam, P., & Ndunguru, B. (1992). Role of *Scutellonema clathricaudatum* in etiology of groundnut growth variability in Niger. *Plant and Soil*, *143*(1), 133–139. doi:10.1007/BF00009138

Siddiqui, M. A. (2007). Seasonal fluctuation in nematode population associated with mango, *Mangifera indica* L. *Archiv für Phytopathologie und Pflanzenschutz*, *40*(6), 389–394. doi:10.1080/03235400600627601

Sikandar, A., Zhang, M., Wang, Y., Zhu, X., Liu, X., Fan, H., Xuan, Y., Chen, L., & Duan, Y. (2020). In vitro evaluation of *Penicillium chrysogenum* Snef1216 against *Meloidogyne incognita* (root-knot nematode). *Scientific Reports, 10*(1), 8342. doi:10.103841598-020-65262-z PMID:32433576

Sorribas, F. J., Verdejo-Lucas, S., Pastor, J., Ornat, C., Pons, J., & Valero, J. (2008). Population densities of *Tylenchulus semipenetrans* related to physicochemical properties of soil and yield of clementine mandarin in Spain. *Plant Disease, 92*(3), 445–450. doi:10.1094/PDIS-92-3-0445 PMID:30769676

Southwood, T. R. E. (1968). Ecological methods. Methuen.

Stirling, G. R. (2021). Surrounding the swollen roots of sweet potato with a decomposing band of an organic amendment enhances nematode-suppressive services and reduces damage caused by root-knot nematode. *Australasian Plant Pathology, 50*(2), 151–168. doi:10.100713313-020-00751-3

Stirling, G. R., Stirling, A. M., & Prichard, M. (2020). Sustainable sweet potato farming systems to improve soil health and reduce losses caused by root-knot nematode. *Australasian Plant Pathology, 49*(6), 591–604. doi:10.100713313-020-00731-7

Thomas, R. J., & Clark, C. A. (1983). Effects of concomitant development on reproduction of *Meloidogyne incognita* and *Rotylenchulus reniformis* on sweet potato. *Journal of Nematology, 15*, 215–221. PMID:19295794

Topalovic, O., Hussain, M., & Heuer, H. (2020). Plants and associated soil microbiota cooperatively suppress plant-parasitic nematodes. *Frontiers in Microbiology, 11*, 313. doi:10.3389/fmicb.2020.00313 PMID:32184773

Treonis, A. M., Unangst, S. K., Kepler, R. M., Buyer, J. S., Cavigelli, M. A., Mirsky, S. B., & Maul, J. E. (2018). Characterization of soil nematode communities in three cropping systems through morphological and DNA metabarcoding approaches. *Scientific Reports, 8*(1), 2004. doi:10.103841598-018-20366-5 PMID:29386563

Van Gundy, S. D., & Martin, G. P. (1962). Soil texture, PH and moisture effect on the development of citrus nematode (*Tylenchulus Semipenetrans*). *Phytopathology, 52*, 31.

Waisen, P., Cheng, Z., Sipes, B. S., DeFrank, J., Marahatta, S. P., & Wang, K. (2020). Effects of biofumigant crop termination methods on suppression of plant-parasitic nematodes. *Applied Soil Ecology, 154*, 103595. doi:10.1016/j.apsoil.2020.103595

Wallace, H. R., & Dongcaster, C. C. (1964). A comparative study of the movement of some microphagus, plant parasitic and animal parasitic nematodes. *Parasitology*, *54*(2), 313–326. doi:10.1017/S0031182000067949 PMID:14170216

Wang, C., Bruening, G., & Williamson, V. M. (2009). Determination of preferred ph for root-knot nematode aggregation using pluronic F-127 gel. *Journal of Chemical Ecology*, *35*(10), 1242–1251. doi:10.100710886-009-9703-8 PMID:19838866

Wesemael, W. M. L., & Moens, M. (2008). Vertical distribution of the plant-parasitic nematode, *Meloidogyne chitwoodi*, under field crops. *European Journal of Plant Pathology*, *120*(3), 249–257. doi:10.100710658-007-9213-x

Winfield, A. L., Coppock, L. J., & Whitlock, A. J. (1970). A survey of brassica cyst eelworm in Bedfordshire. *Plant Pathology*, *19*(2), 91–92. doi:10.1111/j.1365-3059.1970.tb00988.x

Wong, C. L., & Willetts, H. J. (1969). Gall formation in aerial parts of plants inoculated with *Meloidogyne javanica*. *Nematologica*, *15*(3), 425–428. doi:10.1163/187529269X00506

Woolfe, J. A. (1992). *Sweet potato: An untapped food resource*. Cambridge University Press.

Zheng, G. D., Shi, L. B., Wu, H. Y., & Peng, D. L. (2012). Nematode communities in continuous tomato-cropping field soil infested by root-knot nematodes. *Acta Agriculturæ Scandinavica. Section B, Soil and Plant Science*, *62*(3), 216–223. doi:10.1080/09064710.2011.598545

Zhong, S., Zeng, H., & Jin, Z. (2015). Responses of soil nematode abundance and diversity to long-term crop rotations in tropical China. *Pedosphere*, *25*(6), 844–852. doi:10.1016/S1002-0160(15)30065-5

Chapter 3
Plant Pathogenic Nematodes

Wasan Salih Hussain
Mosul University, Iraq

Rayyan Salim Mahmood
Mosul University, Iraq

Hamad Mohammed Hamad
Mosul University, Iraq

ABSTRACT

Plant pathogenic nematode science is a relatively recent science that is related to many applied and research sciences. It is closely related to zoology and botany, which are the origin of all biological sciences. The study of plant-pathogenic nematodes requires a great understanding and knowledge of general zoology, invertebrate zoology, parasitology, and animal behavior science, which helps understand the nutritional behavior of nematodes on their plant hosts and thus the effects of different control methods on nematodes, as well as being familiar with botany and its various branches, especially the functions of organs, appearance, and internal anatomy, especially of the root system, and these are requirements for understanding the different and mutual influences between nematodes and plants.

INTRODUCTION

Nematodes are microscopic, cylindrical that live in the soil. They parasitize plants with a spear that they use to prick and penetrate plant tissues and cause tumors in the roots and ulcers, and some of them cause diseases in the foliage of some plants. Most root plant-parasitic nematodes spend most of their lives either in the soil (Hunt, et al., 2018), in the roots, or in other plant parts in the soil, such as tubers, rhizomes,

DOI: 10.4018/978-1-6684-8083-0.ch003

bulbs, and stems. Others are parasitic on plant parts above the soil surface, such as stems, leaves, buds, flowers, and seeds. infection site depends on the type of nematode and the plant host, and even on the size of the nematode (Bernard et al., 2017).

Identified more than 4000 species of plant nematodes (belonging to about 200 genera in more than 30 families) that attack different parts of their plant hosts. studies that dealt with the role of nematodes in various plant diseases have not exposed participants to more than 150 species. Although no cultivated plant is free from infection with one or more types of nematodes, the number of plant diseases caused by or in which nematodes participate does not exceed more than 100 diseases known (Keçici et al. 2022 & Mesa-Vale et al., 2020).

Some may wonder how nematodes can cause plant disease? The answer is to imagine huge numbers of these nematodes surrounding the root of the plant from every side, constantly absorbing nutrients from the root by means of a spear. There is no doubt that the plant weakens and withers while the worms grow. the matter would have been easier if the damage was limited to what nematode absorbe consume from the plant sap, but most of the damage is caused by what nematode secrete from their saliva, which they inject into the plant cells through the aforementioned spear as continue to feed, the function of saliva is that it works to dilute the contents of the plant cells so that they are easy to eat and to represent with its multiple enzymes that break down cells and lead to the death of plant tissues that appear sores on the root, these enzymes sometimes cause abnormal cell enlargement, such as giant cells caused by root-knot nematode, or salivary enzymes suppress cell division. apical meristem stops growing, roots stop growing; on the contrary, some salivary enzymes may encourage the process of cell division, leading to the formation of root-knots of different sizes and numbers or cells deformation roots or large numbers of lateral roots near infection sites, and other symptoms (Mitiku, 2018).

Moreover, nematodes, by weakening plants and opening gaps in the roots, create a suitable environment for the entry of fungal, bacterial, and viral plant diseases, certain types of nematodes carry disease viruses in their digestive systems and transfer them to healthy plants through their saliva, which they inject into plant cells (Cunha et al., 2018).

HISTORY OF NEMATODES AS PLANT PATHOGENS

Plant parasitic nematodes were discovered a century after the invention of the microscope in the middle of the seventeenth century. because small size of these nematodes, as they are difficult to see with eye compared to other human and animal nematodes. However, their discovery came as a result of their increasing importance as agricultural pests after the development of agriculture and increase pest problems

(Askary & Abd-Elgawad, 2017). Some examples of the discovery of some genera of nematodes (Webster, 2008).

The first global record of plant parasitic nematodes was made in 1743 AD by the English scientist J.T. Needham, who examined infected wheat grains and saw nematode moving in the water called them nematode. These nematode are the second stage juvenile of wheat-warting nematode Anguina tritici.

- 1855 AD, discovered the root-knot nematodes Meloidogyne spp. He noticed the presence of galls on the roots of cucumber plants grown in greenhouses.
- 1857 AD, discovered the stem and bulb nematodes *Ditylenchus dipsaci.*

1859 AD, It was found that the reason for the deterioration of the sugar beet crop in Germany and Europe was due to the presence of nematodes.

1871 AD, Nematodes infesting the roots of sugar beets have been studied and named *Heterodeia Schlachtii* in honor of the scientist Schacht discovered it. as a result of the increasing interest in this pest and the great damage it causes to agricultural crops. This led to development of a new science, plant nematology, which occupied a prominent place among the various agricultural sciences. And the development of this modern science began in the late nineteenth and early twentieth centuries, when the study and development of nematodes were defined and their different types were classified in USA (Mesa-Valle et al, 2020).

The American scientist N. A. Cobb is considered the founder of the modern school of nematology and a pioneer of this science,

as follows:

1. He published his first research in 1913 AD, and he followed with many basic researches (about 125), especially in morphology of nematodes and classification, as well as practical methods for studying them, many of which are still in use at the present time.
2. "nematology" was proposed in 1914 as a distinctive name for this modern branch of agricultural sciences.
3. department establishment for agricultural nematodes at the US Department of Agriculture in 1929 A.D. as a result of its research and educational efforts And in Europe Some of the first books that had a major role in establishing the foundations of this science and its development were written at that time (Barker, 2004).

After that, scientific societies were founded, specialized journals were issued, and centers for research and postgraduate studies were established in universities interested in this science, sponsoring its development and benefiting from its various applications. (Handoo et al., 2013).

NEMATODES ARE IMPORTANT PESTS ON CROPS.

Plant parasitic nematodes represent one of the most important agricultural pests that share our food and greatly affect our agricultural economy. As the cultivation of economic crops on large areas occurs in many countries of the world, their infection with nematodes constitutes heavy losses in terms of the amount of production and the national income of these countries (Jairajpuri, 2019).

Many studies consider that nematodes play a major role in the losses of such crops as sugar beet crops in central Europe, potatoes in England, cotton in Egypt, citrus fruits in the Mediterranean basin, and many vegetable crops in developed and developing countries alike (Zhang et al., 2020). To two spins:

1. Many of the losses caused by nematodes to crops in many countries of the world, especially in developing countries, are still unknown or not specific. This may be due to a lack of complete knowledge of nematodes.

or is completely ignorant of it, or because the farmer is convinced of his crop compared to what farmers get in the neighboring fields.

2. Many of the losses attributed to infection with other pathogens may, in fact, be due to previous infection with nematodes, which leads to weakening the plant and preparing it for infection with other diseases, or it may be the main and effective participant in these losses, and this thus leads to difficulty in obtaining real changes (Bernard et al., 2017). As for the indirect losses, they are numerous and varied.

Most important of which is the increase in production costs. For example, infection with nematodes leads to the inefficiency of the roots in absorbing irrigation water and added fertilizers and making use of them. Symptoms of wilting and a lack of nutrients appear (Abdel Hamid et al., 1988). Thus, the farmer resorts wrongly to increasing irrigation and quantities of fertilizers, but to no avail. Perhaps one of the most important reasons that made nematodes occupy an important and prominent position among many One of the agricultural pests is its wide spread. For example, the root-knot nematode *Meloidogyne* spp. can infect more than 2000 plant species, in addition to the ability of some nematode species to transmit some fungal, bacterial, and viral diseases, and their cooperation and interaction with many pathogens to cause complex diseases (Lamovšek et al., 2013).

CHARACTERISTICS OF PLANT PATHOGENIC NEMATODES

Morphology

plant parasitic nematodes are very small, measuring about 300 to 1000 micrometers, and some of them may reach about 4 mm in length and 15 to 35 micrometers in thickness. The small diameter of the nematode makes it invisible to the naked eye, but it can be easily observed under a microscope (Hunt et al., 2018). Nematodes form Generally cylindrical in shape (with a circular cross-section), it has smooth, undivided bodies that do not have legs or any other appendages. Females in some species can become swollen in the process of maturation and take the form of pear-shaped or spherical bodies (Qing, et al., 2020).

Anatomy

The body of the nematode is almost transparent and covered with a transparent cuticle, which is usually marked with grooves, grooves, or any other signs. cuticle is formed by the hypodermis (Baran M.,2011), which contains living cells. and it extends inside the body cavity in the form of four tendons separating four belts of longitudinal muscles. These muscles enable the nematode to move. In addition, there are specialized muscles on the mouth, along the digestive system, and on the reproductive structures (Palomares-Rius, J. E.et al.,2017).

nematode's body cavity contains a liquid in which blood circulation and respiration occur. As for the digestive system, it is a hollow tube that extends from the mouth and passes through the esophagus, intestines, rectum, and then the anus. The nematode usually has six lips surrounding the mouth. All plant parasitic nematodes have a style or spear, which is used to pierce plant cells (Chitwood, D. J., 2003).

reproductive system is well formed, as the female nematode has one or two ovaries. Following these ovaries are the oviducts and the uterus, which end in the vulva. Also, a pair of intercourse spines are found in the male, which are protruding. Reproduction in nematodes occurs via eggs, and reproduction can be sexual, hermaphrodite, or parthenogenetic. There are many types of nematodes that do not have males (Bird, A. F., & Bird, J., 2012).

PRESENCE OF NEMATODES IN SOIL

Horizontal Distribution

The horizontal distribution of nematodes is characterized as an uneven, irregular distribution, as it is noted that the infection in the fields is in the form of scattered, irregular patches, especially at the beginning of the infection (Brodie, 1976). Therefore, this type of irregular distribution must be taken into consideration when selecting the optimal sampling method in the field. Many factors affect the shape of the horizontal distribution of nematodes. The density of nematodes in the soil at a particular time can change according to the cultivated plant variety (Sharita T. & Sharma, 2020). For example, when cultivating a highly susceptible variety, it can result in a significant increase in nematode density in spots with a relatively low level of initial density (Pi). While in spots with high initial densities, the roots are damaged a lot, which leads to a decrease in the density of nematodes (Norton et al., 2020). In this region, as a result of the lack of adequate food for nematodes, the horizontal distribution is also affected by many interfering factors such as land topography, soil texture, soil use, agricultural operations, as well as the genetic behavior of the nematode type such as type of eggs, method of laying, host range, pathogenicity, survivability, etc. (Nouh, 2022).

Vertical Distribution

The vertical distribution of nematodes in the soil sector is also irregular. It is affected by many environmental factors, but it is generally closely related to the distribution of roots and the surrounding area (Wesemael, 2008). However, there are exceptions to this rule, as it was found that many species prefer a certain section of the soil in which optimal conditions are available for their reproduction (Pudasaini et al., 2006). This optimum zone changes according to the environmental conditions and seasons of the year as the gradient of temperature and moisture levels changes in the soil sector, especially changes in soil temperature, which are usually most severe in the surface layer, and this fluctuation in temperature gradually decreases with the increase in the depth of the soil (Forge et al., 1998).

Most of the plant nematodes are concentrated mainly in the upper 30 cm of the soil, but there are some species that may be found at much greater depths, depending on the depth of the affected roots. For example, the *M. incognita* nematode lives at depths ranging from 150 to 400 cm in some areas of North Carolina. The root-splitting nematode, *T. Christieii*, is found at greater depths than most other nematodes (Dickson & Hewlett, 1988).

NEMATODE DISSEMINATION

Plant nematodes move by themselves (without external interference) over short distances that do not exceed a few centimeters per year, and therefore they depend on different means of transport for their spread. Humans, through their various agricultural activities, are primarily responsible for the spread of nematodes and their arrival in areas that were previously free of them. Since the active stages of nematodes are subject to desiccation, the stages that are resistant to drought or dormant are the most important in the transmission and spread of nematodes over long distances. As mentioned earlier, the degree of nematode settlement and spread in a new area depends on the arrival of a sufficient and active number of nematodes capable of infecting, as well as on the presence of susceptible hosts, in addition to the availability of environmental conditions suitable for nematode reproduction and survival in the area (Lehman, 1994).

Means of spreading plant nematodes

1- Soil and plant tissues

Plant tissues protect nematodes from desiccation during their transport by humans over short and long distances. The soil here is also important, as it is usually transported with plant materials such as seedlings and cuttings, which are contaminated with nematodes from one area to another. Sometimes nematodes are transported with seeds, as they may be found on the surface of seeds, such as sugar beet crop nematodes, or inside infected seeds, such as wheat wart nematodes and stem nematodes. And bulbs. This is in addition to the transmission of nematodes through reproductive vegetative parts such as seedlings, ornamental plants, fruit and bulb rootstocks, tubers, and corms. Sometimes the infected vegetative parts are served as fodder for animals and then spread as the dung spreads (Cunha et al., 2018).

2- Agricultural machinery, tools, and organic fertilizers

Soil contaminated with nematodes or infected plant parts can be transported by agricultural tractors, machinery, and used tools to nematode-free areas outside or inside the farm. It was found that the spread of potato crop nematodes in several farms on Long Island in the New York region was due to the transmission of some contaminated agricultural machinery and tools between these farms. As for the effect of organic fertilizers, it was found that the cause of the spread of the sore nematode *P.brachyurus* in several peanut farms was through the use of bean husks infected with these nematodes, as they were used as a soil conditioner with organic fertilizers (Briar et al., 2016) .

3- Animals

Mud granules contaminated with nematodes or infected plant debris stick to the bodies of birds and other animals and are then carried to where these animals arrive. Drought-tolerant nematodes can be transported over long distances by migratory birds (Wallace, 1973). Also, small animals that live in the soil, such as rodents, insects, and others-such as farm animals can transfer contaminated soil to other healthy areas (Lehman, 1994). It was found that the coconut palm nematode *Rhadinaphelenchus cocophilus*, which causes red ring disease, is spread by the coconut palm weevil, *Rhyncophorus palmarum* (Gerber & Giblin-Davis, 1990). as well as the pine saw beetles *Monochamus alternauts* (Zhao et al., 2007) and the pine wilt nematode *Bursaphelenchus lignicolus*, which causes wilting of pine trees in Japan and the United States of America (Li et al., 2022).

4- Water

Irrigation water plays an important role in transporting nematodes. The distances may be long, depending on the source of this water (Tankam-Chedjou et al., 2008). As for surface water, it is important to transfer it locally (it may be inside the farm or even in the same basin). The degree of nematode spread by water depends on the nematode's resistance to water immersion, and this resistance varies according to the type of nematode as well as its different stages (Lehman, 1994).

5- Wind

There is not enough information about this (Gaur, 1988), but it is believed that the wind can transport nematode crops, which are characterized by their high resistance to drought (Ptatscheck et al., 2018).

SYMPTOMS OF NEMATODE INFECTION

Symptoms are defined as the effects of the disease on the plant as a result of infection with a pathogen, and nematodes are considered one of the pathogens of the plant. The apparent symptoms include what can be seen visible on the root, vegetative or fruiting group of the plant. It is confirmed later that the cause of these pathological symptoms is the nematode when any stage of nematode growth is present in the infected part of the plant, or soil area surrounding the roots (Krusberg,1963).

The pathological symptoms caused by nematodes are divided into above-soil symptoms and sub-soil symptoms.

- Symptoms on plant parts above soil surface

Symptoms of infection are produced on the aerial parts of the plant, as a result of infection of the roots below surface of the soil or the soil as a result of infection of the aerial parts of the plant with nematodes (Endo, 1975).

1- Symptoms resulting from root infection:

When the plant is infected with nematodes parasitic on the roots, as a result, symptoms appear on the plant parts above the soil surface, which are represented in obstructing the plant's growth, and these are similar to those resulting from a lack of nutrients, lack of soil moisture, lack of fertilization, increased soil salinity, high groundwater level, etc. This is one of the unsuitable factors, and the infection usually appears in the form of scattered spots in the field, so infected plants show the following symptoms:

- A general weakness in the growth of the vegetative system, so the plants appear small and dwarfed.
- Yellowing of leaves and pale color.
- Plant wilting, especially in hot times during the day.
- Low production and poor quality.

2- Symptoms resulting from infection of the vegetative system:

These symptoms seem specialized and are due to the parasitism of specialized types of nematodes on the different parts of the shoot (Agrios, 2005).

Leaf spotting and necroses:

Symptoms appear in the form of small pustules, or they may be somewhat large, and the color of these spots eventually changes to a blackish-brown color and is defined by the leaf veins.

Leave-knot

Some types of nematodes, such as *Anguina millefolii*, cause nodules on the leaves of infected plants, especially the Grapes, where the size of these knots reaches a few millimeters.

Leaves curl:

The wheat grain wart nematode *Anguina tritici*, during its stages, causes curling and wrinkling of the leaves during its parasitism on wheat and barley plants, as these stages can be isolated during the milky growth stage of the plant (Gaur,2004).

Stems swelling and deformity:

The stem and bulb nematode *Ditylencus dipsaci* on beans and onions causes various deformities that appear in the form of swellings at the base of the stem, in addition to wrinkling the leaves and wrapping the stem. The inter knots are shortened and the plant dwarfs (Saadi et al., 2019).

Seed gall (warts):

Infection with the wheat grain wart nematode *Anguina tritici* causes seed gall (warts) to form in place of the normal grains in the ears (Ami et al., 1990).

- Roots symptoms

They are the symptoms that appear on the roots as a result of infection of the roots with different types of nematodes. These symptoms are not specialized, as more than one type of nematode can cause the same symptoms, or these symptoms can be caused by other root parasites. These symptoms include:

Root-Knot

Different shapes of knots are formed on the root system as a result of infection with different types of nematodes belonging to the genus *Meloidogyne*. The shape, size, and location of these knots on the roots differ according to the type of nematode (Endo, 1975). The reason for formation of nodules on the root occurs as a result of an increase in the speed of cell division and their numbers (Hyperplasia), as well as in the size of the cells (Hypertrophy) in the surrounding tissue layer (Pericycle), the inner epidermis (Endodermis) and the cortex (Cortex) (Eisenback et al., 2020). The knots arising from infection with the root-knot nematode are distinguished from the bacterial nitrogen-fixing knots in that the nitrogen-fixing knots are lateral and easy to separate and are usually small in size and take a light white or pink color Or lemongrass, while the knots caused by nematodes are caused by swollen root tissues and are rough in texture and cannot be isolated because they are part of the root (Alheyalee & Aljuboori, 2020).

Roots Cankers

Cankering is a localized death of the infected plant tissue and appears in the form of clearly defined dead spots in the root tissues. Ulcers usually form in the inner and outer cortex layers of infected roots. They can also form in different cortex layers. canker form as a result of infection with canker nematode *Pratylenchus spp*. and the burrowing nematode *Rhadopholus similis*, the infection is initially in the form of small spots that enlarge in size with the progression of the infection, and other soil organisms such as bacteria and fungi participate in the decomposition of the infected tissue, which leads to a complete discoloration of the root in a dark brown or black color, and may eventually lead to the death of the plant (Fosu-Nyarko & Jones, 2016).

Root Rot

These symptoms appear in some root plants and bulbs, where some types of nematodes, in association with fungi and other pathogens, destroy plant tissues, such as rotting potato tubers as a result of infection with the stem and bulb nematode *Ditylenchus destructor*. Also, canker caused by canker nematode *Pratylenchus similis* can turn into root rot (Endo, 1975).

Root Breakage

The types of nematodes *Longidorus* spp, *Trichodorus* spp, and *Paratrichodorus* spp cause obstruction in the growth of the roots or stop them completely, as a result of affecting the growing apex where the lateral roots stop growing and thus leads to dwarfing of the roots and the formation of new lateral roots that become infected again (Risède & Simoneau, 2004).

Excessive Root Branching

Crop nematodes usually irritate the growth of roots, especially in young plants, and thus strong secondary roots are formed that resemble the main root so that more than one main root is observed on the plant, and the roots take the reticulate shape, and mature females and crops are observed on the roots (Lilley et al., 2018).

In general, to verify and ensure that these symptoms are caused by nematodes, samples must be taken from the soil and plants, and an appropriate method must be found to extract the nematodes and then diagnose them.

NEMATODES DAMAGE TO PLANTS

Most parasites, including nematodes, cause various damages to the plant. However, in general, the damage caused by the Nematodes does not appear on the plant until several years after the onset of infection, during which the Nematodes has increased its damage and expanded its spread . Because of their microscopic size and slow onset of symptoms, nematodes do not attract farmers' attention, as they usually target environmental and agricultural pathogens and other rapidly evolving pathogens such as fungi, bacteria, and viruses (Seinhorst, 1965).

Nematodes cause direct damage, represented by the wounds they cause by spearing, penetration, and movement through the tissues, which in general have a limited effect compared to the chemical effects that occur by the secretion of enzymes or growth regulators and the consumption of nutrients, which together cause different structural and physiological effects (Ali et al., 2017).

Nematodes play different important roles in transmitting pathogens such as viruses, fungi, and bacteria. They provide the wounds required to penetrate the pathogens, some of which enter directly with them into the plant, and their harmful effects on the plant reduce the plant's resistance to other pathogens (Back et al. 2002).

As a result of structural and physiological Nematodes damage, when accumulated, it will appear in the form of symptoms on the above-ground parts of the plant, such as wilting, yellowing, stunting, regressive death of the limbs, necrosis, ulcers, poor production, etc. (Jones et al., 2013), which share many of them with the damage caused by environmental factors such as drought or other pathogens such as fungi and others. However, there are symptoms that may characterize nematodes, such as deformation of leaves and stems (*Ditylenchus* spp.) (Anonymous,2019), discoloration of leaves (*Aphelenchoides* spp.), nodules on leaves and seeds (*Anguina* spp.), and death and deformation of flowers and growing tops (Wheeler & Crow,2020).

On the underground parts, the Nematodes symptoms are characterized by root sores (*Pratylenchus* spp., *Radophalus* spp.), root nodules (*Meloidogyne spp., Xiphinema* spp., *Hemicycliophora* spp.), excessive root branching (*Paratrichodorus* spp.), and death of the tips of the roots. *Paratrichodorus* spp. and *Xiphinema* spp.) and adhesion of soil masses to the roots as a result of sticky substances produced by the Nematodes *Tylenculus* spp., sores on tubers (*Pratylenchus* spp.), knots (*Meloidogyne* spp.), and soft and dry rot on tubers (*Ditylenchus* spp.). As for field symptoms, they appear in the form of almost circular or irregular patches of affected plants, with dense growth of bushes that expands with time (Visagie et al., 2018).

Plants differ in their sensitivity and resistance to Nematodes infestations. There are sensitive plants that allow free feeding and reproduction of nematodes; tolerant plants that allow feeding and reproduction of nematodes without harming non-host plants; resistant plants that allow limited feeding without allowing nematodes to reproduce; and impervious plants that do not allow feeding or reproduction (Ha et al., 2017).

NEMATODES LIFE CYCLE

life cycle of nematodes is simple and similar among different species. The life cycle consists of six stages: the egg; the first, second, third, and fourth juvenile stages; and the stage of puberty. In most nematodes, first-stage juveniles are formed inside the egg and moulted inside it, where they hatch from second-stage juveniles. the second stage juveniles of the feed before they suffer moulting. In general, all stages of nematodes can feed (Nagachandrabose & Baidoo, 2018). However, some species show that the first and second stages do not feed, depending on the food provided by the egg. Finding the host plant, feeding, and achieving infection are conditions for the survival of the Nematodes in the stages capable of that; otherwise, it will die of starvation, and this may not exceed a few months. However, some stages can go through a period of dormancy, after which they return to activity and achieve infection. In most nematodes, adults are distinguished into males and females, but in some species, there are only females that produce eggs and sperm, and thus they are fertilized eggs. The formation of eggs means the completion of the life cycle (Jonathan, 2022). Nematodes species differ in the number of eggs laid by the female, but in general, it ranges between 50 and 500, and sometimes up to a few thousand (Pudasaini et al., 2006). The duration of the life cycle also varies according to the type of Nematodes, host plant, and environmental conditions, especially temperature. Appropriate temperatures exist, but the life cycle is longer at low temperatures (Bird et al., 2007).

CONTROL OF PLANT PARASITIC NEMATODES

Controlling Nematodes pests that affect various crops is a primary goal for every researcher or specialist in agricultural nematodes to increase agricultural production. Nematodes control requires field and laboratory studies to determine the types of nematodes and the severity of contamination, as well as knowledge of the crop history and agricultural treatments used in the field, as well as knowledge of the economic critical limit for each Nematodes, before giving any recommendation (Oka et al., 2000).

Methods for controlling Nematodes pests parasitic on plants can be divided into:

- Preventive Methods

These methods aim to prevent the arrival or spread of nematodes from one place to another, whether from one country to another or within the region, or even the farm (Cai,1999). These methods include:

1. Agricultural Quarantine

A. International Agricultural Quarantine

It is represented in a set of laws and regulations that are issued to prevent the entry or exit of pests from one country to another, or from one region of the region to another, through the presence of committees of specialists in identifying various pests in ports, airports and international borders to examine seeds, products and plant parts entering the country and grant them a certificate stating They are free from diseases or pests, and if it is proven that they are infected, the authorized party shall control the existing pests, and it may be necessary to destroy these sources of infection.

B. Local Agricultural Quarantine

This type exists between regions within the same country, or even at the level of farms in the same region, when necessary, to prevent the spread of the pest from an infected area to another free of this pest (Ciancio & Mukerji, 2009).

2. Sanitation

They include the necessary measures to prevent the presence of the pest or its transmission from one place to another (Brlansky et al., 2012), especially in nurseries or in the field, following the principle of prevention is better than cure (Chen et al., 2006).

- Cultural Methods

It aims to get rid of nematodes infestations by developing resistant varieties, following the crop rotation system, flooding the soil with water, bringing forward or delaying the planting date, or removing infected plants that may pose a threat to healthy plants (Reddy, 2021). The most important agricultural methods can be summarized as follows:

1. resistant varieties

This method is considered one of the most successful, efficient, and economical methods for controlling plant nematodes. It also saves the costs of chemical control, protects the environment from the dangers of pollution, and aims to develop varieties that are resistant to different types of nematodes. There are a good number of plant varieties that are resistant to certain types of nematodes, but the main problem facing breeders and users of resistant varieties frequently is the emergence of new races capable of breaking the resistance character (Reddy, 2021). This is due to the bad use of the resistant cultivar by planting it several times in a row in the same field as a monoculture. Therefore, it is advised not to plant the resistant cultivar several times in a row in the same field, but rather it must be included in an agricultural cycle that includes susceptible varieties and non-host plants.

2. crop rotation system

It is considered one of the cheapest, most successful, and most effective methods used in controlling plant nematodes. What is meant by the agricultural cycle here is "the cultivation of annual crops that are non-hosts (non-preferred) for nematodes or resistant to them in a specific cascade system in the same field with the main crop that is desired to be cultivated and susceptible to infection with nematodes (Phani et al.,2021). The agricultural rotation aims to deprive the nematode of food and thus reduce its density in the soil to a degree that does not affect its preferred crop when it is grown economically.

3. flooding the soil with water

Although this method is not widely used, early trial results indicate that immersion of 12-22 months completely kills root-knot nematodes in the wild (Somvanshi et al.,2018).

4. time of planting

It is intended to advance or delay the planting date to avoid early infection with nematodes and reduce their damage, and this depends on the type of crop and the Nematodes (Mehl & Kleczewski, 2017). This method is used successfully to control some types of nematodes on some crops

5. host care

Nematodes damage to some crops can be reduced to some extent, by providing agricultural conditions for plant growth such as regular irrigation, good fertilization, and control of other pests such as diseases, insects, and weeds that lead to an increase in the plant's tolerance to damage but are not beneficial for nematodes to multiply on it (El-Sappah et al., 2019).

6. organic matter adding to soil

It was found that the addition of materials (cattle, chicken, and pigeon manure, sewage manure, and green manure) to the soil works to reduce the density of Nematodes.

7. removal or destruction of infected plants

Leaving some annual crops after the end of the season leads to the survival of the root system for several weeks, and this period is sufficient for the development and reproduction of the nematodes present in the roots for an additional generation or two before entering the cold winter season. This is, in fact, a big mistake that many farmers make, as the farmer suffices at the end of the crop harvest season to stop irrigation only and leave the intentions to lemon with time (Kreiter,2018). Therefore, it is advisable to remove the plants as soon as all the crops are finished and turn the soil in preparation for the next season.

- Physical Methods

several methods to control or reduce the density of nematodes to some extent, but they are either used under special circumstances or cannot be relied upon alone to obtain economic control. Generally, these methods are used in combination with other, more efficient methods, to further reduce the density of nematodes in the soil or the tissues of the nuclei.

This group includes many methods, the most important of which are:

1. heat

Heat is one of the most successful ways It is natural and most widely used in the fight against nematodes when it is used either by water vapor or hot water.

2. soil solarization
3. fallowing the ground

It means leaving the land uncultivated and preventing the growth of any plant in it for various periods, by frequent plowing or using herbicides, which leads to the death of the largest types of nematodes present in the soil (Sikora & Roberts, 2018).

- Chemical Control

It refers to the use of chemical compounds known as nematicides to combat plant nematodes, whether present in the soil or within plant tissues (Phani et al., 2021). This control is considered one of the most successful and widely used methods at present, but its use cannot replace the use of resistant varieties and agricultural rotation, it is considered an alternative if resistant varieties are not available, or the agricultural cycle is not possible or economically feasible (Hajihassani, 2018).

They are resistant to nematodes by many chemicals due to the nature of the body structure and the presence of the cuticle envelope surrounding the body and eggs. Therefore, there is difficulty in finding an effective Nematodes that is non-toxic to plants (Mkandawire et al., 2022).

- Biological Control Methods

Biological control is defined as use a living organism, whether an animal or a plant, in combating another living organism. concept of Nematodes biological control is not limited to the use of natural enemies only, but also includes:

- resistant varieties.
- crop rotation system.
- Nematodes-trapping plants.
- anti-Nematodes plants.
- biocides.

In this note, we will focus on the traditional concept of biological control (Abd-Elgawad,2021) which is the use of natural enemies in the fight against plant nematodes, and among these organisms:

Fungi:

There are many fungi in the soil that attack nematodes and feed on them in different ways (Stirling, 2018). These fungi can be divided according to the nature of their intrusion into:

1. Parasitic fungi
2. Predacious fungi
3. Pathogenic fungi

Bacteria

They are types of bacteria that can kill nematodes by sticking to their body and forming a bacterial mycelium that kills nematodes, or by producing toxins that are toxic to nematodes, such as *Pasteuria penetrans*, which is considered a type of obligate parasitic bacteria (Ahmad et al., 2021), which are bacteria that stick to Nematodes bodies and then fix and penetrate the body of nematodes, forming colonies inside nematodes body and kills them. Some scientists have also been able to use some types of *Bacillus* and *Sterptomyces* in combating some types of nematodes, through toxins secreted by these bacteria and considered toxic to nematodes (Moosavi, 2020).

Predacious Nematodes

Some nematode genera *Mononchus*, *Diplogaster*, *Dorylaimus* attack some types of intrusive nematodes and prey on them. These predatory species are often equipped with a wide mouth cavity and a large tooth, such as the genus *Mononchus*, with which it cuts its prey (Devi & George, 2018)., while others are equipped with a spear and secrete neurotoxins that paralyze the prey and then feed on it, such as the genus *Seinura* (Bilgrami, 2008).

Predacious Arthropods

Some of the terrestrial soil mites include the *columbola* mite *Collembolo* (Onychiurus spp.) and many other soil arthropods prey on some plant-parasitic nematodes (Rueda-Ramírez et al., 2023).

REFERENCES

Abd-Elgawad, M. M. (2021). Optimizing safe approaches to manage plant-parasitic nematodes. *Plants*, *10*(9), 1911. doi:10.3390/plants10091911 PMID:34579442

Abdel Hamid, M. E., Ramadan, H. H., Salem, F. M., & Osman, G. Y. (1981). The susceptibility of different field crops to infestation by *Meloidogyne javanica* and *M. incognita acrita. Anzeiger für Schädlingskunde, Pflanzenschutz. Umweltschutz*, *54*, 81–82.

Agrios, G. N. (2005). *Plant pathology*. Elsevier.

Ahmad, G., Khan, A., Khan, A. A., Ali, A., & Mohhamad, H. I. (2021). Biological control: A novel strategy for the control of the plant parasitic nematodes. *Antonie van Leeuwenhoek*, *114*(7), 885–912. doi:10.100710482-021-01577-9 PMID:33893903

Alheyalee, R. S., & Aljuboori, F. K. (2020). The first recording of root-knot nematodes Meloidogyne Javanica on the winter cherry Solanum pseudocapsicum and controlling it in vitro. *EurAsian Journal of BioSciences, 14*(2).

Ali, M. A., Azeem, F., Li, H., & Bohlmann, H. (2017). Smart parasitic nematodes use multifaceted strategies to parasitize plants. *Frontiers in Plant Science*, *8*, 1699. doi:10.3389/fpls.2017.01699 PMID:29046680

Ami, S. N., Taher, I. E., Hussen, F. S., & Ahmed, A. I. (2019). First molecular identification of wheat seed gall nematode races parasitized on wheat in Iraq. *Acta Universitatis Sapientiae. Agriculture and Environment*, *11*(1), 5–15.

Anonymous. (2019). *Floriculture Crops 2018 Summary* (May 2019). National Agricultural Statistics Service.

Askary, T. H., & Abd-Elgawad, M. M. M. (2017). Beneficial nematodes in agroecosystems: a global perspective. In *Biocontrol agents: entomopathogenic and slug parasitic nematodes* (pp. 3–25). CABI. doi:10.1079/9781786390004.0003

Back, M. A., Haydock, P. P. J., & Jenkinson, P. (2002). Disease complexes involving plant parasitic nematodes and soilborne pathogens. *Plant Pathology*, *51*(6), 683–697. doi:10.1046/j.1365-3059.2002.00785.x

Baran, M. (2011). Nematodes reproductive system. *Parasitology*, *7*, 163.

Barker, K. R. (2004). A century of plant nematology. *Nematology Advances and Perspectives, 1*, 1-51.

Bernard, G. C., Egnin, M., & Bonsi, C. (2017). The impact of plant-parasitic nematodes on agriculture and methods of control. *Nematology-Concepts, Diagnosis and Control, 10*, 121-151.

Bilgrami, A. L. (2008). Biological control potentials of predatory nematodes. *Integrated management and biocontrol of vegetable and grain crops nematodes*, 3-28.

Bird, A. F., & Bird, J. (2012). *The structure of nematodes*. Academic Press.

Bird, G., Bishop, B., Grafius, E., Hausbeck, M., Lynnae, J., Williams, K., & Pett, W. (2004). Insect, disease and nematode control for commercial vegetables. *Michigan State University Extension Bulletin, 312*, 81–82.

Briar, S. S., Wichman, D., & Reddy, G. V. (2016). Plant-parasitic nematode problems in organic agriculture. *Organic farming for sustainable agriculture*, 107-122.

Brlansky, R. H., Chung, K. R., & Rogers, M. E. (2012). *2006 Florida citrus pest management guide: Huanglongbing (citrus greening)*. UF/IFAS Extension.

Brodie, B. B. (1976). Vertical distribution of three nematode species in relation to certain soil properties. *Journal of Nematology, 8*, 243–247. PMID:19308230

Cai, W. (1999). Extension theory and its application. *Chinese Science Bulletin, 44*(17), 1538–1548. doi:10.1007/BF02886090

Chen, Z. X., Chen, S. Y., & Dickson, D. W. (Eds.). (2004). Nematology: advances and perspectives. Academic Press.

Chitwood, D. J. (2003). Research on plant-parasitic nematode biology conducted by the United States Department of Agriculture–Agricultural Research Service. *Pest Management Science, 59*(6-7), 748–753. doi:10.1002/ps.684 PMID:12846325

Ciancio, A., & Mukerji, K. G. (Eds.). (2009). *Integrated management of fruit crops and forest nematodes* (Vol. 4). Springer Science & Business Media. doi:10.1007/978-1-4020-9858-1

Cunha, T. G. D., Visôtto, L. E., Lopes, E. A., Oliveira, C. M. G., & God, P. I. V. G. (2018). Diagnostic methods for identification of root-knot nematodes species from Brazil. *Ciência Rural, 48*(2), 48. doi:10.1590/0103-8478cr20170449

Devi, G., & George, J. (2018). Predatory nematodes as bio-control agent against plant-parasitic nematode-A review. *Agricultural Reviews (Karnal), 39*(1), 55–61. doi:10.18805/ag.R-1715

Dickson, D. W., & Hewlett, T. E. (1988). Horizontal and vertical migration of Meloidogyne arenaria in soil columns in the field. *Nematropica, 18*, 4–5.

Eisenback, J. D., & Triantaphyllou, H. H. (2020). Root-knot nematodes: Meloidogyne species and races. In *Manual of agricultural nematology* (pp. 191–274). CRC Press. doi:10.1201/9781003066576-6

El-Sappah, A. H., MM, I., H. El-Awady, H., Yan, S., Qi, S., Liu, J., & Liang, Y. (2019). Tomato natural resistance genes in controlling the root-knot nematode. *Genes*, *10*(11), 925. doi:10.3390/genes10110925 PMID:31739481

Endo, B. Y. (1975). Pathogenesis of nematode-infected plants. *Annual Review of Phytopathology*, *13*(1), 213–238. doi:10.1146/annurev.py.13.090175.001241

Forge, T. A., DeYoung, R., & Vrain, T. C. (1998). Temporal changes in the vertical distribution of Pratylenchus penetrans under raspberry. *Journal of Nematology*, *30*, 179–183. PMID:19274208

Fosu-Nyarko, J., & Jones, M. G. (2016). Advances in understanding the molecular mechanisms of root lesion nematode host interactions. *Annual Review of Phytopathology*, *54*(1), 253–278. doi:10.1146/annurev-phyto-080615-100257 PMID:27296144

Gaur, H. S. (1988). Dissemination and mode of survival of nematodes in dust storms. *Indian Journal of Nematology*, *18*(1), 94–98.

Gaur, H. S. (2004). Nematology: advances and perspectives, volume-i: nematode morphology, physiology and ecology. *Indian Journal of Nematology*, *34*(2), 240–241.

Gerber, K., & Giblin-Davis, R. M. (1990). Association of the red ring nematode and other nematode species with the palm weevil, Rhynchophorus palmarum. *Journal of Nematology*, *22*(2), 143. PMID:19287703

Ha, J., Won, J. C., Jung, Y. H., Yang, J. W., Lee, H. U., Nam, K. J., Park, S.-C., Jeong, J. C., Lee, S.-W., Lee, D. W., Chung, J.-S., Lee, J. J., & Kim, Y. H. (2017). Comparative proteomic analysis of the response of fibrous roots of nematode-resistant and-sensitive sweet potato cultivars to root-knot nematode Meloidogyne incognita. *Acta Physiologiae Plantarum*, *39*(12), 1–11. doi:10.100711738-017-2560-0

Hajihassani, A. (2018). *Chemical nematicides for control of plant-parasitic nematodes in Georgia vegetable crops*. University of Georgia.

Handoo, Z. A., Mowery, J. D., & Chitwood, D. J. (2013, July). The history of the USDA Nematode Collection and its database: valuable resources for taxonomic research and identification. In *Proceedings of the 10th International Nematology Symposium of the Russian Society of Nematologists* (pp. 110-3). Academic Press.

Hunt, D. J., Palomares-Rius, J. E., & Manzanilla-López, R. H. (2018). Identification, morphology and biology of plant parasitic nematodes. In *Plant parasitic nematodes in subtropical and tropical agriculture* (pp. 20–61). CAB International. doi:10.1079/9781786391247.0020

Jairajpuri, M. S. (2019). *Nematode pests of rice*. CRC Press. doi:10.1201/9780429034749

Jonathan, E. I. (2022). *Nematology Fundamentals & Applications (2nd Revised & Enlarged Edition)*. New India Publishing Agency. doi:10.59317/9789395319461

Jones, J. T., Haegeman, A., Danchin, E. G., Gaur, H. S., Helder, J., Jones, M. G., Kikuchi, T., Manzanilla-López, R., Palomares-Rius, J. E., Wesemael, W. M. L., & Perry, R. N. (2013). Top 10 plant-parasitic nematodes in molecular plant pathology. *Molecular Plant Pathology, 14*(9), 946–961. doi:10.1111/mpp.12057 PMID:23809086

Keçici, A. İ., Bozbuğa, R., Öcal, A., Yüksel, E., Özer, G., Yildiz, Ş., Lahlali, R., Slaats, B., Dababat, A. A., & İmren, M. (2022). Diversity and identification of plant-parasitic nematodes in wheat-growing ecosystems. *Microorganisms, 10*(8), 1534. doi:10.3390/microorganisms10081534 PMID:36013952

Kreiter, S. (2018). Pest management in organic grape production. In *Handbook of pest management in organic farming* (pp. 173–217). CAB International. doi:10.1079/9781780644998.0173

Krusberg, L. R. (1963). Host response to nematode infection. *Annual Review of Phytopathology, 1*(1), 219–240. doi:10.1146/annurev.py.01.090163.001251

Lamovšek, J., Urek, G., & Trdan, S. (2013). Biological control of root-knot nematodes (Meloidogyne spp.): Microbes against the pests. *Acta Agriculturae Slovenica, 101*(2), 263–275. doi:10.2478/acas-2013-0022

Lehman, P. S. (1994). Dissemination of phytoparasitic nematodes. *Nematology Circular, 208*.

Li, Z., Tao, J., & Zong, S. (2022). Cold tolerance in pinewood nematode Bursaphelenchus xylophilus promoted multiple invasion events in mid-temperate zone of China. *Forests, 13*(7), 1100. doi:10.3390/f13071100

Lilley, C. J., Maqbool, A., Wu, D., Yusup, H. B., Jones, L. M., Birch, P. R., Banfield, M. J., Urwin, P. E., & Eves-van den Akker, S. (2018). Effector gene birth in plant parasitic nematodes: Neofunctionalization of a housekeeping glutathione synthetase gene. *PLOS Genetics, 14*(4), e1007310. doi:10.1371/journal.pgen.1007310 PMID:29641602

Mehl, H., & Kleczewski, N. (2017). Disease and nematode management in field crops. *Pest Management Guide: Field Crops*, 456-016.

Mesa-Valle, C. M., Garrido-Cardenas, J. A., Cebrian-Carmona, J., Talavera, M., & Manzano-Agugliaro, F. (2020). Global research on plant nematodes. *Agronomy (Basel)*, *10*(8), 1148. doi:10.3390/agronomy10081148

Mesa-Valle, C. M., Garrido-Cardenas, J. A., Cebrian-Carmona, J., Talavera, M., & Manzano-Agugliaro, F. (2020). Global research on plant nematodes. *Agronomy (Basel)*, *10*(8), 1148. doi:10.3390/agronomy10081148

Mitiku, M. (2018). Plant-parasitic nematodes and their management: A review. *Agricultural Research & Technology: Open Access Journal*, *8*(2), 30–38. doi:10.19080/ARTOAJ.2018.16.55580

Mkandawire, T. T., Grencis, R. K., Berriman, M., & Duque-Correa, M. A. (2022). Hatching of parasitic nematode eggs: A crucial step determining infection. *Trends in Parasitology*, *38*(2), 174–187. doi:10.1016/j.pt.2021.08.008 PMID:34538735

Moosavi, M. R. (2020). Efficacy of microbial biocontrol agents in integration with other managing methods against phytoparasitic nematodes. *Management of Phytonematodes: Recent Advances and Future Challenges*, 229-258.

Nagachandrabose, S., & Baidoo, R. (2018). *A guide to introductory nematology*. Nova Science Publishers, Incorporated.

Norton, D. C., & Niblack, T. L. (2020). Biology and ecology of nematodes. In *Manual of agricultural nematology* (pp. 47–72). CRC Press. doi:10.1201/9781003066576-2

Nouh, G. M. (2022). Effect of temperature and soil moisture on the efficacy of indigenous and imported strains of the entomopathogenic nematode, Heterorhabditis sp. against the black cutworm, Agrotis ipsilon (Hufnagel) (Lepidoptera/Noctuidae). *Egyptian Journal of Biological Pest Control, 32*(1), 1-7.

Oka, Y., Koltai, H., Bar-Eyal, M., Mor, M., Sharon, E., Chet, I., & Spiegel, Y. (2000). New strategies for the control of plant-parasitic nematodes. *Pest Management Science: Formerly. Pest Management Science*, *56*(11), 983–988. doi:10.1002/1526-4998(200011)56:11<983::AID-PS233>3.0.CO;2-X

Palomares-Rius, J. E., Escobar, C., Cabrera, J., Vovlas, A., & Castillo, P. (2017). Anatomical alterations in plant tissues induced by plant-parasitic nematodes. *Frontiers in Plant Science*, *8*, 1987. doi:10.3389/fpls.2017.01987 PMID:29201038

Phani, V., Khan, M. R., & Dutta, T. K. (2021). Plant-parasitic nematodes as a potential threat to protected agriculture: Current status and management options. *Crop Protection (Guildford, Surrey)*, *144*, 105573. doi:10.1016/j.cropro.2021.105573

Ptatscheck, C., Gansfort, B., & Traunspurger, W. (2018). The extent of wind-mediated dispersal of small metazoans, focusing nematodes. *Scientific Reports*, *8*(1), 6814. doi:10.103841598-018-24747-8 PMID:29717144

Pudasaini, M. P., Schomaker, C. H., Been, T. H., & Moens, M. (2006). Vertical distribution of the plant-parasitic nematode, Pratylenchus penetrans, under four field crops. *Phytopathology*, *96*(3), 226–233. doi:10.1094/PHYTO-96-0226 PMID:18944436

Qing, X., Wang, M., Karssen, G., Bucki, P., Bert, W., & Braun-Miyara, S. (2020). PPNID: A reference database and molecular identification pipeline for plant-parasitic nematodes. *Bioinformatics (Oxford, England)*, *36*(4), 1052–1056. doi:10.1093/bioinformatics/btz707 PMID:31529041

Reddy, P. P. (2021). *Nematode diseases of crops and their management*. Springer Singapore. doi:10.1007/978-981-16-3242-6

Risède, J. M., & Simoneau, P. (2004). Pathogenic and genetic diversity of soilborne isolates of Cylindrocladium from banana cropping systems. *European Journal of Plant Pathology*, *110*(2), 139–154. doi:10.1023/B:EJPP.0000015337.54178.c0

Rueda-Ramírez, D., Palevsky, E., & Ruess, L. (2023). Soil Nematodes as a Means of Conservation of Soil Predatory Mites for Biocontrol. *Agronomy (Basel)*, *13*(1), 32. doi:10.3390/agronomy13010032

Saadi, I., De Luca, F., Fanelli, E., Troccoli, A., Greco, N., & Benbouza, H. (2019). Morpho-molecular characterization of Ditylenchus gigas and D. oncogenus parasitizing broad bean, Vicia faba, in Algeria. *European Journal of Plant Pathology*, *155*(2), 505–513. doi:10.100710658-019-01784-1

Seinhorst, J. W. (1965). The relation between nematode density and damage to plants. *Nematologica*, *11*(1), 137–154. doi:10.1163/187529265X00582

Sharmishtha, T., & Sharma M. K. (2020). Population Dynamics of Horizontal and Vertical Population of Root-Knot Nematode, Meloidogyne incognita under Protected Cultivation. *International Journal of Current Microbiology and Applied Sciences*, (11), 329-336.

Sikora, R. A., & Roberts, P. A. (2018). Management practices: an overview of integrated nematode management technologie. *Plant parasitic nematodes in subtropical and tropical agriculture*, 795-838.

Somvanshi, V. S., Tathode, M., Shukla, R. N., & Rao, U. (2018). Nematode Genome Announcement: A draft genome for Rice root-knot nematode. *Journal of Nematology*, *50*(2), 111–116. doi:10.21307/jofnem-2018-018 PMID:30451432

Stirling, G. R. (2018). Biological control of plant-parasitic nematodes. In *Diseases of nematodes* (pp. 103–150). CRC Press.

Tankam-Chedjou, I., Touzeau, S., Mailleret, L., Tewa, J. J., & Grognard, F. (2020). Modelling and control of a banana soilborne pest in a multi-seasonal framework. *Mathematical Biosciences*, *322*, 108324. doi:10.1016/j.mbs.2020.108324 PMID:32092468

Visagie, M., Mienie, C. M., Marais, M., Daneel, M., Karssen, G., & Fourie, H. (2018). Identification of Meloidogyne spp. associated with agri-and horticultural crops in South Africa. *Nematology*, *20*(4), 397–401. doi:10.1163/15685411-00003160

Wallace, H. R. (1973). *Nematode ecology and plant disease*. Academic Press.

Webster, J. M. (2008). *An anecdotal history of nematology*. Pensoft Publishers.

Wesemael, W. M., & Moens, M. (2008). Vertical distribution of the plant-parasitic nematode, Meloidogyne chitwoodi, under field crops. *European Journal of Plant Pathology*, *120*(3), 249–257. doi:10.100710658-007-9213-x

Wheeler, L., & Crow, W. T. (2020). Foliar Nematode Aphelenchoides spp. (Nematoda: Aphelenchida: Aphelenchoididae): EENY749/IN1279, 2/2020. *EDIS*, *2020*(3).

Whitehead, A. G. (1997). *Plant nematode control*. CAB International.

Zhang, Y., Li, S., Li, H., Wang, R., Zhang, K. Q., & Xu, J. (2020). Fungi–nematode interactions: Diversity, ecology, and biocontrol prospects in agriculture. *Journal of Fungi (Basel, Switzerland)*, *6*(4), 206. doi:10.3390/jof6040206 PMID:33020457

Zhao, L. L., Wei, W., Kang, L., & Sun, J. H. (2007). Chemotaxis of the pinewood nematode, *Bursaphelenchus xylophilus*, to volatiles associated with host pine, Pinus massoniana, and its vector *Monochamus alternatus*. *Journal of Chemical Ecology*, *33*(6), 1207–1216. doi:10.100710886-007-9289-y PMID:17447123

Chapter 4
Effect of Phytoparasitic Nematodes on Agricultural Yield:
Current Progress and Challenges

Kajol Yadav
Chaudhary Charan Singh University, India

Lovely Bharti
Chaudhary Charan Singh University, India

Ashok Kumar Chaubey
Chaudhary Charan Singh University, India

ABSTRACT

Plant-parasitic nematodes are a high-priced burden in agricultural crop manufacturing. Over 4100 species of plant-parasitic nematodes have been recognized. Together, they represent an envisioned $80–$118 billion dollars annually in damage to crops. Plant-parasitic nematodes are identified as one of the finest threats to vegetation globally. The most damaging plant-parasitic nematodes (PPNs) are taken into consideration to be the root knot nematodes (RKNs), Meloidogyne sp., which are accountable for losses in vegetable plants throughout the world and determine the common use of chemical pesticides. The major genera of phytoparasitic nematodes discovered to cause crop losses were Heterodera, Hoplolaimus, Meloidogyne, Pratylenchus, Rotylenchulus, and Xiphinema. In different reports, M. incognita suppression turned into observed in field soil treated with P. penetrans in contrast to untreated soil. Opportunity means of pest control consisting of the use of biological controls are great interest for crop producers.

DOI: 10.4018/978-1-6684-8083-0.ch004

INTRODUCTION

The coexistence of nematodes and plants over millions of years has led to the emergence of the plant-parasitic nematode. Broadly dispersed pathogens of vascular plants, enormous losses in yields have been ascribed to the nearness of nematodes. The complicated relationship between the parasitic nematode and plant has culminated in an "evolutionary arms race". Phytoparasitic nematodes have advanced procedures to have resistant reactions for the improvement of feeding sites. In turn, plants have created particular molecules to recognize pathogens signaling for immune response. Declining utilize of chemical pesticides has brought incredible consideration to inquire about in elective strategies of nematode control. A successful strategy for nematode management includes the utilization and usage of nematode-resistant cultivars into crop breeding programs. So Genetic sequencing analyses are broadly utilized in the recognizable proof of molecular components of nematode parasitism and are also used to distinguish nematode-resistant and susceptible plant genotypes. These detailed investigations have essentially contributed to our generally understanding of the energetic and complex nature of plant-nematode interactions. Plant-parasitic nematodes are called plant-parasitic since supplements they get come straight forwardly from plants. They have a needle-like structure called a stylet that helps them to penetrate the plant cell wall and reach the juicy substance on which they nourish (Bernard et al., 2017). Correct identification of nematode species is fundamental for choosing the right strategies of control. The morphology based scientific classification of nematodes is complex due to intraspecific variety, but the tools and methodologies based on biochemical and atomic markers can effectively analyze a wide range of nematode species (Carneiro et al., 2017). The economic consequence of crop losses caused by nematodes come in numerous varieties and are associated with a reduction in crops production and yields quality

Plant-parasitic nematodes are recognized as one of the most noteworthy threats to crops globally. It has been found that nematodes, alone or in combination with other soil microorganisms, caused disease almost each part of the plant, consisting roots, stems, fruits, leaves and seeds. Although globally identification of nematodes as critical cause of plant infections did not happen until the mid of the last century, nematodes were examined more than 150 years earlier. The foremost harming plant-parasitic nematodes are considered to be the root-knot nematodes (RKNs), *Meloidogyne* sp. (Almohithef et al., 2018), which are responsible for losses in vegetable crops worldwide and determine the common utilize of chemical pesticides (Sikandar et al., 2020).

The common indications of nematode disease are the root development decrease, expanded shriveling, mineral deficiency-like symptoms, diminished winter toughness, and dieback in perennial plants. Although these indications are related to root disability, a few species of nematodes (*Anguina* spp.) can moreover influence other parts of the plant, leading to the cause of gall on seeds, leaves, and stems (Kumar et al., 2020). By definition, the seed gall nematodes cause seed gall, though the stem gall nematodes cause swelling and twisting of stems, and the leaf nematodes cause leaf discoloration. However, the terrestrial indications of nematode infection are predominantly circuitous and or maybe unpretentious; among them are diminished vitality, hindered development, decreased yield, and leaf chlorosis. India is one of the twelve mega biodiversity regions with 7.7% genetic resources of the world. Nematodes, as one component of the soil ecosystem, interact with biotic and abiotic factors and adapt themselves to their environment even if the environment threatens to change. If they are not able to do so, they become extinct in due course. So, they are considered as important biological indicators because of their tremendous diversity. The plant parasitic nematodes are distributed worldwide and possess a broad host range of economically important crops. They are recognized as causing more economic damage to food crops. Damage due to plant parasitic nematode causes poor growth, decline in the quality and yield of the crop and reduction in the resistance to stressors, such as, drought, diseases, etc. Damaged roots fail to absorb water and nutrients effectively, leading to additional losses.

Life Cycle of Plant Parasitic Nematodes

During its life cycle, a plant-parasitic nematode typically goes through six phases of development *i.e.* the egg, four juvenile and one adult stage (Decraemer & Hunt, 2013). The early stages of development in the life cycle are termed as juvenile stages. The life cycle begins with the laying of eggs, which hatch into first-stage juveniles (J1), then mature into second-stage juveniles (J2). The J2 stage invades the host roots, and it is the only invasive stage in case of sedentary endoparasites. This juvenile stage survives in soil without feeding until it finds a host (Escobar et al., 2015). After finding a suitable host, J2 undergoes moulting to reach 3rd and 4th stage of Juvenile. Thus, the juvenile stages undergo four moulting and at every moult, the size of the juvenile increases. After moulting, juveniles reach maturity (adult stage) (Figure 1) (Decraemer & Hunt, 2013). The life cycle can take only days or over a year, depending upon the species and environmental conditions (Ferris & Ferris, 1998). All nematodes go immature larval stage. At the end of each larval stage, a new cuticle is synthesized and old one is molted off. The ectoparasitic nematode laid their eggs in soil whiles the migratory endoparasites on roots. The egg of sedentary nematode is retained sedentary egg (*Meloidogyne* spp.) or in cyst (*Heterodera* spp. or *Globodera* spp.).

The eggs develop and give rise to the first stage juvenile (J1). The J1 moults in the egg to form the second stage (J2), and the J2 stage emerged out in soil. Nematodes

Figure 1. Life cycle of plant parasitic nematodes (Ayoub, 1980)

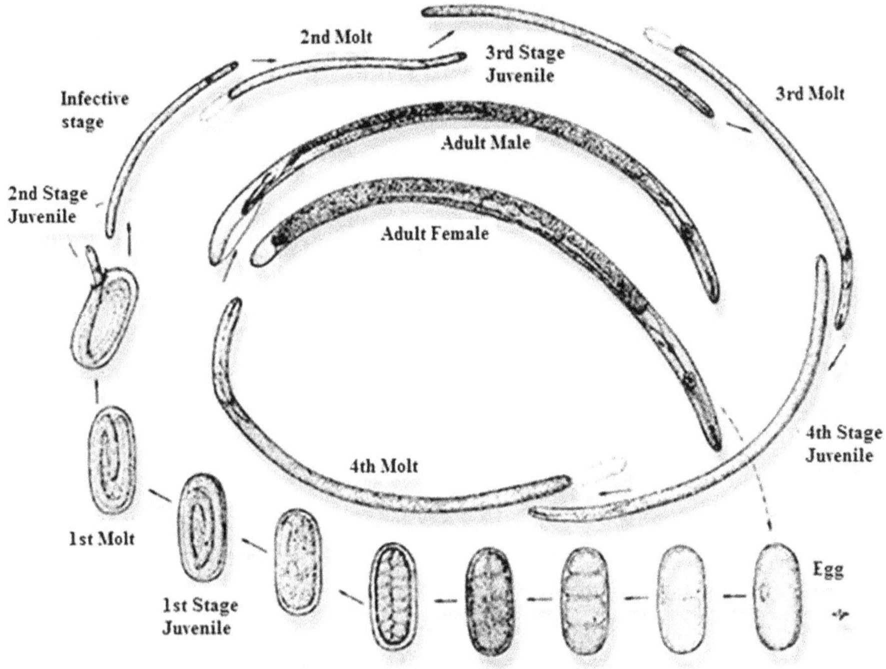

move between the soil particles and move along rhizosphere gradients to find a host root, where they can insert their stylet into the epidermal cells at the root surface. The migratory nematodes withdraw its stylet and move on to other locations after feeding, whereas, endoparasites enter in the root to feed. The second moult resulted in the third stage juvenile (J3) followed by the fourth stage juvenile (J4). After the fourth and final moult, the nematode reaches the adult stage. In contrast to migratory phyto-nematodes, sedentary endoparasites have different life style. As the term proposed after invasion of root tissues, the sedentary endoparasites lose their ability to move after initiating a permanent feeding site. The female is typically located at the feeding site for the rest of her life. When infective J2s reach a host root, they usually penetrate it via the root cells behind the root tip.

Economic Impact of Plant Parasitic Nematodes

There's a basic way to categorize predatory nematodes based on their feeding apparatus and feeding behavior in four major categories as follows: (a) Diplogasterids have a large and solid buccal cavity with a solid claw-like movable dorsal tooth to crush their prey (Bilgrami and Brey, 2005), (b) Mononchids have an exceedingly sclerotised feeding apparatus with a huge pointed dorsal tooth, small teeth, or denticles; (c) Dorilaimids have a penetrating and sucking system and utilize a needle like feeding apparatus, Soil-Borne Nematodes: Affect in Farming and livestock methodologies of Anticipation and Control with Special apparatus to puncture their prey and remove their contents; and) Aphelenchids are predators of fungivorous, parasites of the aerial parts of plant, and insects parasites or predators, and the genus *Seinura* is the only member of the Aphididae group, which is a nematode predator *Seinura* are small worms with a hollow stylet to cut their prey and infuse poison that is produced in an oesophageal gland (Kanwar et al, 2021).

Nematode plagues are a few of the foremost serious issues affecting agricultural yield worldwide and are indeed considered a globally food risk (Coyne et al, 2018). Plant-parasitic nematodes (PPNs) posture a major threat to the quantitative and qualitative production of numerous economic crops around the world. It is assessed that plant parasitic nematodes cause 12.3% of losses in crop yield, which means USD 157 billion every year (Singh et al, 2015). Root-knot nematodes (*Meloidogyne* spp.) are considered the foremost imperative nematodes throughout the world due to their broad and destroying impact on the crops (Sikora et al, 2018). Also, *Meloidogyne* spp. can modify the plants defense, expanding their susceptibility to other pathogens, e.g., microbes and organisms, which results in higher yield losses (Zhou et al, 2016).

Plant-parasitic nematodes are an expensive burden in rural agricultural crop yielding. Over 4100 species of plant-parasitic nematodes have been reported (Decraemer and Hunt, 2006). Collectively, they cause an assessed \$80–\$118 billion dollars annually in damage to crops (Nicol et al, 2011). Including 15% of all reported nematode species, the foremost economically significant species specifically target the major production crops roots and prevent water and nutrient take-up which lead to a reduction in crop yield and quality. Nematodes within the order Tylenchida are pathogens of plants, invertebrates, and fungi and are considered the foremost critical agricultural pests (Sasser and Freckman, 1987). All the essential plant-parasitic nematodes, the foremost successful species are the sedentary groups which build up a feeding site inside the plant host and get nutrients whereas completing their lifecycles. Sedentary nematodes have a characteristic advantage over their migratory relatives due to an interesting and complex strategy of host cell change resulting in the advancement a feasible feeding structure. Interestingly, with over 4000 reported plant-parasitic nematodes, only a small amount produces noteworthy economic losses in crops. In a survey conducted on an assortment of crops in the U.S, the major genera of phytoparasitic nematodes detailed to cause crop losses were *Heterodera, Hoplolaimus, Meloidogyne, Pratylenchus, Rotylenchulus,* and *Xiphinema.* According to Jain et al. (2007), PPNs are responsible for major crop losses of an estimated INR 21.07 billion annually in India. PPNs cause yield losses of about 5% in oilseed crops, 8% in pulse crops, 10% in sugarcane, and 12% in vegetable crops (Sehgal and Gaur, 1999). In many cases, the presence of nematodes makes plant varieties that are resistant to fungus and bacteria susceptible. Moreover, nematodes increase the yield loss worse when they interact with other pathogens.

Economic Losses in Some Major Crops

1. Wheat

Wheat (*Triticum aestivum*) is the most important cereal crop globally. As the main source of food for 40% of the population worldwide, ~783.3 million tons are produced worldwide (FAO, 2023). Wheat production is greatly reduced due to the presence of cereal cyst nematodes (*Heterodera* spp.) in the *Heterodera avenae* group (*H. avenae, Heterodera fi lipjevi* and *Heterodera latipons*), which also affects other cereals such as barley (*Hordeum vulgare*) and oats (*Avena sativa*). Apart from cereal cyst nematodes, other wheat losses are caused by root lesion nematodes *Pratylenchus neglectus* and *Pratylenchus thornei* and seed gall or ear-cockle nematode *Anguina tritici*. The inverse relationship between *H. avenae* and *P. neglectus* was shown on *P. neglectus* resistant and susceptible wheat cultivars infected with *H. avenae* (Lasserre et al, 1994), where a reduction in *P. neglectus* population density was observed in both wheat genotypes. *Anguina tritici* is often a vector of *Rathayibacter tritici*, a gram-positive soil bacterium related to *Clavibacter tritici* that causes seed gall (Riley and Reardon, 1995).

2. Rice

Rice (*Oryza sativa* L.) is the staple food crop for most of the population globally with about 480 million tons currently produced (Childs and Nathan, 2017). Plant parasitic nematodes are one of the major pests of rice and can cause annual yield losses of 10-25% worldwide. More than 100 species of nematodes affect rice production. *Meloidogyne* spp. is distributed globally and are important pathogens of rice and other crops grown in temperate and tropical regions. One of the most important species of *Meloidogyne* is *M. graminicola*, which can reduce rice yield by up to 80% (Soriano et al, 2000). Symptoms of infection are hook-shaped galls, stunting, and reduced number of tillers, poor growth and reproduction. Rice root nematodes *Hirschmanniella oryzae*, *i.e.* rice root nematodes (RRNs), are commonly found in irrigated rice production systems (Kyndt et al, 2014). Widely distributed, *H. oryzae* has been reported from Asian countries such as India, Pakistan, Bangladesh, Sri Lanka, Nepal, Thailand, Vietnam, Indonesia, Philippines, China, Korea and Japan and in the states of Louisiana and Texas of the United States (Bridge et al, 2005).

3. Maize

Maize (*Zea mays*) is widely grown around the world with the three largest productions in North America, Asia and Europe (Nicol et al, 2011). More than 50 species are known to be parasites on maize worldwide, however, the most destructive species include root knot nematodes, *Meloidogyne* spp., root lesion nematodes, *Pratylenchus* spp. and the cyst nematode, *Heterodera* spp. (Nicol et al, 2011). In the United States, the most economically important species are the lesion nematode (*Pratylenchus* spp.) and the corn cyst nematode (*Heterodera zeae*). In most cases, symptoms of infection caused by these nematodes include growth and leaf chlorosis with small nodules (Norton, 1983). The needle nematode *Longidorus breviannulatus* is associated with maize stunting and can cause economic yield losses of up to 60% (Norton and Hoffmann, 1975).

4. Potato

Potato (*Solanum tuberosum*) is the third most important crop in the world with an estimated total global potato production of more than 376 million tons in 2021 (FAOSTAT, 2021), with China and India producing the most (94 million tonnes and 54 million tonnes, respectively. Cyst nematodes are pathogens that proliferate on potatoes causing great yield losses. *Globodera rostochiensis* and *Globodera pallid* are native to South America and are known pests of other members of the Solanaceae family, including the tomato and the woody nightshade (Turner and Evans, 1998). Other major plant parasitic nematodes of potatoes include the root knot nematode (*Meloidogyne* spp.) and the stem nematode *Ditylenchus destructor*. Among the four species of root knot nematodes affecting potato production worldwide, the Columbian root knot nematode (*Meloidogyne chitwoodii*) is considered the most important species (Santo et al, 1980). In addition to potatoes, sweet potato (*Ipomoea batatas* L) is the main host of *D. destructor*, losses up to 100% of yield have occurred in the main producing regions, including the top producer, China (Xu et al, 2015).

5. Sweet Potato

Sweet potato (*Ipomoea Batatas* L) has been considered a plant of great importance throughout human history. Its cultivation dates back to prehistoric times and it has been continuously cultivated as a major food source. The worldwide production of sweet potatoes is estimated at 105 million metric tons (CIP, 2014). Presently the sixth most important food crop, sweet potato production has improved the economic status of communities around the world, especially in developing countries where it ranks as the second most important crop. About 10.2% of sweet potato production is lost every year due to the presence of plant parasitic nematodes (Decraemer and Hun, 2006). The root knot nematode (RKN) is a significant pest of sweet potato, causing infection symptoms including stunted plants, yellowing of leaves, abnormal flowering and gall production on the roots, leading to reduced uptake nutrients and water, as well as necrosis and cracking on the roots. Successful breeding programs for resistance to root knot nematodes in sweet potato involve the identification of resistance genes. Therefore, identification of genetic markers associated with root knot nematode resistance requires broadly molecular studies (Cervantes-Flores, et al, 2008).

6. Soybean

Soybean (*Glycine max* L. Merr) is the most significant oilseed crop in the world. Globally soybean production has increased in the last 20 years, reaching about 210 million tonnes. Additionally, consumption has risen more quickly than production. Soy is used for food and feed, production of raw materials for the industry of vegetable oil and other purposes such as the manufacture of plastics, lubricants, candles, varnishes, soaps, biodiesel and lecithin (Fontana, 2011). Demand for a reliable food supply and food security has increased. With monocultures, intense planting, and the growth of crops into recently opened areas, unsustainable crop production systems have escalated difficulties brought on by new pests and diseases. Plant parasitic nematodes are one of these issues that globally restrict soybean production. Global crop losses from plant parasitic nematodes are estimated to be $78 billion annually, with an average crop yield loss of 10-15%. However, the amount of yield loss caused by nematode parasitism in soybeans varies considerably and is primarily influenced by nematode species, population levels, cultivar susceptibility, cropping systems, temperatures, season, region, and soil factors such as soil texture, pH, and fertility. The yield loss cause by nematodes as high as 30–100% (Schmitt & Barker 1981; Barker, 1998). In relation with soybeans, more than 100 nematode species from 50 genera have been identified. *Meloidogyne javanica, Meloidogyne incognita, Heterodera glycines* (Franzener et al, 2005; Santana et al, 2009), *Pratylenchus brachyurus*, the reniformis nematode *Rotylenchulus reniformis* (Dias et al, 2007), and *Tubixaba tuxaua* (Furlanetto et al, 2010) are the species that harm soybean crop.

7. Faba Bean

The faba bean (*Vicia faba* L.) is widely cultivated in a variety of climates, from temperate to subtropical, and it is host to a diverse range of local, native, and exotic cosmopolitan insect pests, fungi, viruses, and parasitic weeds. The three main varieties of faba beans the Mediterranean-adapted, the cool temperate winter, and the spring are exposed to various diseases and pests at various stages of their growth cycles. Faba bean production was estimated to be 5669185 t/year from 2.67 million hectares of land, with China producing the most (1,723,588 t/year), Ethopia is second (1,070,637 t/year), and followed by United Kingdom, Australia, and Germany producing the majority of the globally faba bean (FAO, 2020). The genera *Ditylenchus*, *Meloidogyne*, and *Pratylenchus* contain the nematodes that are most frequently observed to harm faba beans. The *Ditylenchus dipsaci* (Kühn) Filipjev species complex of stem nematodes affects more than 500 host species and consists of roughly 30 taxa.

Methods to Control Plant Parasitic Nematodes

1. Cultural Control

For many years, crop rotation and cover cropping are frequently utilized in Integrated Pest Management (IPM) conventions to decrease plant-parasitic nematode frequency and replenish soil nutrients levels. Soil nematode numbers have been viably diminished by rotational cultivation of non-host cultivars be that as it may, the wide have run of *Meloidogyne* spp. regularly decreases the adequacy of crop rotation (Sikora and Fernandez, 2005). Planting corn as a rotational crop has been appeared to diminish northern root-knot nematode (*M. hapla*) incidence in any case; population densities of other *Meloidogyne* spp. may increment with persistent. Among them, Plant species with resistance to mixed *Meloidogyne* populations have been recognized. Leguminous cover crops *Mucuna pruriens* L., and *Crotalaria spectabilis* appeared resistance to three species of root-knot nematodes (*Meloidogyne arenaria, M. incognita, M. javanica*) (Osei et al, 2010). In some cases, the exceptionally nature of crop production may suppress the magnitude of infection. Rice is developed under flooding conditions which does not favor the nematode way of life.

2. Plant Extracts

Plant extracts frequently contain a group of compounds which show nematode suppressive properties. Ethanolic extricates of *Azadirachta indica* (neem), *Withania somnifera* (ashwagandha), *Tagetes erecta* (marigold) and *Eucalyptus citriodora* (eucalyptus) were described to show nematicidal activity against *Meloidogyne incognita, Helicotylenchus multicinctus* and *Hoplolaimus* which was comparable to chemical nematicide controls (Khan et al, 2008). In other reports, expanded plant growth and development were appeared in plants propagated with the expansion of a variety of extracts. Root-knot nematode egg hatch and larval development was drastically reduced by leaf extracts from *Hunteria umbellata* and *Mallotus oppositifolius* which coincided with expanded development of cashew seedlings (Okeniyi et al, 2013). Plant height, food production and weights of *M. incognita* contaminated tomato were essentially increased by the addition of ethanol extracts from*Azadirachta indica* leaves, *Capsicum annuum* fruits, *Zingiber officinale* rhizomes and *Parkia biglobosa* seeds in comparison to non-treated controls (Bawa et al, 2014).

3. Biological Control

With rising demands in organic farming and concerns for natural and environmental welfare, the utilization of chemical pesticides has decreased. Another resource of pest management such as the use of biological controls is of immense attention for crop producers. The pathogenicity of nematophagous bacteria and fungi in the control of few nematode pests, consisting cyst and root-knot nematodes have been well-documented (Stirling, 1991). Parasitic bacteria of *Pasteuria* spp. have been described to infect~323 nematode species counting both plant-parasitic nematodes and free-living nematodes (Chen and Dickson, 1998). Three methods of application for *P. Penetrans* were assessed for nematode control including seed, transplant, and post-plant treatments (Kokalis-Burelle, 2015). In greenhouse studies including cucumber, all three *Pasteuria* treatments were appeared to reduce galling caused by *M. incognita* as well as soil nematode numbers and nematode reproductive potential. In other reports, *M. incognita* suppression was shows in field soil treated with *P. penetrans* in comparison to untreated soil (Mankau, 1980; Minton & Sayre, 1989; Weibelzahl-Fulton et al, 1996). Other genera of bacteria including *Bacillus* spp. have appeared great pledge in nematode management. *B. cereus* strain S2 treatment brings about in a mortality of 90.96% to *M. incognita* (Gao et al, 2016). *B. firmus* YBf-10 shown nematicidal activity against *M. incognita*, which was clearly demonstrated by an inhibition of egg hatch and motility (Xiong et al, 2015). Nematophagous organisms *Pochonia chlamydosporia* has potential as a biological control specialist for *M. incognita* in vegetable crops. Among them crop rotational methods; *P. chlamydosporia* was appeared to reduce nematode levels in soil already utilized for root-knot nematode susceptible tomato (Atkins et al, 2003). Nematophagous fungus products including chitinases appear great potential for the development of biopesticides. Certain root-knot nematode species have transparent defensive chitin-containing shells. Filtered chitinase LPCHI1 from *Lecanicillium psalliota* was shown to degrade *M. incognita* eggs (Gan et al, 2007).

4. Nematodes

Nematodes also have a significant role in nature, primarily as essential members of food chains, and they are too in charge of degrading organic matter and recycling nitrogen within the soil (Nisa et al, 2021). They are also significant natural bioregulators of many other populations within the soil, including other nematodes. Most studies related to predatory nematodes have centered on the control of plant-parasitic nematodes, and recently have a few studies begun to investigate their opposing action against ruminant parasitic nematodes (Kanwar et al, 2021). To be considered a great candidate as a predatory nematode, with high desires as a potential agent of control of plant-parasitic nematodes, the predatory nematode should fulfill a few characteristics, such as a good ability to seek for its prey, be specific for its prey, be an productive predatory nematode, have a certain life cycle length and life span, have regenerative potential, and be capable of surviving and adapting to biological conditions (Kanwar et al, 2021). The prey inclination of some predatory nematode confines is a good characteristic because it can focus control on a particular nematode prey (Kanwar et al, 2021). Although some studies about nematode predatory nematodes have appeared successful comes under in vitro conditions, there are limited studies carried out under field conditions. The most restrictions that analysts confront in utilizing predatory nematodes within the control of nematode plagues include mass culture and survival for a certain time after they are discharged in the field. They can caught to be overcome for the management of either plant or animal parasitic nematodes.

Predatory Nematodes

The Mononchida, Dorylaimida, Diplogasterida, and Aphelenchida taxonomic families of nematodes make up the majority of predatory nematodes. Each group has its own type of of food preferences, feeding methods, and feeding apparatuses.

1. Mononchida

The buccal cavity of the mononchids is highly sclerotized and often armed with one tooth, many large puncturing teeth, numerous small gripping teeth, or both. Several commonly found mononchids primarily prey on plant parasitic and other nematodes, though not exclusively (Bilgrami et al., 1986). In a laboratory condition, they can be cultivated on several prey species. If the prey size is small, they may consume it whole, otherwise, they feed by cutting larger preys into pieces. The duration of the life cycle differs between species. *Prionchulus punctatus* requires 45 days to complete one generation, whereas *Mononchus aquaticus* only needs 15 days at 25^{0}C (Maertens, 1975; Grootaert & Maertens, 1976). Cobb (1917) was the first to proposed using predatory nematodes to evaluate populations of plant parasitic nematodes in the soil. Later, feeding by *Clarkus papillatus* on *Meloidogyne* sp. was observed by Cobb (1920) and Steiner & Heinly (1922) and recommended the use of *C. papillatus* to biologically control plant parasitic nematodes in sugar beet fields. Although *C. papillatus* was among the mononchids observed by Thorne (1927) in sugar beet fields, his findings were unremarkable and research revealed that the mononchids population in soil was frequently unstable and sudden changes, sometimes disappearing or drastically declining. *Apicomplexa* infected some species of mononchids, which may have caused to their population decline. However, predatory nematodes might be useful as biocontrol agents against plant parasite nematodes (Cassidy, 1931; Christie 1960; Esser & Sobers, 1964; Mulvey, 1961 & Esser, 1963). Mononchids have been observed feeding on nematodes as well as other types of soil microbes (Banage, 1963). Non-specific predators like mononchids only partially control plant parasitic nematodes (Webster, 1972 & Jones, 1974). While conducting pot trials, Cohn & Mordechai (1974) discovered a consistent correlation between large populations of *Mylonchulus sigmaturus* and low populations of *Tylenchulus semipenetrans*, a citrus nematode. In pot trials, Small (1979) found that the populations of the root-knot nematode *M. incognita* and the potato cyst worm *Globodera rostochiensis* reduced significantly when a predatory nematode *P. punctatus* was present. In field experiments, Ahmad & Jairajpuri (1982) discovered a substantial negative correlation between populations of the plant parasitic nematodes *Trichodorus* sp. and *Hemicycliophora* sp. and the predatory nematode *Parahadronchus shakili*. Similar results were revealed by Azmi (1983), observed a rise in the abundance of *Iotonchus monhystera* and a decline in *Helicotylenchus dihystera*. In mandarin orange orchards, Rama & Dasgupta (1998) found a predator-prey connection between the predator *I. tenuicaudatus* and two plant parasitic nematodes, *T. semipenetrans* and *H. dihystera*.

2. Dorylaimida

Dorylaimids have a hollow stylet, known scientifically as an odontostyle, which they utilise to pierce their prey and ingest through. According to Linford & Oliviera (1937b), when eating other nematodes, the large odontostyle of these predators disorganize the prey's internal organs, quickly immobilizing it. The Nygolaimoidea and *Sectonema* species pierce or slit their prey with huge, narrow, protruding teeth termed a mural tooth. As food is sucked up through the oral orifice, this tooth is not hollow. Predatory nematodes known as dorylaimids are the most prevalent and can be found in all types of soil, climates, and habitats. The predatory potential of dorylaimids is well recognized, but the majority of the research is based on *in vitro* experiments. *Labronema vulvapapillatum* and *Aquatides thornei* predation abilities were originally described in detail by Ways & Grootaert et al. (1977) and Bilgrami et al. (1985), respectively. *L. vulvapapillatum* completes a life cycle in 36 days at 25^{0}C, although dorylaimids usually require 3-6 months (Ferris, 1968;

Wood, 1973). Citrus nematodes have the potential to act as biocontrol agents, as evidenced by the discovery of *Eudorylaimus obtusicaudatus* feeding on eggs inside *Heterodera schachtii* cysts and an increase in the population of *Thornia* species in their presence but a decrease in their absence in pot trials (Boosalis & Mankau, 1965). Dorylaimid predators have an effective rate of predation (Khan et al., 1991; Bilgrami, 1992), a positive correlation between predation and prey trophic groups (Bilgrami, 1993, 1995), a preference for particular prey (Khan et al., 1994, 1995a,b; Khan & Jairajpuri, 1997), an effective ability to find prey, activities that attract and assemble at feeding sites (Bilgrami & Gaugler, 2005).

3. Diplogasterid

The buccal cavity of diplogasterid predators is considerably smaller than that of mononchids, but armed with teeth. They consume bacteria, nematodes, and other soil microorganisms as food. Usually, abundant amounts of diplogasterids can be found in decaying organic manure. They are the most easily cultured nematodes because they can be kept alive on basic nutrient media with bacteria and complete their life cycles in 8–15 days (Yeates, 1969). Diplogasterid predators appear to be more prey-selective than other groups. *Odontopharynx longicaudata* attacked and killed 6 out of 17 prey species provided in the laboratory investigation and 100% of individuals of *Anguina pacificae* were killed (Chitambar & Noffsinger, 1989). They also observed that its ability to completely remove a population of *A. pacificae* is of particular interest because it suggests that a target prey predator relationship exists in the field habitat from where the two species were collected. A strong degree of prey preference was reported for *Mononchoides fortidens, M. longicaudatus* and *M. gaugleri*; they preferred juveniles of endoparasitic over ectoparasitic nematodes (Bilgrami et al., 2005). Bilgrami & Jairajpuri studies on the predatory behavior of diplogasterid predators on plant parasitic nematodes revealed their chemotaxis behavior toward attractants from prey nematodes. In pot tests, Fauzia et al. (1998) showed that *M. longicaudatus* could reduce root galling caused by root-knot nematodes, which enhanced vegetative growth and increased root mass. Diplogaster decreased *M. javanica* and *T. semipenetrans* populations in pots. According to Khan & Kim (2005), a pre-plant application of *M. fortidens* in potted field soil infected with the root-knot nematode *M. arenaria* decreased root galling on tomato plants and suppressed the *M. arenaria* population, and the final population density of predators, *M. fortidens*, increased by approximately two compared to initially added density. Therefore, the diplogasterid predators may show promise as biocontrol agents, but further research is required to validate their commercial viability.

4. Aphelenchid

Predators of aphelenchids include species of the genus *Seinura*. The feeding apparatus is typical aphelenchid style. These predators transfer digestive enzymes into their pray body through the cuticle of nematodes they are feeding on, paralyzing the prey instantly (Hechler, 1963; Wood, 1974). Following that, the contents of the prey body are consumed. These species are usually small in size, yet they have the ability to inject enzymes into larger prey than their own size to feed on it (Linford & Oliviera, 1937b). *Seinura* spp. may be easily and quickly cultivated on fungivorous nematodes in a laboratory setting and have a high reproductive capacity. They also complete their life cycles in 3-6 days (Hechler, 1963; Hechler & Taylor, 1966).

CONCLUSION

The extreme use of chemical pesticides for controlling agricultural pests, especially phytoparasitic nematodes, as well as the utilization of chemical anthelmintic drugs for controlling ruminant parasitic nematodes, are progressively discredited techniques due to a number of undesirable consequences. Additionally, chemical pesticides contaminate the environment with the same destroying impacts on useful organism in nature. There are other options of control that can be utilized in integrated control programs for agricultural pests or ruminant parasitic nematodes. As we have appeared, plagues of significance in the farming or livestock industry are exceedingly complex organisms, and they should be controlled using various strategic tools. In this chapter, diverse procedures of control have been shown and briefly discussed, making extraordinary reference to the utilization of natural nematode antagonists that have been investigated as potential tools of control from the viewpoint of sustainability. Some of the organisms specified in this chapter are still under study, and more data should be generated to consider them as viable measures of control. Other nematode natural enemies have as of now provided great biotechnological tools for the control of nematodes, affecting vital economic crops and against ruminant parasitic nematodes. It is important to consider that research about in this essential zone of information should be encouraged since the seriously use of chemically synthesized molecules and their negative impacts debilitate the environment and public health globally.

REFERENCES

Ahmad, N., & Jairajpuri, M. S. (1982). *Population fluctuations of predatory nematode* Parahadronchus shakili. *Proceedings of Symposium on Ecology and Animal Population in India, 3*, 1–12.

Almohithef, A. H., Al-Yahya, F. A., Al-Hazmi, A. S., Dawabah, A. A. M., & Lafi, H. A. (2018). Prevalence of plant-parasitic nematodes associated with certain greenhouse vegetable crops in Riyadh region. *Journal of the Saudi Society of Agricultural Sciences, 19*(1), 22–25. doi:10.1016/j.jssas.2018.05.001

Atkins, S., Hidalgo-Diaz, L., Kalisz, H., Mauchline, T., Hirsch, P., & Kerry, B. (2003). Development of a new management strategy for the control of root-knot nematodes (*Meloidogyne* spp.) in organic vegetable production. *Pest Management Science, 59*(2), 183–189. doi:10.1002/ps.603 PMID:12587872

Ayoub, S. M. (1980). *Plant Nematology: An agricultural training aid*. Nema Aid Publications.

Azmi, M. I. (1983). Predatory behaviour of nematodes. Biological control of *Helicotylenchus dihystera* through the predaceous nematodes, *Iotonchus monhystera*. *Indian Journal of Nematology, 13*, 1–8.

Banage, W. B. (1963). The ecological importance of free-living soil nematodes with special reference to those of moorland soil. *Journal of Animal Ecology, 32*(1), 133–140. doi:10.2307/2522

Barker, K. R. (1998). Introduction. In K. R. Barker, G. A. Pederson, & G. L. Winhan (Eds.), *Plant and Nematode Interactions* (pp. 1–120). American Society of Agronomy. doi:10.2134/agronmonogr36

Bawa, J., Mohammed, I., & Liadi, S. (2014). Nematicidal effect of some plants extracts on root-knot nematodes (*Meloidogyne incognita*) of tomato (*Lycopersicon esculentum*). *World Journal of Life Science and Medical Research, 3*, 81–87.

Bernard, G. C., Egnin, M., & Bonsi, C. (2017). The Impact of Plant-Parasitic Nematodes on Agriculture and Methods of Control. In Nematology - Concepts, Diagnosis and Control. doi:10.5772/intechopen.68958

Bilgrami, A. L. (1992). Resistance and susceptibility of prey nematodes to predation and strike rate of the predators, *Mononchus aquaticus, Dorylaimus stagnalis*, and *Aquatides thornei*. *Fundamental and Applied Nematology, 15*, 265–270.

Bilgrami, A. L., Ahmad, I., & Jairajpuri, M. S. (1985). Predatory behavior of *Aquatides thornei* (Nygolaimina: Nematoda). *Nematologica, 30*, 457–462.

Bilgrami, A. L., Ahmad, I., & Jairajpuri, M. S. (1986). A study on the intestinal contents of some mononchs. *Review of Nematology*, *17*, 339–340.

Bilgrami, A. L., & Gaugler, R. (2005). Feeding behaviour of the predatory nematodes *Laimydorus baldus* and *Discolaimus major* (Nematoda: Dorylaimida). *Nematology*, *7*(1), 11–20. doi:10.1163/1568541054192207

Bilgrami, A. L., Gaugler, R., & Brey, C. (2005). Prey preference and feeding behaviour of diplogastrid predator *Mononchoides gaugleri* (Nematoda: Diplogasterida). *Nematology*, *7*(3), 333–342. doi:10.1163/156854105774355563

Bilgrami, L. A., & Brey, C. (2005). Potential of predatory nematodes to control plant-parasitic nematodes. In Nematodes as Biocontrol Agents. CAB International. doi:10.1079/9780851990170.0447

Bonfil, D., Dolgin, B., Mufradi, I., & Asido, S. (2004). Bioassay to forecast cereal cyst nematode damage to wheat in fields. *Precision Agriculture*, *5*(4), 329–344. doi:10.1023/B:PRAG.0000040804.97462.02

Bonsi, C., & Phills, B. (1979). Reaction of twelve sweetpotato cultivars and breeding lines to root-knot species with three experimental methods. *Horticultural Science*, *14*, 539–541.

Boosalis, M. G., & Mankau, R. (1965). Parasitism and predation of soil microorganisms. In K. F. Baker & W. C. Snyder (Eds.), *Ecology of Soil-borne Plant Pathogens* (pp. 374–391). Berkeley University of California Press.

Bridge, J., Plowright, R., & Peng, D. (2005). Nematode parasites of rice. In Plant Parasitic Nematodes in Subtropical and Tropical Agriculture (2nd ed.). CABI Bioscience. doi:10.1079/9780851997278.0087

Carneiro, R. M. D. G., de Oliveira Lima, F. S., & Correia, V. R. (2017). *Methods and tools currently used for the identification of plant parasitic nematodes*. Nematology – Concepts, Diagnosis and Control. BoD–Books on Demand. doi:10.5772/intechopen.69403

Cassidy, G. H. (1931). Some mononchs of Hawaii. *Hawaiian Planters' Rec*, *35*(3), 305–339.

Cervantes-Flores, J., Yencho, G., & Davis, E. (2008). Detection of quantitative trait loci and inheritance of root-knot nematode resistance in sweet potato. *Horticultural Science (Prague)*, *133*, 841–851.

Chen, Z., & Dickson, D. (1998). Review of *Pasteuria penetrans*: Biology, ecology, and biological control potential. *Journal of Nematology*, *30*, 313–340. PMID:19274225

Childs& Nathan. (2017). *U.S. 2016/17 Export Forecast Lowered 2.0 Million Cwt to 110.0 Million cwt. USDA-ERS Situation and Outlook*. RCS-17B.

Chitambar, J. J., & Noffsinger, E. M. (1989). Predaceous behaviour and life history of *Odontopharynx longicaudata* (Diplogasterida). *Journal of Nematology, 21*, 284–291. PMID:19287610

Christie, J. R. (1960). Biological control-predaceous nematodes. In J. M. Sasser & W. R. Jenkins (Eds.), *Nematology: Fundamentals and Recent Advances with Emphasis on Plant Parasitic and Soil Forms* (pp. 466–468). University of North Carolina Press.

CIP International Potato Center. (2014). *Sweetpotato Facts and Figures*. Available from: https://cipotato.org/sweetpotato/facts-2/

Cobb, N. A. (1917). The mononchs (Mononchus Bastian, 1865): A genus of free-living predatory nematodes. *Soil Science, 3*(5), 431–486. doi:10.1097/00010694-191705000-00004

Cobb, N. A. (1920). Transfer of nematodes (mononchs) from place to place for economic purposes. *Science, 51*(1330), 640–641. doi:10.1126cience.51.1330.640 PMID:17797737

Cohn, E., & Mordechai, E. (1974). Experiments in suppressing citrus nematode populations by use of marigold and a predacious nematode. *Nematologia Mediterranea, 2*, 43–53.

Coyne, D. L., Dalzell, J., Cortada, L., Claudius-Cole, A. O., Haukeland, S., Luambano, N., & Talwana, H. (2018). Plant-Parasitic Nematodesand Food Security in Sub-Saharan Africa. *Annual Review of Phytopathology, 56*(1), 381–403. doi:10.1146/annurev-phyto-080417-045833 PMID:29958072

Decraemer, W., & Hunt, D. (2006). Structure and classification. In R. Perry & M. Moens (Eds.), *Plant Nematology* (pp. 3–32). CAB International., doi:10.1079/9781845930561.0003

Decraemer, W., & Hunt, D. J. (2013). Structure and classification. In R. N. Perry & M. Moens (Eds.), *Plant nematology* (2nd ed., pp. 3–39). CABI Publishing. doi:10.1079/9781780641515.0003

Dias, W. P., Ribeiro, N. R., Lopes, I. O. N., Garcia, A., Carneiro, G. E. S., & Silva, J. F. V. (2007). Manejo de nematoides na cultura da soja. In *Congresso Brasileiro de Nematologia; 12 de Agosto; Goiânia, Brazil* (pp. 26–30). Sociedade Brasileira de Nematologia.

Escobar, C., Barcala, M., Cabrera, J. & Fenoll, C. (2015). Chapter One - Overview of Root-Knot Nematodes and Giant Cells. *Advances in Botanical Research*, (73), 1-32.

Esser, R. P. (1963). Nematode interactions in plates of no sterile water agar. *Proceedings - Soil and Crop Science Society of Florida*, *23*, 121–128.

Esser, R. P., & Sobers, E. K. (1964). Natural enemies of nematodes. *Proceedings - Soil and Crop Science Society of Florida*, *24*, 326–353.

FAO. Cereal Supply & Demand Brief. (2017). Available from: https://www.fao.org/worldfoodsituation/csdb/en/

FAOSTAT. (2015). *World production of potato*. Available from https://www.nationalpotatocouncil.org/files/1114/4223/8719/Pg._78_World_Potato_Production.pdf

Fauzia, Jairajpuri, M. S. & Khan, Z. (1998). Biocontrol potential of *Mononchoides longicaudatus* on *Meloidogyne incognita* on tomato plants. *International Journal of Nematology*, *8*, 89–91.

Ferris, J. M., & Ferris, V. R. (1998). Biology of Plant-Parasitic Nematodes. In K. R. Barker, G. A. Pederson, & Gl. Windham (Eds.), *Plant and Nematode Interactions* (pp. 21–35). Soil Science Society of America.

Ferris, R. V. (1968). Biometric analysis in the genus *Labronema* (Nematoda: Dorylaimida) with a description of *L. thornei* n. sp. *Nematologica*, *14*(2), 276–284. doi:10.1163/187529268X00480

Fontana, J. D. (2011). *Biodiesel: para leitores de 9 a 90 anos.* UFPR.

Franzener, G., Unfried, J. R., Stangarlin, J. R., & Furlanetto, C. (2005). Nematoides formadores de galhas e de cisto patogênicos a cultura da soja em municípios do Oeste do Paraná. *Nematologia Brasileira*, *29*, 161–165.

Furlanetto, C., Seifert, K. E., Fenterseifer, C. E., Page, E. C., Davi, J. J. S., & Grabouski, M. M. S. (2010). Desenvolvimento das culturas de soja, milho e trigo cultivadas em áreas infestadas com o nematoide Tubixaba tuxaua no Oeste do Paraná. *Tropical Plant Pathology*, *35*, 295–302.

Gan, Z., Yang, J., Tao, N., Liang, L., Mi, Q., Li, J., & Zhang, K. (2007). Cloning of the gene *Lecanicillium psalliotae* chitinase Lpchi1 and identification of its potential role in the biocontrol of root-knot nematode *Meloidogyne incognita. Applied Microbiology and Biotechnology*, *76*(6), 1309–1317. doi:10.100700253-007-1111-9 PMID:17665191

Gao, H., Qi, G., Yin, R., Zhang, H., Li, C., & Zhao, X. (2016). *Bacillus cereus* strain S2 shows high nematicidal activity against *Meloidogyne incognita* by producing sphingosine. *Scientific Reports*, *6*(1), 28756. doi:10.1038rep28756 PMID:27338781

Grootaert, P., & Maertens, D. (1976). Cultivation and life cycle of *Mononchus aquaticus. Nematologica*, *22*(2), 173–181. doi:10.1163/187529276X00265

Hechler, H. C. (1963). Description, developmental biology, and feeding habits of *Seinura tenuicaudata* (De Man), J.B. Goodey, 1960 (Nematoda: Aphelenchoididae), a nematode predator. *Proceedings of the Helminthological Society of Washington*, *30*, 183–195.

Hechler, H. C., & Taylor, D. P. (1966). The life histories of *Seinura celeris, S. olivierae, S. oxura* and *S. steineri* (Nematoda: Aphelenchoididae). *Proceedings of the Helminthological Society of Washington*, *33*, 71–83.

Jain, R., Mathur, K., & Singh, R. (2007). Estimation of losses due to plant parasitic nematodes on different crops in India. *Indian Journal of Nematology*, *37*, 217–219.

Jones, F. G. W. (1974). Control of nematode pest, background and outlook for biological control. In *Proceedings of the 13th Symposium British on Biology in Pest and Disease Control.* Ecology Society Oxford.

Kanwar, R. S., Patil, J. A., & Yadav, S. (2021). Prospects of using predatory nematodes in biological control for plant parasiticnematodes—A review. *Biological Control*, *160*, 104668. doi:10.1016/j.biocontrol.2021.104668

Khan, A., Sayed, M., Shaukat, S., & Handoo, Z. (2008). Efficacy of four plant extracts on nematodes associated with papaya in Sindh, Pakistan. *Nematologia Mediterranea*, *36*, 93–98.

Khan, Z., Bilgrami, A. L., & Jairajpuri, M. S. (1991). Some observations on the predation ability of *Aporcelaimellus nivalis* (Altherr, 1952) Heyns 1966 (Nematoda: Dorylaimida). *Nematologica*, *37*(1-4), 333–342. doi:10.1163/187529291X00330

Khan, Z., Bilgrami, A. L., & Jairajpuri, M. S. (1994). Attraction and food preference behaviour of predatory nematodes, *Allodorylaimus americanus* and *Discolaimus silvicolus* (Nematoda: Dorylaimida). *Indian Journal of Nematology*, *24*, 168–175.

Khan, Z., Bilgrami, A. L., & Jairajpuri, M. S. (1995a). A comparative study on predation by *Allodorylaimus americanus* and *Discolaimus silvicolus* (Nematoda: Dorylaimida) on different species of plant parasitic nematodes. *Indian Journal of Nematology*, *25*, 94–103.

Khan, Z., Bilgrami, A. L., & Jairajpuri, M. S. (1995b). Observations on the predation abilities of *Neoactinolaimus agilis* (Dorylaimida: Actinolaimoidea). *Indian Journal of Nematology, 25*, 129–135.

Khan, Z., & Jairajpuri, M. S. (1997). Predatory behavior of *Paractinolaimus elongatus* on plant parasitic nematodes. *International Journal of Nematology, 8*, 178–181.

Khan, Z., & Kim, Y. H. (2005). The predatory nematode, *Mononchoides fortidens* (Nematoda: Diplogasterida), suppresses the root-knot nematode, *Meloidogyne arenaria*, in potted field soil. *Biological Control, 35*(1), 78–82. doi:10.1016/j.biocontrol.2005.05.015

Kokalis-Burelle, N. (2015). Pasteuria penetrans for control of *Meloidogyne incognita* on tomato and cucumber and *M. arenaria* on snapdragon. *Journal of Nematology, 47*, 207–213. PMID:26527842

Kumar, V., Khan, M. R., & Walia, R. K. (2020). Crop loss estimations due to plant-parasitic nematodes in major crops in India. *National Academy Science Letters, 43*(5), 409–412. doi:10.100740009-020-00895-2

Kyndt, T., Fernandez, D., & Gheysen, G. (2014). Plant-parasitic nematode infections in rice: Molecular and cellular insights. *Annual Review of Phytopathology, 52*(1), 135–153. doi:10.1146/annurev-phyto-102313-050111 PMID:24906129

Lasserre, F., Rivoal, R., & Cook, R. (1994). Interactions between *Heterodera avenae* and *Pratylenchus neglectus* on wheat. *Journal of Nematology, 26*, 336–344. PMID:19279901

Linford, M. B., & Oliviera, J. M. (1937). The feeding of hollow-spear nematodes on other nematodes. *Science, 85*(2203), 295–297. doi:10.1126cience.85.2203.295 PMID:17787453

Maertens, D. (1975). Observations on life cycle of *Prionchulus punctatus* (Cobb, 1917) and culture conditions. *Biologische Jaarboek Dodonaea, 43*, 197–218.

Mankau, R. (1980). Biological control of *Meloidogyne* populations by *Bacillus penetrans* in West Africa. *Journal of Nematology, 12*, 23.

Minton, N. A., & Sayre, R. M. (1989). Suppressive influence of *Pasteuria penetrans* in Georgia soils on reproduction of *Meloidogyne arenaria. Journal of Nematology, 21*, 574–575.

Mulvey, R. H. (1961). The Mononchidae: a family of predaceous nematodes Genus *Mylonchulus* (Enoplida: Mononchidae). *Canadian Journal of Zoology, 39*(5), 665–696. doi:10.1139/z61-070

Nicol, J., Turner, D., Coyne, L., den Nijs, L., Hockland, S., & Maafi, Z. (2011). Current nematode threats to world agriculture. In J. Jones, G. Gheysen, & C. Fenoll (Eds.), *Genomics and Molecular Genetics of Plant-Nematode Interactions* (pp. 21–43). Springer Science Business Media. doi:10.1007/978-94-007-0434-3_2

Nisa, U. R., Tantray, A. Y., Kouser, N., Allie, K. A., Wani, S. M., Alamri, S. A., Alyemeni, M. N., Wijaya, L., & Shah, A. A. (2021). Influence ofecological and edaphic factors on biodiversity of soil nematodes. *Saudi Journal of Biological Sciences, 28*(5), 3049–3059. doi:10.1016/j.sjbs.2021.02.046 PMID:34025181

Norton, D. (1983). Maize nematode problems. *Plant Disease, 67*(3), 253–256. doi:10.1094/PD-67-253

Norton, D., & Hoffmann, J. (1975). *Longidorus breviannulatus* n. sp. (Nematoda: Longidoridae) associated with stunted corn in Iowa. *Journal of Nematology, 7*, 168–171. PMID:19308150

Okeniyi, M., Afolami, S., Fademi, O., & Oduwaye, O. (2013). Effect of botanical extracts on root-knot nematode (*Meloidogyne incognita*) infection and growth of cashew (*Anacardium occidentale*) seedlings. *Academia Journal of Biotechnology, 1*, 81–86.

Osei, K., Gowen, S., Pembroke, B., Brandenburg, R., & Jordan, D. (2010). Potential of leguminous cover crops in management of a mixed population of root-knot nematodes (*Meloidogyne* spp.). *Journal of Nematology, 42*, 173–178. PMID:22736854

Rama, K., & Dasgupta, M. K. (1998). Biocontrol of nematodes associated with mandarin orange decline by the promotion of predatory nematode *Iotonchus tenuicaudatus* (Kreis, 1924). *Indian Journal of Nematology, 28*, 118–124.

Riley, T., & Reardon, T. (1995). Isolation and characterization of *Clavibacter tritici* associated with *Anguina tritici* in wheat from Western Australia. *Plant Pathology, 44*(5), 805–810. doi:10.1111/j.1365-3059.1995.tb02739.x

Santana, H. P. E., Comerlato, A. P., Nasu, E. G. C., & Furlanetto, C. (2009). Variabilidade genética em populações de campo do nematoide de cisto da soja provenientes dos estados do Paraná e Rio Grande do Sul. *Tropical Plant Pathology, 34*(4), 261–264. doi:10.1590/S1982-56762009000400010

Santo, G., O'Bannon, J., Finley, A., & Golden, A. (1980). Occurrence and host range of a new root-knot nematode (*Meloidogyne chitwoodi*) in the Pacific Northwest. *Plant Disease, 64*(10), 951–952. doi:10.1094/PD-64-951

Sasser, J., & Freckman, D. (1987). A world perspective on nematology: The role of the society. In J. Veech & D. Dickson (Eds.), *Vistas on Nematology* (pp. 7–14). Society of Nematologists, Inc.

Schmitt, D. P., & Barker, K. R. (1981). Damage and reproduction potentials of *Pratylenchus brachuyrus* and *P. penetrans* on soybean. *Journal of Nematology, 13*(3), 327–332. PMID:19300771

Sehgal, M., & Gaur, H. S. (1999). *Important nematode problem of crops in India and their management.* National Centre for Integrated Pest Management.

Sikandar, A., Zhang, M. Y., Wang, Y. Y., Zhu, X. F., Liu, X. Y., Fan, H. Y., Xuan, Y. H., Chen, L. J., & Duan, Y. X. (2020). Review article: *Meloidogyne incognita* (root-knot nematode) a risk to agriculture. *Applied Ecology and Environmental Research, 18*(1), 1679–1690. doi:10.15666/aeer/1801_16791690

Sikora, R., Coyne, D., Hallman, J., & Timper, P. (2018). Plant Parasitic Nematodes—The World's Most Important Crop pathogen? *CABIBlog.* Available online: https://blog.cabi.org/2018/07/24/plant-parasitic-nematodes/

Sikora, R., & Fernandez, E. (2005). Nematode parasites of vegetables. In Plant parasitic nematodes in subtropical and tropical. Wallingford: CABI. doi:10.1079/9780851997278.0319

Singh, S., Singh, B., & Singh, A. P. (2015). Nematodes: A threat to sustainability of agriculture. *Procedia Environmental Sciences, 29*, 215–216. doi:10.1016/j.proenv.2015.07.270

Small, R. W. (1979). The effects of predatory nematodes on populations of plant parasitic nematodes in pots. *Nematologica, 25*(1), 94–103. doi:10.1163/187529279X00424

Soriano, I., Prot, J., & Matias, D. (2000). Expression of tolerance for *Meloidogyne graminicola* in rice cultivars as affected by soil type and flooding. *Journal of Nematology, 32*, 309–317. PMID:19270982

Steiner, G., & Heinly, H. (1922). The possibility of control of *Heterodera radicicola* and other plant injurious nematodes by means of predatory nemas, especially *Mononchus aquaticus. Journal of the Washington Academy of Sciences, 12*, 367–385.

Stirling, G. (1991). *Biological Control of Plant-parasitic Nematodes.* CAB International.

Thorne, G. (1927). The life history, habits and economic importance of some mononchs. *Journal of Agricultural Research, 34*, 265–286.

Turner, S., & Evans, K. (1998). The origins, global distribution and biology of potato cyst nematodes (*Globodera rostochiensis* (Woll.) and *G. pallida* Stone. In R. Marks & B. Brodie (Eds.), *Potato Cyst Nematodes: Biology, Distribution, and Control* (pp. 7–26). University Press,pp.

Webster, J. M. (1972). Nematodes and biocontrol. In J. M. Webster (Ed.), *Economic Nematology* (pp. 469–496). Academic Press.

Weibelzahl-Fulton, E., Dickson, D., & Whitty, E. (1996). Suppression of *Meloidogyne incognita* and M. *javanica* by *Pasteuria penetrans* in field soil. *Journal of Nematology*, *28*, 43–49. PMID:19277344

Wood, F. H. (1973). Nematode feeding relationships. Relationships of soil dwelling nematodes. *Soil Biology & Biochemistry*, *5*(5), 593–601. doi:10.1016/0038-0717(73)90049-7

Wyss, U., & Grootaert, P. (1977). Feeding mechanisms in *Labronema* sp. *Med. Fac. Landbouww. Rijskuniv. Gent.*, *42*, 1521–1527.

Xiong, J., Zhou, Q., Luo, H., Xia, L., Li, L., Sun, M., & Yu, Z. (2015). Systemic nematicidal activity and biocontrol efficacy of *Bacillus firmus* against the root-knot nematode *Meloidogyne incognita. World Journal of Microbiology & Biotechnology*, *31*(4), 661–667. doi:10.100711274-015-1820-7 PMID:25672545

Xu, Z., Zhao, Y., Yang, D., Sun, H., Zhang, C., & Xie, Y. (2015). Attractant and repellent effects of sweet potato root exudates on the potato rot nematode, *Ditylenchus destructor. Nematology*, *17*(1), 117–124. doi:10.1163/15685411-00002856

Yeates, G. W. (1969). Predation by *Mononchoides potohikus* (Nematoda: Diplogasteridae) in laboratory culture. *Nematologica*, *15*(1), 1–9. doi:10.1163/187529269X00029

Zhou, L., Yuen, G., Wang, Y., Wei, L., & Ji, G. (2016). Evaluation of bacterial biological control agents for control of root-knot nematode disease on tomato. *Crop Protection (Guildford, Surrey)*, *84*, 8–13. doi:10.1016/j.cropro.2015.12.009

Chapter 5
Entomopathogenic Nematodes for the Control of Insect Pests, as Well as Insect Pathogens for the Control of Plant Pathogenic Nematodes

Hanan Alfy
https://orcid.org/0000-0002-0042-7122
Agricultural Research Center, Plant Protection Institute, Egypt

Marian Malak
https://orcid.org/0000-0002-3468-0716
Faculty of Education, Alexandria University, Egypt

ABSTRACT

Infection by entomopathogen nematodes (EPN) occurs by an infectious juvenile that establishes itself in the body cavity of an insect and suppresses the immune response of the insect. Temperature variations and desiccation are the main abiotic factors affecting EPN survival, and UV light destroys EPNs more quickly than chemical insecticides. Spraying directly on the soil surface is the most powerful strategy for applying EPNs. IJs enter the insect body through natural pores and produce hydrolytic enzymes. Entomopathogenic nematodes (EPNs) have been shown to have a variety of impacts on insect immunology, including the destruction of hemocytes, the destruction of proteins that recognize nematodes, and the disruption of prophenoloxidase activation. The mode of action of bacteria inside the nematodes has been discovered through extensive Bt strain screening and cry gene sequencing. Cry proteins can enter nematodes through the stylet and create lytic holes in their cell membranes. LPS has an antagonistic action against nematodes.
DOI: 10.4018/978-1-6684-8083-0.ch005

INSECT-PARASITIC NEMATODE EFFICACY

One of the animal kingdoms with the most diversity is the nematode. Nematodes can be free-living or parasitic due to their resilient cuticle, which allows them to adapt to drastic environmental changes and colonize a variety of settings. (Brivio & Mastore, 2018). Nematodes are classified into two groups: beneficial and non-beneficial (plan-parasitic) nematodes (Van & Malan, 2014). Entomopathogenic nematodes (EPNs) are beneficial nematodes and can be distinguished from other parasitic nematodes by their capacity to kill insect hosts by partnering with insect-pathogenic bacteria (Brivio & Mastore, 2018). EPNs are tiny, non-segmented, colorless, elongated, insect-parasitic roundworms that exist naturally in soil (Van & Malan, 2014). EPNs are considered valuable organisms since they are harmless to humans and are typically specific to their target insect, making them ideal for integrated pest control strategies (Shapiro-Ilan et al., 2012). Recent developments have resulted in a considerable number of biocontrol products that are currently on the market. Several of the widely regarded commercial stocks contain the microorganisms *Bacillus firmus, Pasteuria penetrans,* and *Purpureocillium lilacinus*.

EPNs are excellent animal models for biological insect management and studying interspecies interactions such as mutualism, symbiosis, and parasitism. EPNs include two nematode genera, *Steinernema* and *Heterorhabditis* (Castillo et al., 2011; Silva et al., 2021). *Steinernema* and *Heterorhabditis* are identified by their accompanying Gram-negative symbiotic bacteria, *Xenorhabdus* and *Photorhabdus*, which are directly responsible for insect mortality (Van & Malan, 2014). The presence of symbiotic bacteria in the specific intestinal vesicle of *Steinernema*, as well as in the initial portion of the intestine and throughout the intestinal lumen of *Heterorhabditis* (Ozakman & Eleftherianos, 2021). Infection by nematodes occurs when an infectious juvenile (IJ) establishes itself in the body cavity of an insect and suppresses the insect's immune response (Van & Malan, 2014). The bacterial cells then replicate and produce severe toxins with a high insecticidal capability to cause the death of the host in less than two days (Tomar et al., 2022), whereas most biological agents need days or weeks to kill the insect (Gozel et al., 2016).

The ideal nematode must be able to escape an insect's defenses and cause pathogenesis in order for EPNs to be effective at suppressing insects. (Shapiro-Ilan et al., 2014). Furthermore, the stage of the target insect is necessary to achieve successful control by EPNs. Even though adult insects are frequently more mobile, insect larvae and pupae are more vulnerable to nematodes. Of that kind, variations in susceptibility might emerge from a variety of factors, including higher activity levels and avoidance behaviors in adults, immune system differences, or physical barriers to nematode invasion (Labaude & Griffin, 2018). In addition, the low sensitivity of some pupae may be related to a lack of natural entrance pathways for nematodes, as well as a thicker cuticle (Garriga et al., 2018). IJs have the ability to infect pupae and then kill adults once the host undergoes metamorphosis. (Williams et al., 2015). EPNs can live for several weeks to months in the infectious stage and are straightforward, affordable animals to culture. (Gozel et al., 2016).

EPNs are strictly limited in that they belong to live microorganisms that need specific conditions to be successful in murdering the insect host, like some biological control agents. Temperature variations and desiccation are the main abiotic factors affecting EPN survival. (Vashisth et al., 2013). In temperatures between 20 °C and 30 °C, EPNs can be more effective than pesticides and are more influenced by poor soil type, depth, and irrigation frequency (Shapiro-Ilan et al., 2006). The soil in the treated area should then be kept moist for at least two weeks (Vashisth et al., 2013). Similarly, UV light destroys EPNs more quickly than chemical insecticides. Hence, applications should be carried out in a fashion that avoids direct sunlight, such as early morning or midnight treatments, depending on the biology of the nematodes (Vashisth et al., 2013). Application to aboveground target regions is difficult because of the worm's susceptibility to desiccation and UV radiation; nevertheless, superior formulations have increased nematode efficacy against aboveground target pests (Shapiro-Ilan et al., 2010). Spraying directly on the soil surface is the most powerful strategy for applying entomopathogenic nematodes (Vashisth et al., 2013).

Additionally, IJs use a variety of cues and signals to navigate different types of soil and locate hosts (Koppenhöfer et al., 2020). Alarm signals are produced by EPNs in response to volatile cues from their host (Baiocchi et al., 2017) or plants attacked by herbivorous insects (Turlings et al., 2012). IJs can differentiate between target hosts as well as infected and uninfected hosts (Koppenhöfer et al., 2020). Significantly, IJ cues seem to have a negative impact on the performance of root insect herbivores. (Helms et al., 2019), highlighting the dual benefits of employing EPNs in pest management strategies. Important contributions to the rapidly growing area emphasize the significance of a thorough knowledge of these finely tuned interactions when considering EPNs as a pest control method (Koppenhöfer et al., 2020).

The Mode of Action

The infectious juvenile (IJ) stage of EPN infections initiates the infection cycle. (Shapiro-Ilan et al., 2014) (Figure 1). The third-stage juveniles' range in length from 0.4 to 1.1 mm and are free-living, non-feeding, and developmentally stopped. (Devi & Nath, 2017). Because of its greater susceptibility to abiotic stressors and reliance on nutrients gained from its prior host, the IJ is the only stage that can survive outside the host (Koppenhöfer et al., 2020). These IJs aggressively search for insect hosts using well-developed chemoreception and substantial mechanoreception to detect insect movement (Chaisson & Hallem, 2012).

EPN infection results when IJ enters the insect body through natural pores such as the mouth, anus, spiracles, or, in rare cases, the cuticle of the insect (Shapiro-Ilan et al., 2014). After entering the body cavity, *Steinernema* IJs produce hydrolytic enzymes such as serine proteases (serine carboxypeptidases, trypsin, aspartyl proteases and zinc carboxypeptidases) and protease inhibitors (Lu et al., 2017). These proteases may aid in the transmission of IJs from the gut of insects to establish themselves in the insect hemolymph (Kenney & Eleftherianos, 2016; Lewis & Clarke, 2012). Various of these proteases have been isolated, and their contribution to pathogenesis seems to be via immunological inhibition, such as suppressing phenoloxidase activity and/or influencing hemocyte performance. In *Heterorhabditid* IJ, however, penetration through the cuticle may be aided by a specific "tooth" structure across the intersegmental membranes of the insect. *Heterorhabditis* does not release proteases and is not pathogenic in the absence of its bacterial symbionts (Lewis & Clarke, 2012). *Heterorhabditis* IJs induce a lower immunological response than colonized nematodes or *Photorhabdus* on their own, indicating that they are not quickly recognized by the immune response of insects (Eleftherianos et al., 2010). Furthermore, *Heterorhabditis* appears to interact with the cellular immune response by inhibiting hemocyte phagocytic capability during the early stages of infection, leading to hemocyte death subsequently (Castillo et al., 2011).

Following insect host penetration, the IJ is induced to discharge the bacterial symbionts that it carries into the hemolymph of the insect host (Lewis & Clarke, 2012). Then, IJ develops into an adult hermaphrodite nematode (in the case of *Heterorhabditis*) or either an adult male or female nematode (in the case of *Steinernema* except for *S. hermaphroditum*) (Lewis & Clarke, 2012). However, the discharge of nematode mutualistic bacteria is delayed by 30 minutes in the case of *Heterorhabditis* and 4 to 6 hours in the case of *Steinernema* (Kenney & Eleftherianos, 2016). The pigments formed by the monoculture of mutualistic bacteria living in the host are responsible for the insect's body color (Gozel et al., 2016). Insect cadavers infected with nematodes do not smell awful, and the insect cuticle remains undamaged until late in the infection period (Paschapur et al., 2018). The EPN and its associated bacteria have a mutualistic relationship: the nematode enables the transport of the bacteria through the soil to the hemolymph of an insect to multiply rapidly. The bacteria protect the dead insect (through various toxins and antimicrobial agents) from invasion by undesirable and contaminating soil microbes (Gozel et al., 2016; Van & Malan, 2014). In turn, the nematode depends on the bacteria's pathogenic capacity to kill the insect host shortly after through septicemia (Tomar et al., 2022).

The infective nematodes live in the host for one-to-many generations, relying on the bacteria and nutrients in the dead host (Shan et al., 2019). When the host cadaver's resources are expended, a new cohort of IJs with symbiotic bacteria cells in their malfunctioning digestive system leaves to seek new insect hosts (Koppenhöfer et al., 2020). A generation from egg to egg takes at least 4 to 7 days. Across most circumstances, there are approximately at least two generations within a host before new juveniles emerge looking for a new host, hence the duration between the first infection by juveniles and the appearance of "new" juveniles can range from 8 to 14 days (Van & Malan, 2014). The length of time is influenced by the temperature of the soil, the host size, and the type of nematode involved (Gozel et al., 2016).

The Immunological Response of Insects Towards EPNs

Some insects have evolved behavioral defenses against EPNs infections, which include intense aggressive and grooming behavior when in contact with IJs to avoid infection or even ignoring an area with many IJs (Alvandi et al., 2017; Koppenhöfer et al., 2020). Insects may have also developed physical barriers to keep IJs from accessing their hemolymph, including the larval cuticle, adult exoskeleton, chitinous lining of the tracheae, and peritrophic membrane of the midgut (Koppenhöfer et al., 2020; Tafesh-Edwards & Eleftherianos, 2022).

IJs must overcome the host's immunological response if they are to defeat insect behavioral and physical barriers (Shapiro-Ilan et al., 2017). EPNs often avoid the camouflage response by shedding the second-stage juvenile cuticle that still surrounds the IJs and depositing de-novo generated or host-sequestered immune components as camouflage to prevent detection by the immune system of the insect. In the absence of the bacterial symbiont, *Steinernema* spp. can interact with the immune response of insects through the synthesis and release of proteases, but they typically cannot complete their life cycle (Koppenhöfer et al., 2020). *Heterorhabditis* seems to actively modify the immune system of *Manduca sexta* by diminishing the phagocytic potential of circulating hemocytes (Lewis & Clarke, 2012). Symbionts in both EPN genera have complementary functions in bypassing insect defenses and effectively inhibiting the immunological response. Several toxins fatal to the insect are produced and released by *Xenorhabdus* and *Photorhabdus* species, as well as antibiotics to avoid subsequent infections of the corpse by other pathogens or scavengers (Koppenhöfer et al., 2020).

Cellular and humoral immunity is a physiological second line of defense. Phagocytosis, nodulation, and encapsulation are all part of the cellular immune recognition system. Inducible antimicrobial peptides, lysozyme, lectins, reactive oxygen radicals, the phenoloxidase cascade, and clotting are all components of the humoral system (Alvandi et al., 2017; Tafesh-Edwards & Eleftherianos, 2022). It has also been found that EPNs reduce phenoloxidase activity and hemocyte encapsulation (Ono & Yoshiga, 2019) through developed methods to prevent the encapsulation reaction by EPNs cuticles (Kenney & Eleftherianos, 2016). Furthermore, EPNs have been shown to have a variety of impacts on insect immunology, including the destruction of hemocytes, the destruction of proteins that recognize nematodes, and the disruption of prophenoloxidase activation (Labaude & Griffin, 2018; Ono & Yoshiga, 2019). In addition, the infection with *Steinernema carpocapsae* (Weiser) in the last instar larvae of Colorado potato beetles could display adult discoloration, and wing distortion (Ebrahimi et al., 2014) or cause a delayed metamorphosis in the case of the infection by *Steinernema feltiae* (Filgueiras & Willett, 2021).

Entomopathogenic Nematodes' Involvement as Biological Control Agents

Current food production systems nowadays involve pest control products and technologies that were developed as the result of research on biopesticides that have been shown commercially effective. At the present time, 20 species of *Heterorhabditis* and 90 species of *Steinernema* are acknowledged as potential biological control agents for a variety of economically relevant insect pests worldwide (Labaude & Griffin, 2018; Tomar et al., 2022). Because of the development of application procedures and the capacity to make them in vast numbers under laboratory conditions, EPNs are efficient biological control agents (Filgueiras & Willett, 2021). Furthermore, IJs are frequently fatal to kill their insect hosts within 24-48 hours. Because of the poisonous chemicals and proteins generated as secondary metabolites, EPN-associated bacterial endosymbionts have had a great potential for agricultural pest management in recent times and at least 13 EPN species have been marketed as biopesticides (Koppenhöfer et al., 2020; Thakur et al., 2022a; Thakur et al., 2022b). Different species of Heteroptera, Lepidoptera, Coleoptera, Diptera, and Hymenoptera have been successfully controlled by commercial products comprising *Heterorhabditis bacteriophora*, *Heterorhabditis megidis, S. feltiae,* and *S. carpocapsae* (Tarasco et al., 2023). Meanwhile, combinations of EPN species are widely utilized as an effective method to manage different insect pests and are increasingly supplied in commercial products (Wu et al., 2023).

Many chemical and biological insecticides, fungicides, herbicides, fertilizers, and growth regulators can be tank-mixed and sprayed simultaneously. Combining chemical treatments with nematodes can result in synergistic levels of insect death. Some compounds should be handled with caution or avoided, such as aldicarb, carbofuran, diazinon, iodine, methomyl, and other nematicides, since they have no compatible interactions. A successful quality control method is built on maintaining high viability and virulence during manufacture, formulation, and storage. This provides a cost-effective alternative to the pest control method and allows nematodes to be used in integrated pest management systems easily and quickly (Vashisth et al., 2013). EPNs have a significant role in the management of insect pests in a variety of agricultural crops, especially when the use of insecticides is prohibited, such as just before harvest (Nxitywa and Malan, 2023).

The EPN is mass-produced and does not require specialist application equipment since it is compatible with regular agrochemical equipment such as sprayers and irrigation systems (Shapiro-Ilan & Gaugler, 2002). With EPNs currently produced in water-soluble powder, readily dispersed granules, sponge, and water-based gel forms, there has been significant advancement in EPN formulation techniques (Nxitywa and Malan, 2023). These developments are complemented by studies into the biology and ecology of EPNs, such as the mechanisms of infection and environmental factors affecting their survival and behavior, which also helps to improve their field performance (Labaude & Griffin, 2018).

Bioinsecticide Potentials of Nematodes

Root-knot nematodes (RKN) are important parasites of a variety of crops, especially vegetables, and they significantly reduce crop yields, mostly in tropical and subtropical areas (Sikora & Fernandez, 2005). Certain biological insects, such as Bts, are nematicidal to the eggs and juvenile root knot nematodes *Meloidogyne javanica* that have just been hatched (Mohammed et al., 2008). A helpful bacterium called *Bacillus thuringiensis* (Bt) creates huge crystal proteins that are poisonous to several invertebrates during their sporulation phase, primarily insects and nematodes (Wei et al., 2003). *Bacillus thuringiensis* subsp. *mexicanensis* Btm treatments produced high efficacy against J2 *M. incognita* after 24, 48, and 72 hours of exposure in vivo. When compared to the control, the results showed a highly significant difference in the percentage of mortality compared with the untreated control, which was the same for two concentrations (20% and 30%) after 24, 48 hours and was respectively 87.5% and 90.6% after 72 hours. Also, there were notable variations in the mean number of the fresh weight of the roots, the number of galls per root, the number of egg masses per root, and the number of eggs per egg mass between the two Btm concentrations and the untreated control as well as between each other (Alfy et al., 2020). The root galls on tomato plants significantly lowered the hatching, motility, penetration, development, and reproduction of RKN by applying an effective Bt isolate (Peng et al., 2011).

To assess their effectiveness in controlling *M. javanica*, two types of bacterial isolates, bacterial suspension (BS) and spore/crystal mixture (SCM)—were examined under laboratory and field circumstances. In vitro, the BS of ToIr65 and ToIr67 had 70% more nematicidal activity than SCM. Two variants of ToIr65 considerably (by 51%) reduced the number of galls on tomato plants in pot trials, while ToIr67 had no such effect (Ravari & Moghaddam, 2015).

The Mode of Action of Bacteria Inside the Nematodes

More than 700 cry gene sequences have been discovered because of extensive Bt strain screening and cry gene sequencing (Crickmore et al., 2011). These sequences were classified according to how similar their amino acid sequences were in at least 70 different cry gene groups, where Cry5, Cry6, Cry12, Cry13, Cry14, and Cry21 showed nematicidal activity (Bravo et al., 2012). According to a recent study (Zhang et al., 2012), some crystal proteins can enter nematodes through the stylet. The cry protein affects gut epithelial cells by creating lytic holes in their cell membranes. The crystals disintegrate inside the nematode's gut once the target nematode larvae intake the toxin, which is followed by proteolytic activation (Crickmore, 2005). However, depending on the toxin type and nematode species, the poisoning mechanism may vary.

Reitz et al. (2002) demonstrated that lipopolysaccharides, or LPS (lipid.A), which can be isolated from bacterial cultures and is characterized as an essential component of the outer membrane of the cell, have an antagonistic action against the nematode *M. incognita*. This is in harmony with the endotoxin called LPS (lipid.A), which is released when a bacterial cell ends up dead.

The actual cause of *M. incognita* J2's death may have been due to the antagonist, particularly *P. fluorescens* (Khan et al., 2012), which released hydrogen cyanide, ammonia, phenazine, pyoleutorin, pyrrolnintrin, and 2, 4-diacetyl phloroglucinol (Whistler et al., 2000; Schoonbeek et al., 2003). In the study of native antagonists' nematicidal toxicity towards *M. incognita*, in-vitro tests were conducted using *Pseudomonas fluorescens, Bacillus megaterium, Azotobacter sp., Rhizobium sp., Purpureocillium lilacinum,* and *Pochonia chlamydosporia*. The association between antagonists, concentration, and time demonstrated that all tested antagonists were capable of significantly increasing *M. incognita* J2 mortality with increases in filtrate concentrations and exposure times (Bharali et al., 2019). The wide range of Bt strains and the toxins they produce enable the production of bioinsecticides utilizing the bacteria themselves as well as the use of the toxin genes to create transgenic plants. Nematode J2-mortality was significantly reduced in *M. incognita* treated with Btm. Also, significant differences in the mean number of all plant parameter measurements were noted in all studies.

REFERENCES

Alfy, H., Ghareeb, R. Y., & Aziz, W. Z. (2020). Semi-field Application of a New Formulation Based on *Spodoptera litura* Nuclear Polyhedrosis Virus *and Bacillus thuringiensis* subsp. mexicanensis Against Cotton Leaf Worm, *Spodoptera littoralis* and Root Knot Nematode, *Meloidogyne incognita* in Egypt. *Current Applied Science and Technology*, 512-527.

Alvandi, J., Karimi, J., Ghadamyari, M., Sharifi, M., & Asoodeh, A. (2017). Physiological defense of the white grub, *Polyphylla adspersa* Motschulsky (Col., Scarabaeidae) against entomopathogenic nematodes. *Journal of Asia-Pacific Entomology, 20*(3), 878–885. doi:10.1016/j.aspen.2017.06.003

Baiocchi, T., Lee, G., Choe, D.-H., & Dillman, A. R. (2017). Host-seeking parasitic nematodes use specific odors to assess host resources. *Scientific Reports, 7*(1), 1. Advance online publication. doi:10.103841598-017-06620-2 PMID:28740104

Bharali, A., Bhagawati, B., & Kurulkar, U. (2019). Nematicidal toxicity of native antagonists against *Meloidogyne incognita. International Journal of Current Microbiology and Applied Sciences, 8*(2), 1385–1398. doi:10.20546/ijcmas.2019.802.163

Bravo, A., Gómez, I., Porta, H., García-Gómez, B. I., Rodriguez-Almazan, C., Pardo, L., & Soberón, M. (2012). Evolution of *Bacillus thuringiensis* Cry toxins insecticidal activity. *Microbial Biotechnology, 6*(1), 17–26. doi:10.1111/j.1751-7915.2012.00342.x PMID:22463726

Brivio, M. F., & Mastore, M. (2018). Nematobacterial Complexes and Insect Hosts: Different Weapons for the Same War. *Insects, 9*(3), 117. doi:10.3390/insects9030117

Castillo, J. C., Reynolds, S. E., & Eleftherianos, I. (2011). Insect immune responses to nematode parasites. *Trends in Parasitology, 27*(12), 537–547. doi:10.1016/j.pt.2011.09.001 PMID:21982477

Chaisson, K., & Hallem, E. (2012). Chemosensory behaviors of parasites. *Trends in Parasitology, 28*(10), 427–436. doi:10.1016/j.pt.2012.07.004 PMID:22921895

Crickmore, N. (2005). Using worms to better understand how *Bacillus thuringiensis* kills insects. *Trends in Microbiology, 13*(8), 347–350. doi:10.1016/j.tim.2005.06.002 PMID:15967665

Crickmore, N., Zeigler, D. R., Schnepf, E., Van Rie, J., Lereclus, D., Baum, J., Bravo, A., & Dean, D. H. (2011). *Bacillus thuringiensis toxin nomenclature*. Available at http://www. lifesci. sussex.ac.uk/Home/Neil_Crickmore/Bt/

Devi, G., & Nath, D. (2017). Entomopathogenic nematodes: A tool in biocontrol of insect pests of vegetables-A review. *Agricultural Reviews (Karnal)*, *38*(2). Advance online publication. doi:10.18805/ag.v38i02.7945

Ebrahimi, L., Niknam, G., Dunphy, G. B., & Toorchi, M. (2014). Side effects of immune response of Colorado potato beetle, *Leptinotarsa decemlineata* against the entomopathogenic nematode, *Steinernema carpocapsae* infection. *Invertebrate Survival Journal : ISJ*, *11*(1), 132–142.

Eleftherianos, I., Joyce, S., Ffrench-Constant, R. H., Clarke, D. J., & Reynolds, S. E. (2010). Probing the tri-trophic interaction between insects, nematodes, and *Photorhabdus*. *Parasitology*, *137*(11), 1695–1706. doi:10.1017/S0031182010000508 PMID:20500922

Filgueiras, C. C., & Willett, D. S. (2021). Non-lethal effects of entomopathogenic nematode infection. *Scientific Reports*, *11*(1), 1–8. doi:10.103841598-021-96270-2 PMID:34429457

Garriga, A., Morton, A., & Garcia-del-Pino, F. (2018). Is *Drosophila suzukii* as susceptible to entomopathogenic nematodes as *Drosophila melanogaster? Journal of Pest Science*, *91*(2), 789–798. doi:10.100710340-017-0920-6

Gozel, U., & Gozel, C. (2016). Entomopathogenic Nematodes in Pest Management. In Integrated Pest Management (IPM): Environmentally Sound Pest Management. IntechOpen. https://doi.org/ doi:10.5772/63894

Helms, A. M., Ray, S., Matulis, N. L., Kuzemchak, M. C., Grisales, W., Tooker, J. F., & Ali, J. G. (2019). Chemical cues linked to risk: Cues from below-ground natural enemies enhance plant defences and influence herbivore behaviour and performance. *Functional Ecology*, *33*(5), 798–808. doi:10.1111/1365-2435.13297

Kenney, E., & Eleftherianos, I. (2016). Entomopathogenic and plant pathogenic nematodes as opposing forces in agriculture. *International Journal for Parasitology*, *46*(1), 13–19. doi:10.1016/j.ijpara.2015.09.005 PMID:26527129

Khan, M. R., Khan, M. M., Anwer, M. A., & Haque, Z. (2012). Laboratory and field performance of some soil bacteria used as seed treatments on Meloidogyne incognita in chickpea. *Nematologia Mediterranea*, *40*, 143–151.

Koppenhöfer, A. M., Shapiro-Ilan, D. I., & Hiltpold, I. (2020). Advances in the use of entomopathogenic nematodes (EPNs) as biopesticides in suppressing crop insect pests. In *Biopesticides for sustainable agriculture*. Burleigh Dodds Science Publishing. doi:10.19103/AS.2020.0073.10

Labaude, S., & Griffin, C. T. (2018). Transmission Success of Entomopathogenic Nematodes Used in Pest Control. *Insects*, *9*(2), 72. doi:10.3390/insects9020072 PMID:29925806

Lewis, E. E., & Clarke, D. J. (2012). *Nematode parasites and entomopathogens*. Academic Press.

Lu, D., Macchietto, M., Chang, D., Barros, M. M., Baldwin, J., Mortazavi, A., & Dillman, A. R. (2017). Activated entomopathogenic nematode infective juveniles release lethal venom proteins. *PLoS Pathogens*, *13*(4), e1006302. doi:10.1371/journal.ppat.1006302 PMID:28426766

Mohammed, S.H., El Saedy, M.A., Enan, M.R., Ibrahim, N.E., Ghareeb, A. and Moustafa, S.A., (2008). Biocontrol efficiency of *Bacillus thuringiensis* toxins against root-knot nematode, *Meloidogyne incognita*. *Journal of Cell and Molecular Biology, 7*(1), pp.57-66.

Nxitywa, A., & Malan, A. P. (2023). Formulation of *Steinernema yirgalemense* in gel for long-term storage at room temperature. *Journal of Plant Diseases and Protection*, *130*(4), 809–816. doi:10.100741348-023-00764-2

Ono, M., & Yoshiga, T. (2019). Cellular immunity in the insect *Galleria mellonella* against insect non-parasitic nematodes. *Parasitology*, *146*(6), 708–715. doi:10.1017/S003118201800210X PMID:30567609

Ozakman, Y., & Eleftherianos, I. (2021). Nematode infection and antinematode immunity in Drosophila. *Trends in Parasitology*, *37*(11), 1002–1013. doi:10.1016/j.pt.2021.06.001 PMID:34154933

Paschapur, A. U., Lakshmi, K. V., & Pawar, B. S. V. (2018). Survey for the presence of entomopathogenic nematodes (EPNs) in sugarcane growing regions of Telangana state. *Journal of Entomology and Zoology Studies*, *6*(6), 320–326.

Peng, D., Chai, L., Wang, F., Zhang, F., Ruan, L., & Sun, M. (2011). Synergistic activity between Bacillus thuringiensis Cry6Aa and Cry55Aa toxins against *Meloidogyne incognita*. *Microbial Biotechnology*, *4*(6), 794–798. doi:10.1111/j.1751-7915.2011.00295.x PMID:21923640

Ravari, S. B., & Moghaddam, E. M. (2015). Efficacy of Bacillus thuringiensis Cry14 toxin against root knot nematode, *Meloidogyne javanica*. *Plant Protection Science*, *51*(1), 46–51. doi:10.17221/93/2013-PPS

Reitz, M., Oger, P., Meyer, A., Niehaus, K., Farrand, S. K., Hallman, J., Sikora, R., & Richard, S. A. (2002). Importance of the Oantigen, core-region and lipid A of rhizobial induction of systemic resistance in potato to *Globodera pallida*. *Nematology*, *4*(1), 73–79. doi:10.1163/156854102760082221

Schoonbeek, H., van Nistelrooy, J. G., & de Waard, M. A. (2003). Functional Analysis of ABC Transporter Genes from *Botrytis cinerea* Identifies BcatrB as a Transporter of Eugenol. *European Journal of Plant Pathology*, *109*(9), 1003–1011. doi:10.1023/B:EJPP.0000003936.61182.14

Shan, S., Wang, W., Song, C., Wang, M., Sun, B., Li, Y., Fu, Y., Gu, X., Ruan, W., & Rasmann, S. (2019). The symbiotic bacteria Alcaligenes faecalis of the entomopathogenic nematodes *Oscheius* spp. Exhibit potential biocontrol of plant- and entomopathogenic fungi. *Microbial Biotechnology*, *12*(3), 459–471. doi:10.1111/1751-7915.13365 PMID:30618110

Shapiro-Ilan, D. I., Bruck, D. J., & Lacey, L. A. (2012). Principles of epizootiology and microbial control. *Insect Pathology*, *2*, 29–72. doi:10.1016/B978-0-12-384984-7.00003-8

Shapiro-Ilan, D. I., Cottrell, T. E., Mizell, R. F. III, Horton, D. L., Behle, R. W., & Dunlap, C. A. (2010). Efficacy of *Steinernema carpocapsae* for control of the lesser peachtree borer, *Synanthedon pictipes*: Improved aboveground suppression with a novel gel application. *Biological Control*, *54*(1), 23–28. doi:10.1016/j.biocontrol.2009.11.009

Shapiro-Ilan, D. I., & Gaugler, R. (2002). Production technology for entomopathogenic nematodes and their bacterial symbionts. *Journal of Industrial Microbiology & Biotechnology*, *28*(3), 137–146. doi:10.1038j.jim.7000230 PMID:12074087

Shapiro-Ilan, D. I., Gouge, D. H., Piggott, S. J., & Fife, J. P. (2006). Application technology and environmental considerations for use of entomopathogenic nematodes in biological control. *Biological Control*, *38*(1), 124–133. doi:10.1016/j.biocontrol.2005.09.005

Shapiro-Ilan, D. I., Han, R., & Qiu, X. (2014). Production of Entomopathogenic Nematodes. In J. A. Morales-Ramos, M. G. Rojas, & D. I. Shapiro-Ilan (Eds.), *Mass Production of Beneficial Organisms* (pp. 321–355). Academic Press. doi:10.1016/B978-0-12-391453-8.00010-8

Shapiro-Ilan, D. I., Hiltpold, I., & Lewis, E. E. (2017). Nematodes. In Ecology of Invertebrate Diseases (pp. 415–440). John Wiley & Sons, Ltd. doi:10.1002/9781119256106.ch11

Sikora, R. A., & Fernandez, E. (2005). Nematode parasites of vegetables. In *Plant parasitic nematodes in subtropical and tropical agriculture* (pp. 319–392). CABI Publishing. doi:10.1079/9780851997278.0319

Silva, M. S. O., Cardoso, J. F. M., Ferreira, M. E. P., Baldo, F. B., Silva, R. S. A., Chacon-Orozco, J. G., Shapiro-Ilan, D. I., Hazir, S., Bueno, C. J., & Leite, L. G. (2021). An Assessment of *Steinernema rarum* as a Biocontrol Agent in Sugarcane with Focus on *Sphenophorus levis*, Host-Finding Ability, Compatibility with Vinasse and Field Efficacy. *Agriculture*, *11*(6), 6. Advance online publication. doi:10.3390/agriculture11060500

Tafesh-Edwards, G., & Eleftherianos, I. (2022). Entomopathogenic Nematodes as a Model for Probing the Insect Immune System. *Nematodes as Model Organisms*, 42–57. https://doi.org/ doi:10.1079/9781789248814.0003

Tarasco, E., Fanelli, E., Salvemini, C., El-Khoury, Y., Troccoli, A., Vovlas, A., & De Luca, F. (2023). Entomopathogenic nematodes and their symbiotic bacteria: From genes to field uses. *Front. Insect Sci.*, *3*, 1195254. doi:10.3389/finsc.2023.1195254

Thakur, N., Tomar, P., Kaur, S., & Kumari, P. (2022a). Virulence of native entomopathogenic nematodes against major lepidopteran insect species of tomato (*Solanum lycopersicum* L.). *Journal of Applied Biology and Biotechnology*, *10*(1), 6–14. doi:10.7324/JABB.2022.10s102

Thakur, N., Tomar, P., Sharma, S., Kaur, S., Sharma, S., Yadav, A. N., & Hesham, A. E.-L. (2022b). Synergistic effect of entomopathogens against *Spodoptera litura* (Fabricius) under laboratory and greenhouse conditions. *Egyptian Journal of Biological Pest Control*, *32*(1), 1–10. doi:10.118641938-022-00537-3

Tomar, P., Thakur, N., & Yadav, A. N. (2022). Endosymbiotic microbes from entomopathogenic nematode (EPNs) and their applications as biocontrol agents for agro-environmental sustainability. *Egyptian Journal of Biological Pest Control*, *32*(1), 80. doi:10.118641938-022-00579-7

Turlings, T. C. J., Hiltpold, I., & Rasmann, S. (2012). The importance of root-produced volatiles as foraging cues for entomopathogenic nematodes. *Plant and Soil*, *358*(1), 51–60. doi:10.100711104-012-1295-3

Van, Z. C., & Malan, A. P. (2014). The role of entomopathogenic nematodes as biological control agents of insect pests, with emphasis on the history of their mass culturing and in vivo production: Review article. *African Entomology*, *22*(2), 235–249. doi:10.4001/003.022.0222

Vashisth, S., Chandel, Y. S., & Sharma, P. K. (2013). Entomopathogenic nematodes—A review. *Agricultural Reviews (Karnal)*, *34*(3), 163. Advance online publication. doi :10.5958/j.0976-0741.34.3.001

Wei, J., Hale, K., Carta, L., Platzer, E., Wong, C., Fang, S., & Aroian, V. (2003). *Bacillus thurigiensis* crystal proteins that target nematode. *Microbiology*, *100*, 2760–2765. PMID:12598644

Whistler, C. A., Stockwell, V. O., & Loper, J. E. (2000). Lon protease influences antibiotic production and UV tolerance of Pseudomonas fluorescens Pf-5. *Applied and Environmental Microbiology*, *66*(7), 2718–2725. doi:10.1128/ΛEM.66.7.2718-2725.2000 PMID:10877760

Williams, C. D., Dillon, A. B., Ennis, D., Hennessy, R., & Griffin, C. T. (2015). Differential susceptibility of pine weevil, *Hylobius abietis* (Coleoptera: Curculionidae), larvae and pupae to entomopathogenic nematodes and death of adults infected as pupae. *BioControl*, *60*(4), 537–546. doi:10.100710526-015-9658-3

Wu, S. Y., Tang, H., Zhang, C., Tang, F., Lin, J., Wang, Y., Chen, L., & Hou, Y. (2023). Potential of entomopathogenic nematode-infected insect cadavers for the biocontrol of the red imported fire ant *Solenopsis invicta*. *Pest Management Science*, ps.7637. Advance online publication. doi:10.1002/ps.7637 PMID:37384581

Zhang, F., Peng, D., Ye, P., Yu, Z., Hu, Z., Ruan, L., & Sun, M. (2012). *In vitro* uptake of 140 kDa *Bacillus thuringiensis* nematicidal crystal proteins by the second stage juvenile of *Meloidogyne hapla*. *PLoS One*, *7*(6), e38534. doi:10.1371/journal.pone.0038534 PMID:22737212

APPENDIX

Figure 1. Diagram representing the entomopathogenic nematode cycle, with the Photorhabdus and Xenorhabdus bacteria demonstrating their symbiosis. Infectious juveniles from plant soil enter the insect body cavity via the mouth, cuticle, anus, or spiracles. When nematodes gain access to the hemolymph, they release symbiotic bacteria into the insect host, suppressing the insect immune response and resulting in rapid septicemia. The nematodes and their symbiotic bacteria replicate within the insect, where they complete their life cycles before emerging from the insect body as a complex in search of new suitable insect hosts.

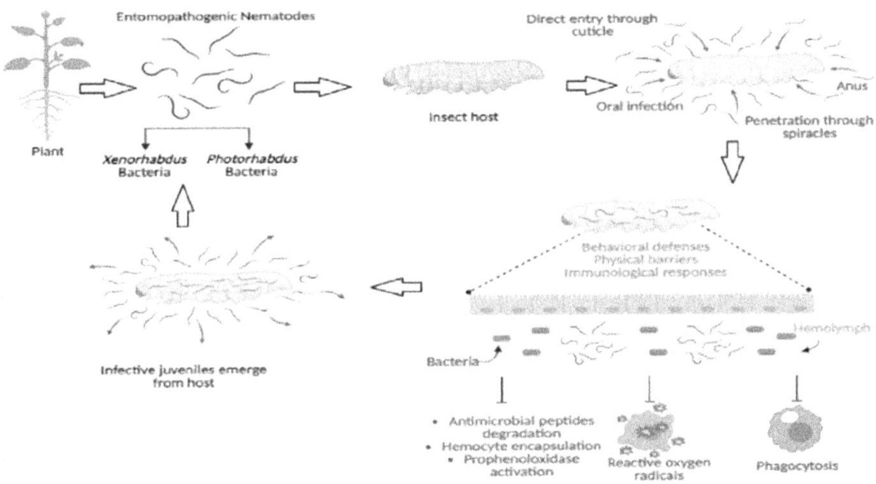

Chapter 6
Relationship Between Fruit Tree Growth Conditions in Orchards and the Attack by Nematodes

Magdi Ismaiel Saif
Faculty of Agriculture, Suez Canal University, Egypt

ABSTRACT

The variation in different fruit plant groups morphologically, anatomically, and physiologically beside its type in addition to the preferred place of cultivation all make a difference in the response to the infection by nematodes. In addition, the conditions in the orchard around the fruit plants affect the infection by nematodes. Some plant fruits are big trees (pears), but others are small (grapes). Some of them have a woody stem and others don't (banana). Deciduous fruits can grow in a cold climate, but evergreen fruits grow in a worm places. Orchards of pome and stone fruits are found in cold areas, but bananas grow in hot places and need a lot of water while olives prefer sandy soil with small amounts of water. The nematodes could be affected, and the amount could be decreased in sunny places with the decrease of water and by using mulch. Starting with clean and healthy seedlings or by grafting on tolerated rootstocks could minimize the nematode infection in orchards.

Plant parasitic nematodes are economically damaging fruit orchards. Their impact crop yield losses estimated to be billions of dollars per year. Most of the damage is caused by a relatively small number of nematode genera that attack fruits like Northern root-knot nematode (RKN) (*Meloidogyne hapla),* Root-Lesion Nematodes

DOI: 10.4018/978-1-6684-8083-0.ch006

(*Pratylenchus* spp.), Dagger Nematode (*Xiphinema* spp.), and Rinspot Virus (ToRSV) (*Criconemoides xenoplax*).

Whilst, pears could be affected by root lesion nematodes, grapes could be affected by root knot, root lesion and ring nematodes.

The symptoms of the infection by nematodes varied from one or more of the following symptoms on the tree parts: hot spots, stunted, short internodes, small roots, galls, and tree death. In addition to the symptoms on plants, there are some symptoms in the orchard due to Tomato ringspot virus and dangerous bacteria as the infection by bacterial spot, bacterial canker, and ring nematode connection. The management of nematodes has to start before planting fruit trees. Before starting in the procedures of nematode management, there is a need to evaluate the nematode genera and its populations in the orchard to choose the best tools for controlling it. Quarantines and regulations choosing the best rootstocks of the fruit trees and treating the orchard soil before planting trees in the orchard decrease the chance of affecting by nematodes.

The condition in orchard (climate and orchard management) could affect not only the plants but also affect its resistance to enemies like bacteria, fungi and nematodes.

Climate is the most important factor on which choice of the crop for a region depends. Climate is defined as the whole of average atmospheric phenomena for a certain region calculated for a period of thirty years. These phenomena are light, heat, water and air. The growth of the plants depends primarily on temperature. Availability of heat units decide the crop for a given place and the average temperature of a place gives an idea about heat units available on the basis of which crop can be decided.

The temperate fruit is suitable for growing in cold to temperate climate regions. Many of the world's favorite and well-known temperate fruits, such as apples, pears, peaches, nectarine, plums, apricot, cherries and grapes, are adapted to the climate of mid-latitudes with distinctly cold winters. They become dormant due to short day conditions in the region and need chilling of various lengths to break dormancy (Yahia, 2019).

Subtropical fruits require warm to moderate temperatures all year round, but can survive light frosts. The most common subtropical fruits are citrus fruits such as oranges, grapefruits, lemons and limes. Subtropical fruits include dates, figs, olives and pomegranates. This fruits are sensitive to chilling injury.

Tropical fruits are fruits that grow on trees native to the tropics. Frost and chilling are harmful for tropical and subtropical plants. Some examples of these fruits include bananas, plantains, mangos, papayas, Coconut, Cashew, and guava. All tropical fruits are sensitive to chilling injury, and therefore they should be maintained at high temperatures (Yahia, 2019). Banana is among the most popular tropical fruits. Fruit plants in this region require strong sunlight, a warm, humid climate, very mild winters, and cannot tolerate frost.

The orchard management starting with the cultivation of plants. In cultivation could be plant the trees in orchard alone or use the space between the row of trees to grow other plants (green mulching) or cover the ground under trees by mulch as pest and weed control. In clean cultivation distance among plants is kept clean by tillage and elimination of weeds. In regions in which soil is eroded for the duration of rains and drainage is poor, soil is cultivated and cover crops are grown among the rows for the duration of rains. As cover crops, legumes must be favored due to the fact they upload more N in soil via fixation of Atmospheric-N of their nodules. They additionally enhance soil condition, soil fertility and will increase water retention potential of soil and the organic complexes of soil.

Mulching is the covering of the soil around plants to create more favorable conditions for growth and to retain usable soil moisture. In this management system, open soil is covered with a loose mulch of straw, hay, crop residues, leaves, sawdust, and plastic. It is essentially a surface barrier to evaporation and prevents weeds from growing, reducing competition for nutrients. This is one of the important soil management practices adopted in some countries in the garden. It inhibits weed growth, regulates soil temperature, conserves moisture by protecting from sun and wind, improves soil structure by increasing soil organic matter levels and controlling erosion (Naik and Thippesh, 2014)

Among all the mulching materials, plastic mulches are becoming popular especially black polythene mulch, where weed control is desired. Mulching is common in tropics especially in crops like banana (polythene, straw mulch, banana trash and sugarcane trash), grapes (black polythene), pineapple.

Fruit tree planting spacing is often affected by a variety of factors, including climate and soil, variety selection, tree cultivation method, rootstock used, nature of irrigation and pruning techniques used. The distances between trees may vary according to the different plant systems. One of the most important aspects of successful fruit growing is optimal fruit tree spacing. The optimal distance regulates the correct use of the sun's rays, avoids competition for the absorption of nutrients by root collision and favors correct watering. Increasingly the latest in high density planting system technology, trees are planted at critical distances to maximize space. It is difficult to recommend exact spacing for fruit trees for each site or soil (Pollock, 2012). The spacing could be 1x1 to 3x4 m for grapes (depending in training method) while it reaches 2x2 to 3x3m with banana plants. It is recommended to cultivate apple, plum and peach on 4x4 m while increasing this space to be 5x5 or 6x6 m with apricot and pears.

There are some details, could be declared for the nematodes symptoms and its management in the orchards of grapes, stone fruits, pome fruits and subtropical or tropical fruits but it depends on variation in the morphology, anatomy and physiology of fruit trees and the orchard management in which cause difference response to the attack by different species of nematodes. For that it's needed to know some information on the biology of fruit trees as the following.

GRAPES

A grape is a fruit, botanically a berry, of the deciduous woody vines of the flowering plant genus Vitis. Most domesticated grapes come from cultivars of *Vitis vinifera*, a grapevine native to the Mediterranean and Central Asia. Minor amounts of fruit and wine come from American and Asian species (Creasy and Creasy, 2018). The grapevine is a woody perennial plant, which is differentiated into the above ground shoot system (canopy) and the underground root system.

The underground part consists of an underground trunk with the root system. The root system is the interface between aboveground part of the grapevine and the soil. The primary root, producing lateral roots that branch into smaller lateral roots. The number, depth and pattern of the lateral roots depend on rootstock, variety, climate and soil properties as they grow towards water and nutrient areas. Lateral roots produce many short, fine feeder roots which responsible for water and nutrient uptake. The roots also play a significant role as storage organs for carbohydrates, water and nutrients, which support the initial growth of the grapevines after the winter. Any vineyard grower wants to grow a healthy root system because the grapevine roots have many functions and thus directly affecting grape quality and quantity.

The Majority of grapevine's roots are found in the top 100 cm of soil, while individual roots can grow to down to 9 m depth. The fine roots are at depths of 30 cm of soil (Rijal, 2016). The distribution of roots depends on soil characteristics, rootstock, and cultural practices used in cultivation. Various factors affect root growth, such as irrigation, fertilization and pruning technique. Roots don't grow near harvest time or at winter (dormancy time). Little growth occurs in early spring. During the dormant season, grapevine roots can grow if the soil is warm enough (Ollat *et al.*, 2016).

Healthy root is the key to produce the best yield of grapes with high quality and quantity values. Knowing the timing of root growth and development is important in order to implement appropriate management practices, such as irrigation, fertilization and other floor managements. With appropriate irrigation management practice, the vine vigour could be controlled. In severe water deficit condition, flow and root growth stop because the grapevine root system can't uptake mineral nutrients from the soil. Nitrogen and calcium must be dissolved in the water then the roots can even dry out. On the other hand, too much water in the soil decreases or even prevents breathing and growth of the roots, due to an inhibition oxygen supply that cause disorder in grapevine growth that finally ending with the plant death (Hunter *et al.*, 2016).

Stone Fruits

Stone fruits are a group of fruits have thin skin and a "stone" in the center, which contains the seed. They're in season starting in early spring and continue through early fall. So many deciduous fruits like peaches, plums, apricots, and cherries fall into the stone fruit category (Hill, and Perry, 2012).

The domesticated peach tree (*Prunus persica*), is a deciduous tree which produces tasty, juicy fruits with wide variety characteristics and could be produce fruits with glossy skin (nectarines). It belongs to the rose family's Prunus genus, which also includes the cherry, apricot, almond, and plum (Sinha *et al.*, 2012). Peach trees can reach heights of 7 to 10 meters, but when they are pruned, they typically only reach 3 to 4 meters. The lanceolate leaves measure 8–15 cm in length and 2-3 cm in width. The flower has five petals with a diameter of 2.5 to 3 cm, produced in the early spring before the leaves. Depending on the cultivar, the fruit's skin can be velvety (like peaches) or smooth (like nectarines), and its flesh can be yellow or pale. The flesh is very delicate and easily bruised in some cultivars, but is fairly firm in some commercial varieties, especially when green. The seed is enclosed in a husk that resembles wood and is reddish-brown in color, oval in form, and 1.3-2 cm long. Depending on their resilience to diseases, pests, and other damages, peach trees have a shorter lifespan (10 to 25 years) than some other fruit trees (Layne and Bassi, 2008).

The apricot tree is a modest, 10–20 m tall tree with a trunk that can reach a diameter of 40–50 cm and a dense canopy. The ovate leaves have a rounded base, a pointed tip, and a coarsely serrated border. They are 5–9 cm long and 4–8 cm wide. The flowers have five white to pinkish petals and a diameter of 2–4.5 cm; they are produced singly or in pairs in the early spring before the leaves. The fruit is a drupe (stonefruit) about the size of a small peach, 1.5–2.5 cm in diameter, and can range in color from yellow to orange, frequently with a reddish tinge on the side that receives the most sun. The fruit's surface can be smooth (referred to botanically as glabrous) or velvety with very short hairs (botanically: pubescent). The flesh is usually juicy (Xi and Lei, 2020), but dry in some species such as *P. sibirica*. Its taste ranges from sweet to sour. A single seed or "kernel" is encased in a hard shell, often called a "stone", that is granular and smooth in texture except for three ridges running along one side (Roussos *et al.*, 2016).

Plums are a diverse species but commercially important plum trees are medium-sized, usually pruned to a height of 5 to 6 meters. Without pruning, the tree can reach 15 meters high and 10 meters wide. They bloom in different months in different parts of the world, from January in some places to early April in others. The fruit is usually medium in size, 2 to 7 cm in diameter, and globular to elliptical in shape. The pulp is firm and juicy. The skin of the fruit is smooth and has a natural waxy surface that adheres to the flesh. Prunes are drupes. That is, the fleshy fruit of the plum surrounds a single hard pit that surrounds the seed of the fruit (Erez, 2013).

A cherry tree belonging to the genus *Prunus* (Rosaceae). Commercial production includes cherries (*Prunus cerasus*), frozen or preserved and used in sauces and baked foods, and cherries (*P. avium*), are usually consumed fresh and preserved mostly in real or imitation maraschino liqueurs. Many species are grown as ornamentals for their prolific spring flowers, and their dark red wood is used to make fine furniture. Most cherry species are native to the northern hemisphere. The native range of the species from which the cultivated cherry is derived is believed to be in Western Asia and Eastern Europe, from the Caspian Sea to the Balkans. Cherries need the cold of winter to bloom in the spring. The tree blooms very early in spring, just after the peaches and earlier than the apple trees. Because of their fruits, mainly three types of cherries are cultivated. Cherries, sour cherries and, to a much lesser extent, sweet cherries, which are hybrids between sweet cherries and cherries. The cherry trees are large and fairly erect, reaching heights of up to 11 meters. The fruit is a fleshy drupe (stone), usually heart-shaped or nearly spherical, about 2 cm in diameter, ranging in color from yellow to red to almost black. Cherries are low in acidity. The higher the acidity of the cherry, the more distinctive it is. Cherry trees are smaller and rarely exceed 5 meters in height. The fruit is round or flat, usually dark red, and too high in acid, making it unsuitable for fresh consumption. Duke cherry is intermediate in both wood character and fruit character.

Almond is a deciduous tree, 4 to 12.2 m tall, with a trunk diameter of up to 30 cm. Young branches are green at first, then purple in sunlight, then turn gray after 2 years. Leaves are 8-13cm long, with saw teeth on the edge, petiole length is 2.5cm, flowers are white to light pink, 3-5cm in diameter, 5 petals, 1 or 2 pairs, bloom before leaves in early spring. Almonds grow best in Mediterranean climates with warm, dry summers and mild, wet winters. The optimal temperature for growth is 15-30°C, and tree buds require 200-700 hours of refrigerated storage at 7.2°C or lower to break dormancy. Almonds begin to bear fruit 3 years after planting. The tree bears full fruit 5-6 years after planting. Fruits ripen in autumn 7-8 months after flowering. The almond fruit is drupe, not a nut with 3.5 to 6 cm long. The outer integument or hull covering consisting of the skin and the fleshy mesocarp or pulp in other Prunus members such as plums and cherries is instead a thick, leathery gray-green rind (which has a fluffy appearance). It's called a corpus. Inside the body is a woody endocarp that forms a reticulated hard shell (similar to the outer shell of a peach pit) called a pyrene. Inside the shell is an edible seed, commonly referred to as a nut. There is usually one seed, but sometimes two are found. After the fruit ripens, the skin cracks and separates from the husk, forming a deciduous layer between the stem and fruit, which can cause the fruit to fall off the tree (Westwood, 1993).

Pome Fruits

Pome fruits are members of the plant family Rosaceae, sub-family pomoideae. They are fruits that have a "core" of several small seeds, surrounded by a tough membrane. The membrane is encased in an edible layer of flesh. Trees are deciduous and have a dormant winter period that requires cold temperatures for the tree to properly break dormancy in spring. Apples, pears, quince are pome fruits, grown from spring blossom and harvested from late summer through to late autumn (Hill and Perry, 2012).

Pome fruit is primarily attacked by root-lesion, dagger and stubby root nematodes. It is of critical importance to minimize the number of nematodes before planting new trees. The infection could be controlled by the integrated nematode management.

The apple tree is a deciduous tree, usually growing to a height of 2 to 4.5 m in cultivation and up to 9 m in the wild. In cultivation, the size, shape and density of branches are determined by the rootstock selection and pruning method. The leaves are alternately arranged, simple elliptical, dark green with serrated edges and slightly pubescent at the base. The flowers form in spring at the same time as the leaves open on spurs and long shoots. The flowers are 3-4 cm in size, white with gradually fading pink, 5 petals, and inflorescences with 4-6 flowers on raceme. The central flower of the inflorescence is called the "royal flower". It opens first and can develop a larger fruit. The fruit is a pome that ripens in late summer or fall, and varieties vary in size. Commercial growers tend to produce apples between 7 and 8.5 cm in diameter, depending on market preference. The skin of a ripe apple is usually red, yellow, green, pink, or maroon, but there are also many two- or three-color variants. The skin can also be wholly or partially brown, i.e. rough and brown. The skin is covered with a protective layer of epicuticular wax. The exocarp (pulp) are usually pale off-white, but pink, yellow, or green sheaths are also found (Ferree and Warrington, 2003).

Pears are native to the coastal and temperate regions of the Old World, from Western Europe and North Africa eastward through Asia. It is a medium-sized tree, reaching 10–17 m tall, often with a tall, narrow crown. several types of shrubs. Leaves alternate, simple, 2–12 cm long, glossy green in some species, densely hairy in others. The shape of the leaf varies from broad oval to narrow lanceolate. Most pears are deciduous, but one or two species in Southeast Asia are evergreen. Most are cold tolerant, able to withstand temperatures of −25 to −40 °C in winter, with evergreen species only tolerant down to −15 °C. The flowers are white, rarely tinted yellow or pink, 2–4 cm diameter, and have five petals. Like that of the related apple, the pear fruit is a pome, in wildest species 1–4 cm diameter, but in some cultivated forms up to 18 cm long and 8 cm broad; the shape varies in most species from oblate or globose, to the classic pyriform "pear shape" of the European pear with an elongated basal portion and a bulbous end. The fruit is composed of the receptacle or upper end of the flower stalk (the so-called calyx tube) greatly dilated. Enclosed within its cellular flesh is the true fruit: five 'cartilaginous' carpels, known colloquially as the "core". From the upper rim of the receptacle are given off the five sepals, the five petals, and the very numerous stamens. Pears and apples cannot always be distinguished by the form of the fruit. Some pears look very much like some apples, e.g. the nashi pear. One of the main differences is that the pulp of pear fruits contains stone cells (Erez, 2013).

Subtropical and Tropical Fruits

Olive (*Olea europaea*), is a species of small tree or shrub in the family Oleaceae, found traditionally in the Mediterranean Basin. Olives are cultivated in many regions of the world with Mediterranean climates, such as South Africa, Chile, Peru, Pakistan, Australia, Oregon, and California, and in areas with temperate climates such as New Zealand which has a temperate climate with rainy summers and dry winters Olives like hot weather and sunny positions without any shade, while temperatures below −10 C may injure even a mature tree. They tolerate drought well because of their sturdy and extensive root systems. Olive trees can remain productive for centuries as long as they are pruned correctly and regularly. The olive tree, *Olea europaea*, is an evergreen tree or shrub native to the Mediterranean, Europe, Asia, and Africa. It is short and squat, rarely exceeding 8-15 m in height. The leaves are silvery green, elliptical, 4–10 cm long and 1–3 cm wide. The trunk is usually lumpy and twisted. Small white feathery flowers with 10-lobed sepals and corollas, 2 stamens and forked stigmas appear on last year's trees as racemes, usually from the axils. The fruit is a small drupe with a length of 1 to 2.5 cm when ripe, and the flesh of wild plants is thinner and smaller than that of horticultural varieties. Olives are harvested in the green to purple stage (Therios, 2009, Pollock, 2012 and Jaiswal, 2020).

Bananas (*Musa* spp.) are herbaceous perennials with short underground rhizomes from which grow up to 3 m laterally from the rhizome (corm) and are mostly in the upper 40 cm of soil. Fewer roots grow vertically or deeper although rooting density and distribution are influenced by the texture and depth of the topsoil. Bananas thrive in the lowland tropical regions where rainfall is in excess of 1250 mm/year and there is a mean minimum temperature above 15°C. Banana plants are the largest herbaceous flowering plants. All aboveground parts of a banana plant grow in structures commonly referred to as "corm". Plants are usually tall, fairly sturdy, and tree-like, but what appears to be a stem is actually a "false stem" or pseudostem. Bananas grow in a variety of soils as long as the soil is at least 60 cm deep, well drained and not compacted. Banana plants grow the fastest of all plants, with an increase of 1.4 to 1.6 square meters of surface area per day. The leaves of banana plants consist of a stem (petiole) and a plate (plate). The base of the petiole expands to form a leaf sheath. The densely packed sheaths constitute the pseudo stems that support the plant. The edges of the shell converge when first created, giving it a tubular shape. When new growth appears in the center of the pseudostem, the edges move apart. Cultivated banana plants vary in height depending on the variety and growing conditions. Most are about 5 m tall, from the 'Dwarf Cavendish' plant, which is about 3 m tall, to the 'Gros Michel' plant, which is more than 7 m tall. The leaves are spirally arranged and can reach 2.7 m in length and 60 cm in width. They tear easily in the wind, creating the familiar leaf shape. When a banana plant matures, the corm stops producing new leaves and begins to form a peduncle or inflorescence. A growing stem develops inside the pseudostem, carrying immature inflorescences that eventually emerge at the top. Each pseudostem produces one inflorescence, commonly referred to as a "banana heart". In afforestation, only one of the branches is allowed to develop to maintain spacing. The inflorescence has many bracts (sometimes erroneously called petals) between the rows of flowers. Female flowers (which may develop into fruit) appear in rows down the stem (closer to the leaves) in male flower rows. The ovary is inferior, so that small petals and other parts of the flower appear at the end of the ovary. The banana fruit grows in large hanging bundles made up of layers (called "arms") from the centre of the banana, with up to 20 fruits per layer. The hanging bunches are 3–20 tiered bunches or commercially known as "banana stems" and weigh 30–50 kg. An individual banana fruit (popularly known as a banana or "finger") weighs an average of 125 g, of which about 75% water and 25% dry matter (nutrient table, bottom right). The fruit has been described as a "skinny berry" (Robinson and Saúco, 2010).

1. NEMATODE SYMPTOMS AND MANAGEMENT IN FRUIT ORCHARDS

Grapes

Grapes (*Vitis vinifera*) are attacked by numerous parasitic nematodes such as dagger nematode (*Xiphinema* spp.), root-knot nematode (*Meloidogyne* spp.), root-lesion nematode (*Pratylenchus* spp.), needle nematode (*Longidorus* spp.), ring nematode (*Criconemoides xenoplax*) and stunt nematode (*Tylenchorhynchus* spp.). The annual global loss due to grape nematodes is 12.5%. Meloidogyns are sedentary endoparasites that form galls on the roots of vines. Pratylenchus - a migratory endoparasite that causes root damage; *Tylenchulus semipenetrans* is a root-infecting endoparasite, adhering to the roots. Criconemoides, Tylenchorhynchus, feed ectoparasitically on roots. Other nematodes such as *Xiphinema* spp., *Longidorus* spp. and *Trichodorus* spp., in addition to ectoparasitic feeding on roots, act as vectors to transmit viruses to plants.

Due to the synergistic effect of nematodes and viruses, fruit set can be drastically reduced, reaching up to 80%, and even cause the vines to die. *Meloidogyne Arenaria*, *M. incognita* and *M. javanica* are major problems in most viticultural regions of California, the Mediterranean Basin and South Africa. *M. hapla* in South Australia; middle. Ethiopia in Chile.

Nematode infestation is manifested by slight yellowing of the leaves, deterioration of plant health and vine vigor, and reduction of crop quality and quantity. In addition, damage by nematodes to plant root tissues makes them susceptible to further infection by other pathogens present in the soil, such as bacteria and fungi. Ground symptoms in plants caused by plant parasitic nematodes are not very specific in most cases and are often confused with other problems such as lack of moisture, malnutrition, and fungal or bacterial infections (Askary *et al.*, 2018).

Dagger Nematodes

Dagger nematodes are root ectoparasites of grapevines by *Xiphinema* genus that ectoparasites armed with long stylets which they insert deep into root tissues normally at the growing tips of the fine feeder roots. It is results in a swollen root tip which are often confused with the galls caused by root-knot nematodes but the galls are produced only at the root tip. Suppressed or clubbed roots from which fine feeder roots arise, gives the appearance of a "witches broom", the resulting abnormalities in heavily parasitized roots. Due the infection, the result is significant reduction in top and root weight, inflorescence number and fruit size of grapes. Moreover, Xiphinema nematode acts as vector and is capable of transmitting virus causing fan-leaf virus disease in grapevine. Vine shoots have abnormal branching, double nodes, short internodes, zigzag growth, develop yellow mosaics on leaves, and show yellow stripes along the midribs of mature leaves. The disease can cause up to 80% fruit loss. Symptoms can be confused with herbicide and mite infestation.

Root-Knot Nematode

Meloidogyne root-knot nematodes are sedentary endoparasites that embed themselves in roots. *M. arenaria*, M. incognita and *M. javanica* are a major problem in most vineyards grown in moderate temperature sandy soils in California, the Mediterranean Basin and South Africa. Gall formation occurs in the root due to cortical hyperplasia, rostral hyperplasia, and may be associated with the formation of giant cells that are feeding sites for nematodes. The infection causes the leaves to turn yellow, reducing fruit size and grape yield.

Root-Lesion Nematode

The root nematode, Pratylenchus, is a migratory endoparasite that travels internally and intercellularly inside the root, feeding on cells in the cortex and causing apoptosis and destruction.

Nematodes move intercellularly and intracellularly in cortical tissues, where they feed and cause cell death and decay. During crop maturation, nematodes leave the roots and move into the soil, where they again increase soil populations (Walia and Bajaj 2003). As a result of this movement and nutrition, dark brown necrotic lesions appear in the epidermis, cortex, and endoderm cells of the root. The lesions initially appear as small, oblong, water-soaked spots. The necrotic spots coalesce over time, causing necrosis of the entire root. Severe nematode infestations can destroy the entire root system. The degree of damage depends on the phenol content in the cells, as the lesions darken due to the oxidation of phenol. Damage to the root surface opens the way for other pathogens, especially fungi, to enter the root system. Stunting, yellowing or tawny leaves, and plant dying in hot weather are symptoms seen in airborne plants (Askary *et al.*, 2012).

Needle Nematode

Longidorus needle nematodes are migratory ectoparasites that cause significant damage to many plant species. Nematodes enter the root cells with the help of the odontostyle, the anterior part of the stylet, and the posterior part of the odontophore, which help to differentiate deep areas within the plant roots. The root system is greatly reduced. The lateral, sometimes straight roots are very stunted. Bile has large cells with dense cytoplasm. Eating Longidorus directly into the vine root results in lower yields. Damage mainly occurs to seedlings and young branches. However, if the nematode transfers the virus (nepovirus) to the roots of the grapevine, the synergistic action of the nematode and the virus causes severe damage and greatly reduces the yield.

Ring Nematode

The ringed nematode *Mesocriconema xenoplax* is a sedentary ectoparasitic nematode. It has a roughly striated cuticle, giving it a segmented appearance, like a single row of rings on its body. The nematodes feed on the cells of the root bark of plants, puncturing or damaging them, weakening the vine. Top dressing leads to rapid darkening and destruction of the root system. The root system is stunted with few vegetative roots, and the vine is stunted, chlorotic, and weak in vigor.

Stubby Root Nematode

Feeding root nematodes can cause atrophied or stocky root systems, which is why these nematodes are commonly referred to as short-root nematodes. They feed on root hairs and epidermal cells adjacent to the elongation zone at the tip of the root. *Trichodorus* spp. is an economically important plant parasite that serves as a vector of viruses for plants.

Stunt Nematode

The stunt nematode *Tylenchorhynchus* spp. is a soil ectoparasite has many economically important agricultural crops, including grapes. They are found in a variety of soils but generally prefer sandy or loamy sandy soils for their activity. They are migratory ectoparasites that feed on cells of the epidermis and outer cortex, often at the tip of the root. Common symptoms exhibited by plants are stunting, leaf chlorosis and stunting of the root system.

Management

The planting material of seedlings should be selected from nematode free plants at nurseries. Crop rotation with non-host crops is sufficient when nematode numbers are high for three years. However, grapevine fan leaf virus can still be detected in nematodes stored for 4 years in rootless soil. A gap of at least one year between removing the vines and replanting will help reduce the nematode population as the nematodes will starve during this time. Soil solarization and mulching of wet area soil with obvious polyethylene plastic sheet for 4-6 weeks in warm summer. This will reason quick wave radiation from solar to effortlessly penetrate the polythene sheet and warmth the soil however the lengthy wave radiation leaving the soil not able to do so, subsequently retained below the sheet. The temperature will growth with inside the top soil layer through approximately 5-10 °C diploma over regular temperature for numerous hours every day thereby killing or inflicting metabolic exhaustion of many nematodes. Soil solarization is powerful in opposition to root-knot nematodes throughout tropical and subtropical elements of the world. The nematode populace may be decreased considerably through this approach however can't be completely eradicated. Using neem, castor bean, compost, or manure as soil fertilizer before planting can be an effective means of reducing nematode populations in the soil.

Planting cover crops, including mustard, arugula, and Sudan grass in the vineyard area can help reduce nematode populations in the area. Strawberry clover, Salina, is a poor host for *M. incognita*. Barley, Columbia, Sudangrass, SS-222, Vetch, and Cahaba White are *X. americanum* antagonists that are not x hosts. index and *M. hapla*. Rootstocks 420A and 101-14 are very resistant to ring nematode. Soil fumigation is one of the best ways to control plant parasitic nematodes. Fumigants such as metam sodium and 1,3-dichloropropene can be used in vineyards, but are expensive and require special equipment to apply. Other issues relate to technical knowledge and security issues. Biofumigation, which uses plant poisons to suppress the presence of nematodes in the soil. Cruciferous plants are used for biofumigation because they contain glucosinolates with maximum concentrations in the flowering stage. At this stage, the aerial parts are ground as finely as possible to destroy all plant cells to release the maximum amount of glucosinolates. Subsequent crops are planted at least 2 weeks after the start of biofumigation. This helps prevent damage to staple crops by nematodes.

Biological control using fungal nematode biological control agents such as *Paecilomyces lilacinus*, *Pochonia chlamydosporia*, *Trichoderma harzianum* and *Aspergillus niger* have be proven effective against various species of root-knot nematodes. These biological fungicides are environmentally friendly, economically viable, commercialized worldwide under a variety of trade names, and used by plant growers to control root-knot nematodes (Askary, 2017).

Any damage of root system gives chances of grapevine infections from the soil by bacteria, fungus and nematodes. To grow healthy grapevine orchard, it is important to use grafted vines on suitable rootstocks for the area of growing plants. It is very important to choose the rootstocks that facing the infection and bad conditions that could be start in the grapevine's orchard. Selection of rootstock variety is an important factor in vineyard development since rootstock can influence on vine growth, drought tolerance, pest resistance as well as nutrient uptake efficiency. And therefore help grapevine to growth in unfavourable environmental conditions, such as poor soil, low water or soil infection (Hunter *et al*, 2016).

Among the top ten rootstocks for 42 cultivars and accessions used in Australia, only Ramsey shows a unanimous root-knot nematode resistance rating to *M. incognita*, as well as to *M. javanicapt* 1103P. Ramsey consider a durable rootstock resistance to root-knot nematode pathotypes present in Australian vineyards (Smith *et al.*, 2016).

Grapevine rootstocks differ in their susceptibility to nematodes, either endoparasites as Meloidogyne or transmitting viruses as Xiphinema. There is a high variability among Vitus species and the related genus *Muscadinia* for their interaction with nematodes. *Meloidogyne* spp. among commercial rootstocks, the *V. champinii* cultivars Ramsey and Dog Ridge appear to exhibit a sustainable resistance to root-knot nematodes. Harmony and Freedom, bred later on from complex parentage including *V. champinii*, were firstly considered as controlling these nematodes. However, the emergence of virulent pathotypes of *M. incognita* and *M. arenaria* for both rootstocks has been described. In addition, 110R, 140Ru, 420A, 5BB, *V. riparia* 'Gloire de Montpellier', Borner and 1616C appeared resistant to moderately resistant to some virulent *Meloidogyne* spp. strains. 1613C showed resistance to *M. incognita* and *M. javanica*. Recently several breeding programmes conducted in the United States have resulted in the release of rootstocks that are resistant to several dagger nematodes. The USDA series and the RS series appear to express sustainable resistance towards a range of nematode populations (Schurig *et al.*, 2021).

SO_4 was used as a highly susceptible control and the rootstock cultivar Nemadex A.B. served as a reference partially resistant to nematodes. The rootstock cultivars Börner, Rici, and Cina were the candidates for evaluation. Generally, distinct root gall development was found on the roots of SO_4 after X. index inoculation, while galls were hardly found on roots of Nemadex A.B.

The rootstocks Salt Creek, Freedom, Harmony, St. George, Riparia Gloire, 101-14 Mgt, 3309C, 110R, 420A, and Matador were poor hosts for *M. hapla*. Populations of *M. hapla* varied in reproductive potential and virulence on own-rooted Chardonnay (Zasada *et al.*, 2018).

Grapevines either grafted to 'O39-16′ (RKN-susceptible) or 'Freedom' (RKN-resistant) in 3 gallon pots were inoculated with a RKN, *Meloidogyne incognita* (Kofoid & White) Chitwood. Nematode counts in the roots, were greater in the susceptible 'O39-16′ rootstocks than resistant 'Freedom' (Wallis, 2020).

Resistance to *Meloidogyne* spp. of Ramsey rootstock for several table grape varieties have been obtained with sandy soils. Ramsey and 99 Richter, would appear to be susceptible to *C. xenoplax*. Some cover crops resulted in reduction of nematode population levels in grapevine's orchards. The use of shortened irrigation cycles and shallower rootstocks of grapevines as well as the use of mulches and cover crops impacted on the presence of nematode pests in the soil (Storey *et al.*, 2017).

Use of certified rootstocks that are free from nematodes. The use of resistant rootstocks can be helpful in solving the grapevine infection. Nemadex, O39-16 and GRN1-5 are resistant rootstocks to X. index. Rootstock of Dog Ridge, Freedom, 101-14, 110R, 3309C, 420A, Ramsey, Harmony are resistant to *M. hapla* and *M. arenaria*. RS 2 and RS 9 have shown resistance to root-knot nematodes. GRN1-5 is resistant to root-knot and resistance to other nematode species. Host plant, UCD GRN1 and VR O39-16 are resistance to *M. rotundifolia*.

Stone Fruits

The major nematodes of stone fruits are root-knot nematode (*Meloidogyne* spp.), root-lesion nematode (*Pratylenchus* spp.), ring nematode (*Criconemella* spp.), dagger nematodes (*Xiphinema* spp.), spiral nematode (*Helicotylenchus* spp.), and pin nematode (*Paratylenchus* spp.).

Root-Knot Nematode (*Meloidogyne* spp.)

Root-knot (Gall) nematodes are considered a potential threat to peach production. These are sedentary endoparasites that cause root irritation. Affected plants produce small yields of reduced quality. Infectious stages that affect the plant's roots lead to gallbladder formation and the affected roots do not elongate. Affected root systems appear densely pubescent. Leaf yellowing, stunting, and early plant senescence are ground symptoms that eventually lead to low yield and fruit production (Jonathan 2010).

Ring Nematode (*Criconemella* spp.)

Ring nematodes are ectoparasitic and have long probes that help them reach the bark cells beneath the root cuticle. This nematode can cause bacterial cancer and short-lived peach tree syndrome. Transplanted trees in particular shorten the tree's lifespan. The roots of plants infected with nematodes often show longitudinal cracks and darken. Severe infection leads to microscopic root death. Affected plants show stunted growth, mineral deficiency syndrome and susceptibility to water stress.

Stone fruit trees could be die due to the infection of nematodes especially root-knot nematodes. The damage caused when stone fruit is reestablished on the same soils or when establish new trees between old trees that are scheduled to be removed. There possible infestation sources are water, plant material and soil. Water originating from rivers with significant agricultural activity can act as a nematode infestation source even it is a minor source of infestation compared to infected soil. Planting in relatively clean soil cause the trees to resist nematodes. It is important to remove old roots of the previous crop as possible. It is particularly the endo-parasitic root-lesion and root-knot nematodes that find shelter in the roots.

Management

Strategic management practices are advised to be adopted to overcome the menace of these tiny microorganisms. These include physical, cultural, biological, and chemical methods. Producers should be encouraged to use resistant rootstocks in areas with nematode problems (Ozarslandan and Tanriver 2018). Various nematode control tactics must be combined, such as deep plowing, hot water treatment of seedlings, seedbed nematode treatment, use of potential biological control agents and resistant stocking. The selected planting material must not be infested with nematodes.

planting material requires slightly higher temperatures to kill nematodes (50-52 °C hot water treatment for 5-10 minutes). Treatment of rootstocks with carbofuran prior to planting in the nursery helps reduce nematode infestation. Solarisation (temperature above 50°C) is a method to control soil-borne organisms and pathogens through the use of raised soil temperatures that reduces the nematode population. Most resistant rootstocks of stone fruit are exceptionally susceptiblc to root-knot nematodes and the trees will still be attacked by other nematodes. Peach rootstock Greenpac, Mantianhong, Okinawa, R-15-2, Tsukaba-4, Tsukuba-5, GF-31, GxN No. 15, Torinel, AD101, Monopol, Nemaguard, and Cadamum were found to be resistant to root nodule nematodes. P2175, Pixy, San Julian 655-2, Marianna 2624, Damsons PSM 101, P2032, Torinel and Red glow are rootstocks of plums that resistant to root-knot nematodes. Cherry, Morello, Mazarda, Montmorency and Stockton-Morello rootstocks have been found to be resistant to root knot nematodes.

Fumigation with methyl bromide, carbofuran or metam sodium may reduce nematode populations in soil. It is sometimes essential to fumigate the soil before the new trees are established especially when the nematode population is exceptionally high and no fallow period is planned. Biological fumigation (biofumigation) is a technique that uses certain plants own protection functions to control a range of organisms and pathogens, including nematodes and other pathogeneses as fungi, bacteria, insects and some weeds. The plants produce special volatile compounds that reduce the population of nematodes. Planting apricot trees with marigolds and oats helped control nematode populations (Sharma and Kashyap 2009). Promoting root health is essential that result in fewer nematode problems (Stirling, 2014).

Pome Fruits

Apple (*Malus domestica*) is considered as one of the most important deciduous tree fruit in the world which are propagated by budding or grafting the desired scion onto the seedling rootstock in the nursery. A large number of plant parasitic nematodes belonging to different genera have been reported to attack apple trees. Among them *Pratylenchus*, *Meloidogyne*, *Paratylenchus*, *Xiphinema* and *Longidorus* are of major economic importance as they cause pronounced deleterious effects on plant growth and productivity. Plant parasitic nematodes present in the soil parasitize the roots of apple plant and thus the disease acquired in the nursery later on introduce into the orchard. Besides, some nematode groups belonging to *Xiphinema* and *Longidorus* have also been reported to act as vectors for transmission of virus in apple trees. Pest nematodes (*Pratylenchus* spp.) is the cause of "soil sickness" in apple nurseries and orchards.

Symptoms of *Pratylenchus* spp. a migratory endoparasitic form of the nematode that causes marked necrotic lesions and discoloration of the roots, which are initially usually reddish-brown but later darken and eventually black. The lesions enlarge and coalesce, eventually leading to a complete band of roots and loss of cortical tissue as a result of rejection (Nazeer Udin, 2014). The shape of the feeding roots may be significantly reduced. A strong infection can destroy the entire root system. Severely affected plants can be easily pulled out of the soil. Symptoms of gout in plants include stunted growth, yellowing or tawny leaves, and wilting of the plant during hours of hot sunlight.

Pear

Several types of nematodes have been reported to be associated with pears (*Pyrus communis*), but only a few show signs of parasitism. The needle nematode *Longidorus elongatus* and the lesion nematode *Pratylenchus penetrans* are considered to be of great economic importance in Europe, while in Japan the root-knot nematodes *M. hapla* and *M. incognita* parasitize pears. Apple growers must be able to use certified nematode-free seedlings. Polychlorinated petroleum hydrocarbons significantly reduce the population density of the nematode *Pratylenchus* spp. in the apple orchard. Apple growers must have access to certified rootstock. Plant roots should be immersed in water above 40°C. Dagger nematode (*Xiphinema* spp.) *Xiphinema* spp. Because of their long, dagger-like spears, they are generally considered dagger nematodes. They are migratory ectoparasites and feed on newly emerging roots. These nematodes are of great economic importance in apple-growing regions around the world. During feeding, the nematodes throw their probes, causing rupture and death of epidermal and cortical cells, resulting in cell necrosis. Vegetative growth slows down, and at a later stage complete destruction of the root system occurs with signs of severe necrosis, with twisting and swelling of the roots near the root tips and shriveling of the proximal part of the root. It contains or necrotic cells produced by *X. diversicaudatum* and is located at the distal end of the root. Needle nematode (*Longidorus* spp.) It is a large ectoparasite with long spear-shaped needles in its mouth and parasitizes on roots. They feed deep into the root tip. During feeding, the virus is later transmitted by nematodes. Symptoms Stunning, branching, swelling, curling of root tips, and necrosis are some of the common symptoms caused by *L. elongatus* on apple plants (Askary *et al.,* 2021).

SUBTROPICAL AND TROPICAL FRUITS

Olive

Two root knot species, *M. incognita* and *M. javanica*, can infect olives (*Olea europaea*) and have been shown to significantly reduce seedling growth in grafting experiments and have been identified as factors to be considered in olive nurseries. *M. arenaria* race 2, *M. javanica* and *M. incognita* race 1 caused growth retardation, yellowing and leaf fall in seedlings of two common olive rootstocks. *M. incognita* infection exacerbates the fungus *Verticillium daliae* that causes wilt. Proposals for uncovered root plunges of seedlings in suspensions of nematicidal chemicals, earlier to exchange into groves, have too been advertised for decreasing root knot nematode infestation. Soil solarization strategies were as compelling as soil fumigation to disinfest olive nursery soils of *T. semipenetrans*, *P. vulnus* and *Criconemoides xenoplax*, in Mediterranean locales (Ali *et al.*, 2014).

Banana

The nematode species that have proven most detrimental to banana (*Musa* spp.) are those that engage in destruction of primary roots, disrupt fixation systems, and resulting in plant fall. The most common and important are some species of *Radopholus similis*, *Pratylenchus* spp. and *Helicotylenchus multicinctus*. It is also very common to find some sedentary parasites such as *Meloidogyne* spp. and *Rotylenchulus reniformis*, parasitic to the root system (Almeida *et al.*, 2018). For most banana propagation, the traditional method of reproduction is by cutting off the side shoots (suckers) that are scattered around the mother plant. These sprouts are then sown directly in the field, where they may be infected with nematodes. The development of meristem culture (micropropagation), in which tissues produce large numbers of new shoots in a sterile culture medium, revolutionized banana propagation. The ability to prevent the spread of certain pests and diseases, including nematodes, is one of the many advantages of plants produced using this system. Most commercial growers have adopted the use of laboratory plants when replanting is required. Commercial tissue culture laboratories capable of producing millions of plants are now being established worldwide (Jones, 2018).

2. RELATIONSHIP BETWEEN BIOLOGY OF FRUITS (MORPHOLOGY, ANATOMY AND PHYSIOLOGY) WITH THE ORCHARD CONDITIONS AND THE NEMATODE INFECTION

Increased gall and egg mass formation in *M. incognita* must be related to increased penetration rates in sandy soil conditions, which provide sufficient aeration due to the coarse nature of the soil for nematodes, resulting in increased mobility for improved root penetration. For *M. hapla*, sandy soil conditions are expected to favor healthy plant growth with low accumulation of inhibitory substances and adequate aeration, enhancing nematode growth and feeding activity (Kim *et al.*, 2017).

Mulching using the remaining stems at harvest had a significant impact on the production and respiration of the fungivores, and the yields were higher by 71.8%, 62.7%, and 71.5%, respectively with 100% coverage compared to no coverage, 33% and 67% (Luo *et al.*, 2021).

After mulching with straw, the number of soil nematodes was lower than that of the control. However, metabolic and structural traces of omnivorous predator nematodes showed higher values than those of the control group (Luo *et al.*, 2021). The abundance of plant parasitic nematodes and omnivorous predators was negatively correlated with ammonium nitrogen and dissolved organic nitrogen, while soil moisture was negatively correlated with the abundance of all nematodes (Song *et al.*, 2020).

Parasitic loads were highest in summer to the prevalence of Heterakis gallinarum by 22.5% and the mean intensity was 26.83. It was found that parasitic load occurred the most in summer. The overall prevalence of *Capillaria* spp., was 11.75% and the mean intensity was 4.59. Fall had the highest parasitic load. The most numerous species was identified as *A. galli*. A significant relationship has been shown between seasonality and the prevalence of parasitic parasites (Ara *et al.*, 2021).

The grape's cultivars Harmony, Freedom and UCD-GRN rootstocks are resistant to nematodes (Ferris *et al.*, 2012).

Nematode numbers can be minimized by using new Cavendish farms established from tissue culture plants during fallow periods or after crop rotations (Robinson and Saúco, 2010).

VR039-16, Schwartzmann and Freedom grape rootstocks indicate resistance to nematodes (Ferris *et al.*, 2012).

The nematode life cycle is completed in 4-6 weeks depending on soil temperature. Nematodes are active in warm, moist soils that support the growth of host plants and can produce several generations in a single season, increasing population density by several folds (verdejo-Lucas and Talavera, 2009).

Prunus seedlings have traditionally been used as rootstocks because they adapt well to the local ecological and edaphic conditions prevailing in a region. Therefore, sweet and sour cherries are usually grafted onto seedlings of *P. mahaleb* or *P. avium* while peaches, plums and apricots are grafting on *P. persica*, *P. cerasifera*, *P. domestica*, *P. insititia*, or *P. salicina*, or their interspecies hybrids for protection against nematodes. Peach-almond hybrids 'Nemaguard', 'Nemared', 'Flordaguard', 'Okinawa', 'Cadaman' and 'Guardian' rootstocks were resistant to nematodes *M. incognita*, *M. javanica*. In California, 85 percent of almond, peach, plum, and nectarine plantations were replanted on Nemaguard (*P. persica* × *P. davidiana*) because it conferred field resistance to all types of root-knot nematodes. In plum, root-knot nematode resistance was present in 'Mariana' (*P. cerasifera* × *P. munsoniana*) and 'Myrobalan' (*P. cerasifera*). Cloning 'Myrobalan' completely inhibits root nematode strains and provides a stable, high level full spectrum. *M. incognita*, *M. arenaria* and *M. javanica* (Ciancio and Mukerji, 2009). Exposure of soil to sunlight is an effective way to control olive nematodes (Stirling, 2014).

The use of pathogen-free plant material and nematode-free soil during olive propagation is essential to reduce or minimize the effects of pathogens (especially *V. dahliae* nematodes) from infection single or concomitant during early development of the olive trees to prevent or avoid its spread (Therios, 2009).

In nurseries, clear polyethylene film and black polyethylene film are recommended for orchard growers to reduce nematode population and prevent excessive heat damage to the trees (Luo *et al.*, 2021).

Ascolana and Moraiolo, and a selection of wild olive DA 12I as rooted explants, showed interesting results in the reproduction of M. incognita. In the most popular Italian cultivar, cv. Coratin appears to have a resistance effect to root-knot nematodes (*M. incognita* and *M. javanica*) and cv. Leccino is only resistance for *M. javanica* (Ciancio and Mukerji, 2009).

The lowest average number of larval parasitic nematodes was observed at the lowest applied water level (60% evapotranspiration), and reached the highest average (3697.3) using the surface drip irrigation system while, it recorded (3007.4) using the higher levels of applied water (100% evapotranspiration) in the drip irrigation under the subsurface system (Samak and Abo-Korah, 2021).

3. CONCLUSION

The variation in different fruit plant groups morphologically, anatomically and physiologically beside its type (deciduous or evergreen) in addition to the preferred place of cultivation all make difference to response to the infection by nematodes. In addition, the conditions in the orchard around the fruit plants affect the infection by nematodes. Some plant fruits are big trees (pears and plums) but others are small (grapes and peaches). Some of them have a woody stem (pome and stone fruits) and others haven't (banana). Deciduous fruits can grow in cold climate but evergreen fruits grow in a worm places. Orchard of pome and stone fruits found in cold area but banana grow in hot places and need a lot of water while olive prefer the sandy soil with small amount of water. The nematodes could be affected and its amount could be decreased in sunny places with the decrease of water and by using mulch. The starting with clean and healthy seedlings by the use of tissue culture or by grafting on tolerated rootstocks could minimize the nematode infection in orchard.

REFERENCES

Ali, N., Chapuis, E., Tavoillot, J., & Mateille, T. (2014). Plant-parasitic nematodes associated with olive tree (Olea europaea L.) with a focus on the Mediterranean Basin: A review. *Comptes Rendus Biologies*, *337*(7-8), 423–442. doi:10.1016/j.crvi.2014.05.006 PMID:25103828

Almeida, N. O., Teixeira, R. A., Carneiro, F. A., de Oliveira, C. M., Ribeiro, V. A., Lobo Júnior, M., & da Rocha, M. R. (2018). Occurrence and correlations of nematodes, Fusarium oxysporum and edaphic factors on banana plantations. *Journal of Phytopathology*, *166*(4), 265–272. doi:10.1111/jph.12683

Ara, I., Khan, H., Syed, T., & Bhat, B. (2021). Prevalence and seasonal dynamics of gastrointestinal nematodes of domestic fowls (Gallus gallus domesticus) in Kashmir, India. *Journal of Advanced Veterinary and Animal Research*, *8*(3), 448. doi:10.5455/javar.2021.h533 PMID:34722743

Askary, T. H. (2017). Diversity of plant parasitic nematodes in pulses. In A. A. Ansari, S. S. Gill, Z. K. Abbas, & M. Naeem (Eds.), *Plant diversity: present situation and future scenario* (pp. 232–274). CAB International. doi:10.1079/9781780646947.0239

Askary, T. H., Banday, S. A., Iqbal, U., Khan, A. A., Mir, M. M., & Waliullah, M. I. S. (2012). Plant Parasitic Nematode Diversity in Pome, Stone and Nut Fruits. In E. Lichtfouse (Ed.), *Agroecology and Strategies for Climate Change. Sustainable Agriculture Reviews* (Vol. 8). Springer. doi:10.1007/978-94-007-1905-7_10

Askary, T. H., Gani, M., & Wani, A. R. (2021). Nematodes Associated with Stone Fruits and Their Management Strategies. Production Technology of Stone Fruits, 423-436. doi:10.1007/978-981-15-8920-1_16

Askary, T. H., Khalil, A., Nazir, N., Khan, A. A., & Banday, S. A. (2018). Nematode Parasites of Grapevines. In E. Lichtfouse (Ed.), *Sustainable Agriculture Reviews 31. Sustainable Agriculture Reviews* (Vol. 31). Springer. doi:10.1007/978-3-319-94232-2_7

Ciancio, A., & Mukerji, K. G. (Eds.). (2009). Integrated management of fruit crops and forest nematodes (Vol. 4). Springer Science & Business Media. doi:10.1007/978-1-4020-9858-1

Creasy, G. L., & Creasy, L. L. (2018). *Grapes* (Vol. 27). Cabi. doi:10.1079/9781786391360.0000

Erez, A. (Ed.). (2013). *Temperate fruit crops in warm climates*. Springer Science & Business Media.

Ferree, D. C., & Warrington, I. J. (Eds.). (2003). *Apples: botany, production, and uses*. CABI. doi:10.1079/9780851995922.0000

Ferris, H., Zheng, L., & Walker, M. A. (2012). Resistance of Grape Rootstocks to Plant-parasitic Nematodes. *Journal of Nematology*, *44*(4), 377–386. PMID:23482972

Hill, L., & Perry, L. (2012). *The Fruit Gardener's Bible: A Complete Guide to Growing Fruits and Nuts in the Home Garden*. Storey Publishing, LLC.

Hunter, J. J., Archer, E., Van Schalkwyk, D., Strever, A. E., & Volschenk, C. G. (2016). Grapevine roots: Interaction with natural factors and agronomic practices. *Acta Horticulturae*, (1136), 63–80. doi:10.17660/ActaHortic.2016.1136.10

Jaiswal, A. K. (Ed.). (2020). *Nutritional composition and antioxidant properties of fruits and vegetables*. Elsevier Academic Press. doi:10.1016/C2016-0-04117-7

Jonathan, E. I. (2010). Nematology: Fundamentals and applications. Pitam Pura: New India Publishing Agency.

Jones, D. R. (Ed.). (2018). *Handbook of diseases of banana, abaca and enset*. CABI.

Kim, E., Seo, Y., Kim, Y., Park, Y., & Kim, Y. H. (2017). Effects of Soil Textures on Infectivity of Root-Knot Nematodes on Carrot. *The Plant Pathology Journal*, *33*(1), 66–74. doi:10.5423/PPJ.OA.07.2016.0155 PMID:28167889

Layne, D., & Bassi, D. (Eds.). (2008). *The peach: botany, production and uses*. Cabi. doi:10.1079/9781845933869.0000

Luo, J., Zhang, X., Kou, X., Xie, H., Bao, X., Mahamood, M., & Liang, W. (2021). Effects of residue mulching amounts on metabolic footprints based on production and respiration of soil nematodes in a long-term no-tillage system. *Land Degradation & Development*, *32*(7), 2383–2392. doi:10.1002/ldr.3918

Naik, B. H., & Thippesh, D. (2014). *Fundamentals of Horticulture & Production Technology of Fruit Crops*. UAHS.

Ollat, N., Peccoux, A., Papura, D., Esmenjaud, D., Marguerit, E., Tandonnet, J.-P., … Delrot, S. (2016). Rootstocks as a component of adaptation to environment. *Grapevine in a Changing Environment*, 68–108. doi:10.1002/9781118735985.ch4

Ozarslandan, A., & Tanriver, E. (2018). Evaluation of some stone fruit rootstocks against resistance to root knot nematode (Meloidogyne incognita). *International Journal of Agricultural and Natural Science, 1*, 137–141.

Pollock, M. (Ed.). (2012). *Fruit & Vegetable Gardening*. DK Publishing.

Rijal, J., & Bergh, J. (2016). Food-Finding Capability of Grape Root Borer (Lepidoptera: Sesiidae) Neonates in Soil Column Bioassays. *Journal of Entomological Science, 51*. . doi:10.18474/JES15-21.1

Robinson, J. C., & Saúco, V. G. (2010). *Bananas and plantains* (Vol. 19). Cabi. doi:10.1079/9781845936587.0000

Roussos, P. A., Denaxa, N.-K., Tsafouros, A., Efstathios, N., & Intidhar, B. (2016). Apricot (Prunus armeniaca L.). *Nutritional Composition of Fruit Cultivars*, 19–48. doi:10.1016/B978-0-12-408117-8.00002-7

Samak, A. A., & Abo-Korah, M. S. (2021). Impact of Irrigation Systems and Applied Water on Crop Yield, WUE, Certain Nematicides Efficiency, and Plant-Parasitic Nematodes Activity Infected Tomato Plants. *Egyptian Academic Journal of Biological Sciences. F, Toxicology and Pest Control, 13*(2), 301–316. doi:10.21608/eajbsf.2021.244427

Schurig, J., Ipach, U., Hahn, M., & Winterhagen, P. (2021). Evaluating nematode resistance of grapevine rootstocks based on Xiphinema index reproduction rates in a fast screening assay. *European Journal of Plant Pathology, 160*(1), 233–238. doi:10.100710658-021-02227-6

Sharma, G. C., & Kashyap, A. S. (2009). Effect of different intercrops on the nematode populations and yield of apricot var. New Castale. *Indian Journal of Horticulture, 66*, 420–421.

Sinha, N. K., Sidhu, J., Barta, J., Wu, J., & Cano, M. P. (Eds.). (2012). *Handbook of fruits and fruit processing* (2nd ed.). John Wiley & Sons. doi:10.1002/9781118352533

Smith, B. P., Morales, N. B., Thomas, M. R., Smith, H. M., & Clingeleffer, P. R. (2016). Grapevine rootstocks resistant to the root-knot nematodeMeloidogyne javanica. *Australian Journal of Grape and Wine Research, 23*(1), 125–131. doi:10.1111/ajgw.12242

Song, D., Tariq, A., Pan, K., Chen, W., Zhang, A., Sun, X., Ran, Y., & Zeng, F. (2020). Effects of straw mulching practices on soil nematode communities under walnut plantation. *Scientific Reports*, *10*(1), 1–10. doi:10.103841598-020-72530-5 PMID:32948816

Stirling, G. R. (2014). *Biological control of plant-parasitic nematodes: soil ecosystem management in sustainable agriculture*. CABI. doi:10.1079/9781780644158.0000

Storey, S. G., Malan, A. P., & Hugo, H. J. (2017). Nematode Pests of Grapevine. In H. Fourie, V. Spaull, R. Jones, M. Daneel, & D. De Waele (Eds.), *Nematology in South Africa: A View from the 21st Century*. Springer. doi:10.1007/978-3-319-44210-5_14

Therios, I. (2009). *Crop Production Science in Horticulture 18, OLIVES* (1st ed.). CABI.

Udin, N. (2014). Organic Farming Impact on Sustainable Livelihoods of Marginal Farmers in Shimoga District of Karnataka. *American Journal of Rural Development.*, *2*(4), 81–88. doi:10.12691/ajrd-2-4-4

Verdejo-Lucas, S., & Talavera, M. (2009). Integrated Managemenr of Nematodes Parasitic on *Prunus SPP*. *Integrated Management of Fruit Crops Nematodes*, 177-194.

Walia, R. K., & Bajaj, H. K. (2003). *Text book on introductory plant nematology. Directorate of Information and Publications of Agriculture*. ICAR.

Wallis, C. M. (2020). Grapevine (*Vitis* spp.) rootstock stilbenoid associations with host resistance to and induction by root knot nematodes, Meloidogyne incognita. *BMC Research Notes*, *13*(1), 360. doi:10.118613104-020-05201-3 PMID:32727572

Westwood, M. N. (1993). *Temperate-zone pomology: Physiology and Culture*. Timber press.

Xi, W., & Lei, Y. (2020). Apricot. *Nutritional Composition and Antioxidant Properties of Fruits and Vegetables*, 613–629. doi:10.1016/B978-0-12-812780-3.00038-6

Yahia, E. M. (2019). Classification of Horticultural Commodities. *Postharvest Technology of Perishable Horticultural Commodities*, 71–97. doi:10.1016/B978-0-12-813276-0.00003-1

Zasada, I. A., Howland, A. D., Peetz, A. B., East, K., & Moyer, M. (2018). *Vitis* spp. Rootstocks Are Poor Hosts for Meloidogyne hapla, a Nematode Commonly Found in Washington Winegrape Vineyards. *American Journal of Enology and Viticulture*. doi:10.5344/ajev.2018.18027

Chapter 7
Managing Nematode Infection in Fruit Orchards

Waleed Fouad Abobatta
Horticulture Research Institute, Agriculture Research Center, Egypt

Mohamed Abdel-Raheem
Pests & Plant Protection Department, Agricultural and Biological Research Institute, Cairo, Egypt

ABSTRACT

Plant-parasitic nematodes are one of the biggest threats to the production of various fruit crops worldwide. Nematodes cause adverse impacts on various crops, such as declining plant growth, and may lead to the destruction of large areas, like in banana, peach, and citrus orchards. Yellowing, stunting, and wilting are common symptoms of nematode infection in fruit orchards, and they are all associated with tree decline and decreasing yield. The efficiency of controlling nematodes varies due to many biotic and abiotic factors, such as soil type, nematode species, host plant physiology, and climate conditions, which affect nematode dynamics and infestation in the soil. Therefore, using integrated strategies with new management practices for controlling nematode infection is required to reduce economic loss for fruit producers. Hence, using biological control field management practices such as fallow, cover crops, and mulching, could be an effective way to inhibit nematodes' growth. Thus, they sustain the production of various crops, protect the environment, and preserve human health.

DOI: 10.4018/978-1-6684-8083-0.ch007

1. INTRODUCTION

Plant-parasitic nematodes (PPNs) are microscopic roundworms that live in the soil and feed on plant roots or foliage. Nematode feeding can cause a reduction in root volume and mass, but above-ground symptoms such as stunting, yellowing, and wilting are often misdiagnosed as nutrient deficiencies. PPNs are one of the biggest threats to various crop production globally; they live in wide habitats and cause economic loss every year. PPNs that are detected in orchards and small fruit fields can transmit viruses that quickly decline trees. Nematodes are serious threats across fruit-growing regions globally, while their severity is higher in sandy soils in arid climates.

Damage from plant parasitic nematodes costs fruit growers millions of dollars annually (Hajihassani, et al., 2018). Generally, the harmful effect of all kinds of nematodes is related to the population density in the soil. Therefore, prior to establishing a new planting, it is highly recommended to analyze the soil for nematode infection. So, monitoring nematode populations should continue after the establishment of orchards, or every one or two years after cultivation, and then every three to five years.

Plant-parasitic nematodes have different lifestyles, as some of them invade plant cells and live inside them until they complete their lifecycle (Endoparasitic) without killing the cells around them, such as the root-knot nematode. While other nematodes (Ectoparasitic) obtain food from the outside organs of host plants, where they feed on plant roots and thus remain outside the cells of the host plant, also, there are semi-endoparasitic nematodes that able to penetrate plant cell partially without moving in plant tissue (Smant, et al., 2018). The majority of PPNs feed by using needle-like structures called stylets to pierce and kill cells. The lesion, lance, needle, sting, stunt, and sting nematodes all use this method of feeding (Kumar & Yadav, 2020).

Xiphinema (California dagger nematode) is an ectoparasitic nematode that transmits the Grapevine fanleaf virus, which causes severe economic losses in viticulture around the world (Malik et al., 2022).

Endoparasitic nematodes are further classified into two groups migratory and sedentary. Migratory endoparasitic nematodes move within the root tissue and remove cytoplasm killing the host cell while sedentary nematodes become static after the forming of a feeding site within the host tissue (Smiley, 2021). Lesion nematodes (Pratylenchus spp.), burrowing nematodes (Radopholus spp.), and rice root nematodes (Hirschmanniella) are a few examples of migratory endoparasitic nematodes that cause economic loss (Khan, 2015).

Nematodes infesting the newly planted region via infected seedlings, organic fertilizers, plant materials, irrigation, and machinery, pose a threat to citrus development and yield. Furthermore, in some cases, transferring fertile soil from cultivated areas that could be nematode-infested to virgin desert soil as part of a reclamation practice to improve soil properties increases the prevalence of nematodes infesting trees. Over time, nematodes cause economic losses on farms due to the rapid reproduction of nematodes (Abd-Elgawad et al., 2016).

Post-plant control options for nematodes are limited, so it is preferable to manage nematodes before cultivated bushes, seedlings, trees, or vines are planted. Therefore, there is a need for new management/cultural practices and alternative chemicals for controlling nematode pathogens that will provide the producer with optimum productivity of fruit orchards.

This chapter explores various approaches for the management of nematode infection, which will be reviewed as specific measures due to cultivation systems that are used in different crops, in addition to the most nonchemical strategies and future approaches for controlling nematodes.

2. NEMATODES AND FRUIT ORCHARDS

Plant-parasitic nematodes are widespread and are among the most damaging pests of fruit tree varieties. Infection by nematodes causes not only significant crop losses on a commercial scale but also limits tree productivity and viability.

Currently, it is more interesting to use integrated strategies for the control of nematodes in fruit orchards. Integrated management for controlling nematodes includes agriculture practice management like fallow, cover crops, crop rotation, and mulching, in addition to biological control.

The efficiency of controlling nematodes varies due to many biotic and abiotic factors such as soil type, nematode species, host plant physiology, and climate conditions, which affect nematode dynamics and infestation in the soil (Lazarova, et al., 2021).

Hereunder, we explore management and control nematode techniques that improve the growth and productivity of some fruit varieties.

2.1. Banana and Nematodes

In general, all banana varieties are considered hosts for parasitic nematodes, but a few of them cause economic losses, while, the root-knot nematode, which belongs to the genus Meloidogyne, is one of the most dangerous nematode species and causes dwarfing and deformation of the roots, causing significant economic losses to banana growers worldwide (Bahadur, 2021).

Infected plants with nematodes are less able to absorb water and nutrients, which leads to stunted plants and an extended vegetative growth stage and delays the time to complete punch growth. The damage varies according to the type of nematode, plant vigor, and other environmental factors. The damage ranges from a slight extension of the vegetative growth stage to a complete collapse of plants, which is the most obvious symptom (Hamouda & El-Ansary, 2017).

- The Nematode Problem in Banana Cultivation

Losses due to nematode infestations of different crops are estimated based on improved production after managing the infestation. From a mechanical perspective, three levels of nematode damage can be identified as follows:

1. Prolonging the vegetative stage without major effects on yield: Nematode infection of banana cultivations causes a prolongation of the different phenological stages (the period between planting and flowering, the growth till maturity of the bunches, harvesting, etc.) as in banana cultivation in Latin America.
2. Lengthening the vegetative phase with a reduction in the total yield:

There are two sub-levels according to the reduction in the number of harvestable bunches.

a. poor quality bunches or immature delayed fruit).
b. the reduction in the average fruit size and bunch weight.

This type of damage is often observed in commercial plantations in West Africa.

3. Extending of the vegetative stage, with a decrease in the total harvest bunch or the longevity of the cultivation: This condition occurs when hazardous strains of nematodes infect the plantation such as R. similis, cause severe growth inhibition and dwarfing of the plants until they are completely destroyed and the farm is uprooted completely. This situation of damage is observed in dessert banana cultivation worldwide.

The Main Parasitic Nematode Species of Bananas

It is well documented that there are various nematode species have been associated with banana plants (Musa spp.) throughout the world, according to local conditions. Moreover, there are five major species worldwide considered the main serious pests for banana cultivation, which are:

1. Migratory endoparasites (Radopholus similis).
2. Lesion nematodes (Pratylenchus coffeae).
3. Lesion nematodes (Pratylenchus goodeyi).
4. Endoparasite (Helicotylenchus multicinctus).
5. Sedentary parasite (Meloidogyne spp.)

Other species of nematodes that infect banana cultivation with minor incidence include Rotylenchulus reniformis, Hoplolaimus pararobustus, H. seinhorsti, Heterodera oryzicola, and Zygotylenchus taomasinae (Kamira, et al., 2013).

The problem of nematode infection was discovered in banana plantations, especially in sandy and loamy soils, from a very early age. Since Ashby (1915) from Jamaica described nematode infection symptoms in banana rhizomes and called it a 'Black head disease of bananas'.

While, Loos (1962) is the first author to describe the full life cycle of banana parasitic nematodes from eggs to eggs in 20-25 days under a temperature of 24-32°C, and explained that eggs hatch after 8-10 days and the juvenile stage develops completely within 10-13 days (Ramclam & Araya, 2006).

Later, parasitic nematode received great attention from researchers in major production areas such as Latin America and South Asia, where it causes a decrease in productivity exceeding 20% in the case of infection with the root-knot nematode, or P. goodeyi or M. incognita (Ciancio & Mukerji, 2009).

In general, the root-knot nematodes are more active in subtropical conditions, whereas the large populations in banana plantations, such as in the Middle East and South Asian regions (Sikora, et al., 2018).

a. Radopholus similis

Some species of nematodes could have habitats different hosts, such as R. similis which could be parasitic various crops including Musa Spp., Citrus spp., and weeds, while this nematode can't survive in the soil without live hosts for more than 6 months (Tranier, *et al.*, 2014).

Due to the widespread of R. similis nematode throughout the world, it could habituate in the roots of different hosts, such as tea, coffee, pepper, woody trees, ornamental plants, and herbs. Therefore, R. similis considered the second dangerous nematode of banana cultivations, due to their easily spread and infected banana plantations, consequently, causing great economic damage (Ravichandra & Ravichandra, 2014).

b. *Helicotylenchus multicinctus*

The damage of spiral nematode *Helicotylenchus multicinctus may be* the most damaging nematode on bananas after *R. similis*, and root-knot spp. (Daneel et al., 2015).

While, *Pratylenchus coffeae* and *P. goodeyi* are documented as damaging species for banana cultivations among the lesion nematodes, furthermore, they have the similar symptoms of the burrowing nematode on banana plants (Mgonja, et al., 2020).

There is association between environment conditions and spread nematode, for instance spreading of Pratylenchus goodeyi in bananas in tropical conditions is in relation with the altitude and the latitude or in other words with soil temperature (Luambano, et al., 2019).

The P. coffeae nematode is one of the main pests in tropical climates and infects many economic fruit crops such as coffee and bananas, as well as tuber crops and ornamental plants (Coyne, 2021).

- Damage and Economic Importance

The nematode has a very adverse impact on various crops, for instance in banana cultivation it causes excessive losses and may lead to the uprooting of large areas (Daneel & De Waele, 2017). As happened in Jamaica in the fifties of the last century, where R. similis destroyed its production of bananas during this period, as it caused "Black head toppling disease"

Nematodes affect negatively the growth and production of bananas because of destroying the root system, which is accompanied by an increase in the number of nematode species in the soil around roots (such as burrowing and lesion nematodes), which increases root damage, reducing the biomass of roots and plant death by the end.

2.2. Nematodes in Peach Orchards

Peach (Prunus persica (L.) Batsch) is one of the principal stone fruits, which has promising potential in a temperate climate, but the production of peaches is nevertheless hampered by a number of difficulties and constraints (Elsadr, 2016).

Climate change is associated with edaphic stress, such as increased soil salinity and alkalinity and affecting the sustainability of peach production regions, in addition to variables biotic stress such as bacteria, fungi, and nematodes (Miller, et al., 2017).

Nematodes are one of the major constrain to peach cultivation, thus peach growers are mainly concerned with the main genera of plant-parasitic nematodes that are known to be harmful to peaches, which include:

- Ring nematode (*Mesocriconema xenoplax* (Raski) Loof & de Grisse (= *C. xenoplax* (Raski) Luc and Raski).
- Root-knot nematode [*Meloidogyne incognita* (Kofoid and White) Chitwood, *M. javanica* (Treub) Chitwood, and *M. floridensis (*Handoo et al., 2004).

- Nematodes of root-lesion (*Pratylenchus vulnus* Allen and Jensen) (Nyczepir and Esmenjaud, 2008).

1. Ring nematode

Infection of the ring nematode, M. xenoplax is one of the main threats to peach plants due to its association with pathogens. Nyczepir, (2011) reported that infection with the ring nematode is directly accompanied by Peach Tree Short Life (PTSL) disease.

Furthermore, young peach trees (3-5 years) that have been infested with ring nematodes are more susceptible to various threats like PTSL and canker bacteria (Pseudomonas syringae pv. syringae van Hall) as well as a combination of diseases (Shirley, et al., 2019).

Infection with ring nematode in peach trees causes various symptoms like wilting of young leaves and discolored cambial tissue, particularly at the end of growth seasons. In the next season, the growth of new shoots will be very weak, resulting in decline of vegetative growth (Luo, et al., 2022).

2. Root-knot nematode

The root-knot nematodes are obligate endoparasitic nematodes that cause root-galling symptoms because of locking vascular parenchyma cells. Therefore, early defoliation and a deterioration in tree vigor are signs of strong infection, which ultimately lead to yield loss, particularly, the problems are further exacerbated in sandy soils after exposed trees to environmental stress like cold or drought. (Adaskaveg, et al., 2008).

The root-knot nematodes (Meloidogyne spp.) are the most common parasitoid species in the world and are responsible for noteworthy economic losses since they can survive in a variety of plant production regions and diverse climatic circumstances ranging from tropical to temperate (Teketel, 2019). In Peach orchards, infection with root-knot and root-lesion nematodes (P. vulnus) causes adverse effects on tree growth and productivity (Nyczepir, 2011).

To reduce the risks posed by nematodes in peach orchards, resistant peach rootstocks could be used. As an illustration, Nemaguard is reported to be resistant to M. javanica and M. incognita (Shirley, et al., 2019). Whereas Nemaguard is vulnerable of ring nematode (Lesmes-Vesga et al., 2022).

2.3. Nematodes and Citrus Production

Citrus fruits are one of the most popular fruits that are widely used in the world and rank third globally in fruit production, as they are grown in about 150 countries under various conditions of climate and soil. Citrus fruits have a high nutritional value because they contain numerous minerals, fibers, and vitamins, especially vitamin C, which is an important part of a healthy diet for humans (Abobatta, 2019).

Nematodes are one of the main restricted factors for citrus cultivation, which constrains the expansion of citrus cultivation, particularly in sandy soil. Tylenchulus semipenetrans is considered the main threat species to citrus trees and is known as the citrus nematode (Elkelany, et al., 2021).

Main Hazards of Nematodes

Parasitic nematodes spread all over the world, infect most economic crops, cause more burdens on crop production, and represent one of the main threats to global food security.

Yellowing, stunting, and wilting are common nematode signs, and they are all associated with tree decline and decreasing yield. Root-knot nematodes (Meloidogyne spp.) cyst nematodes (Heterodera and Globodera spp.) and lesion nematodes (Pratylenchus spp.) are considered the most dangerous species that cause huge levels of damage because of their ability to infect many economic plants in different climates, which causes significant losses to different crops and negatively affects food production worldwide.

Alternate nematode control techniques are becoming more popular because of the hazards of chemical nematicides. Identification and implementation of host resistance for nematode species is considered a useful technique in controlling nematode infection.

The development of genetic control techniques such as gene silencing and RNA interference is one of the alternative methods for fighting nematode infection that targets important nematode genes involved in parasitism (Zotti, et al., 2018).

Transcriptome profiling analysis have recently been used to discriminate nematode-resistance and susceptible genotypes and to pinpoint the precise molecular elements and pathways activated during the plant immunological response to nematode infection (Banerjee, et al., 2017).

Plant-parasitic nematodes develop specific feeding structures and secretory substances that increase the infection of the host and increase the absorption of nutrients from the host plant through the stylet and subventral and dorsal esophageal glands to increase the reproduction of the nematode, which quickly damages host tissue and leads to the collapse of the infected plants (Hewezi & Baum, 2017). Parasitic nematodes use a needle-like appendage to penetrate plant tissues and utilize endometrial and dorsal glands to release various proteinaceous secretions that affect host plant cells and allow nematode entry (Reddy & Reddy, 2021).

Beta-1,4-endoglucanase genes have been isolated from plant-parasitic cyst nematodes with catalytic domains belonging to family 5 of the glycosyl hydrolases (Xu, et al., 2012).

3. NEMATODES CONTROL

Plant-parasitic nematodes represent a major threat to the agricultural production of different crops worldwide. Due to the severity of the hazards of chemical nematicides and their toxicity to humans and the environment. Thus, using alternative strategies to control nematodes is necessary to sustain the production of various crops, protect the environment, and preserve human health.

There are other strategies that could be used to control nematodes and reduce their hazards, including management practices such as fallow, cover crops, and mulching. Furthermore, using biological control for nematodes by microorganisms like fungi and bacteria.

3.1. Integrated Nematode Management

Previously, nematode control management in banana cultivation was applied at determined times annually, over the last few decades, the control treatment has been done according to the rates of infection and the population in the root zone, in order to reduce the use of pesticides.

Various agricultural practices have been tried in banana-growing areas to get rid of the R. similis nematode from the soil (De Brida, et al., 2017), and among these practices:

- Continuous flooding of the soil before planting.
- bare soil for enough time before planting.
- Cultivation of non-hosting plants of nematodes such as pangolin and Sudan grass.
- Using crop rotation with other crops such as sugarcane, cassava, or pineapple.
- Mulching

3.2. Practices Management to Control Nematode Infection

a. Fallow

Fallow is one of preplant techniques to control nematodes that is used as a tool to avoid nematode infection (Zasada, et al., 2010).

Fallow acts to reduce available plant materials by keeping soil without cultivation for a time and using some agricultural practices such as plowing to starve the nematode, reduce population, and control it, in addition to using herbicides to avoid plant growth, which is considered an efficient technique to control nematode before planting (Abd-Elgawad, 2019).

Nematodes plant-parasitic is obligate parasites that depend on living plants for their food requirements to complete their life cycle. Most of the plant-parasitic nematodes die after the first six months, while the other species cannot tolerate starving for more than From 12 to 18 months (Khoja, et al., 2021).

Therefore, the fallow technique is one of the efficient control methods for fighting nematodes (Sasanelli, et al., 2021).

Furthermore, fallow reduces nematodes' population through a prolonged drought. Most types of nematodes cannot survive when exposed to heat and drought for a long time, with some exceptions (Sumner, 2018).

b. Cover Crops

Cover crops represent one of the solutions used to control the infection of nematodes. There is no specific plant that can be used against all nematodes, but nematode infestations can be successfully controlled by using specific cover crops before planting. Cover crops are used for different reasons, depending on the type of nematode, the crop, and the specific nematode crop composition. In addition, cover crops could be repellents for nematodes (Hooks, et al., 2010).

There are "antagonistic" plants that inhibit nematodes spreading by exuding toxic substances that affect nematodes' growth. As the nematodes are parasitic on certain plants, using certain cover plants that are not hosts for nematodes, such as marigold (Tagetes spp.), asparagus (Asparagus officinalis), and onion (Allium cepa), starves nematodes and reduces the population, similar to the action of the fallow technique.

There are cover plants that act as traps for nematodes, such as Crotalaria spp., that withstand the invasion of nematodes, but they prevent the development of larvae into full animals, which limits the reproduction of nematodes. For instance, Crotalaria spp. is used as a trap plant for root-knot nematodes (Rech, et al., 2022).

c. Organic Manuring and Soil Amendments

The addition of large quantities of organic manure to the soil could decrease the population of plant parasitic nematodes and increase crop productivity. This decrease in nematodes may be due to the increase in the availability of their natural enemies, in addition, decomposing organic matter improves the growth of plants and their tolerance to nematode infection.

The positive effect of the decomposition of organic fertilizers and plant residues on nematode infection may be due to the formation of various substances such as butyric acid during the decomposition of organic matter and the remains of cover crops, in addition to rising soil temperatures during decomposition.

Moreover, green manure crops and soil additives are considered effective substances to reduce populations of parasitic nematodes, such as crushed sesame stems mixed with crabmeal. Furthermore, mixing urea with agricultural residues before decomposition in the soil releases ammonia, which could be an effective factor in killing nematodes (Devi, et al., 2020).

d. Crop Rotation

One of the natural methods used to fight nematodes is crop rotation, which is considered an effective control practice that is widely used in different regions of the world. While using nonhost plants or antagonism for nematodes in the sequence of rotation is necessary, Crop rotation is considered a short-term strategy for controlling nematodes in field crops and vegetables, while long-term rotation could be an effective method for fruit orchards (Sumner, 2018).

e. Mulching

Straw mulching technique is mainly used in food crops to achieve several goals such as weed control and reducing nematode infestation (Figure 1), as it depends

Figure 1. Field image of mulching with rice straw
Source: Dr. Abobatta (2019)

on the physical, chemical, and biological effects of soil and the physiological and environmental responses of the mulched tree species (Song, et al., 2020).

Moreover, stimulate beneficial nematode in soil and reducing population of pathogenic nematodes, in addition, straw mulching using to improve the sustainable development of citrus orchards and reduce weed growth (Liu, et al., 2020)

f. Biological Control

Due to the adverse effects of nematicides on human health and their damaging effects on the environment. Using microorganisms like fungi and bacteria agents is considered an efficient technique to fight nematodes and reduce negative impacts on various crops.

There are numerous species used as biocontrol Nematode agents (BNAs) while, the Trichoderma, mycorrhizal, and Bacillus spp. are considered the main species that are used practically in controlling nematodes in fruit orchards (Ferreira & Musumeci, 2021).

Trichoderma and mycorrhizal are the most important fungi used as biological Nematode agents (BNAs), these fungi affect both directly by parasitizing or paralyzing nematodes or indirectly by competing for nutrients and secreting enzymes such as lytic enzymes that reduce nematode activity, in addition, to increase nutrients and water availability for plants which improve plant growth (Abd-Elgawad, 2020).

The filamentous fungi induce plant-defense mechanisms against nematode infestation by secreting substances like salicylic and jasmonic acid, strigolactone among others. Moreover, altering the transport pathway of chemical defense substances through the plant or the synthesis of plant secondary metabolites and various enzymes contributes to enhancing plant defenses (Poveda, et al., 2020).

Using of filamentous fungi to control nematodes is one of the promising solutions within the strategy of biological control of nematodes parasitic in fruit orchards (Abdel-Raheem, et al., 2022).

4. CONCLUSION

Parasitic nematodes are one of the most dangerous biotic stressors for plant species and cause economic loss for fruit growers all over the world, as they can live in different environments and under varying environmental conditions.

Yellowing, stunting, and wilting are common symptoms of nematode infection, consequently causing weak vegetative growth and low productivity, which lead to the deterioration of the whole tree, such as in bananas, peaches, and citrus cultivations.

There are different types of nematodes, some of which live inside plant tissues (endoparasitic), such as the root-knot nematode, and some of which live outside of plants in the root zone (ectoparasitic) and feed on cells and damage the root system, like Xiphinema nematode. There are also sime-endoparasitic nematodes that penetrate plant cells and live inside the cell without mobility.

The hazards of nematode infestation vary due to many biotic and abiotic factors, such as soil type, nematode species, host plant physiology, and climate conditions, which affect nematode dynamics and spreading in the soil. Therefore, using integrated strategies that include agricultural management practices with resistant rootstocks and proper nematicides for controlling nematode infection is required to reduce the economic loss for fruit producers.

REFERENCES

Abd-Elgawad, M. M. (2016). Comments on the use of biocontrol agents against plant-parasitic nematodes. *International Journal of Pharm Tech Research*, *9*(12), 352–359.

Abd-Elgawad, M. M. (2019). Plant-parasitic nematodes of strawberry in Egypt: A review. *Bulletin of the National Research Center*, *43*(1), 1–13. doi:10.118642269-019-0049-2

Abd-Elgawad, M. M. (2020). Biological control agents in the integrated nematode management of potato in Egypt. *Egyptian Journal of Biological Pest Control*, *30*(1), 1–13. doi:10.118641938-020-00325-x

Abdel-Raheem, M., Sedhom, H., Ibrahim, D. S. S., & Bhattacharyya, J. (2022) Entomopathogenic Nematodes and Insect Pests. Mittal Publication.

Abobatta, W. F. (2020). Healthy Benefits of Fruits. *EC Nutrition, 15*(3), 1-8.

Adaskaveg, J. E., Schnabel, G., & Förster, H. (2008). *Diseases of peach caused by fungi and fungal-like organisms: biology, epidemiology and management. In The peach: Botany, production and uses*. CABI.

Bahadur, A. (2021). *Nematodes diseases of fruits and vegetables crops in India*. IntechOpen.

Banerjee, S., Banerjee, A., Gill, S. S., Gupta, O. P., Dahuja, A., Jain, P. K., & Sirohi, A. (2017). RNA interference: A novel source of resistance to combat plant parasitic nematodes. *Frontiers in Plant Science*, *8*, 834. doi:10.3389/fpls.2017.00834 PMID:28580003

Ciancio, A., & Mukerji, K. G. (Eds.). (2009). *Integrated management of fruit crops and forest nematodes* (Vol. 4). Springer Science & Business Media. doi:10.1007/978-1-4020-9858-1

Coyne, D. (2021). The shifting sands of banana nematode communities under mixed cropping conditions. In *Integrated Nematode Management: State-of-the-art and visions for the future* (pp. 167–173). CABI. doi:10.1079/9781789247541.0024

Daneel, M., DeJager, K., Van den Bergh, I., De Smet, M., & De Waele, D. (2015). Occurrence and pathogenicity of plant-parasitic nematodes on commonly grown banana cultivars in South Africa. *Nematropica, 45*(1), 118–127.

Daneel, M. S., & De Waele, D. (2017). Nematode pests of banana. *Nematology in South Africa: A view from the 21st century*, 359-371.

De Brida, A. L., Rosa, J. M. O., Oliveira, C. M. G. D., Serrão, J. E., Zanuncio, J. C., Leite, L. G., & Wilcken, S. R. S. (2017). Entomopathogenic nematodes in agricultural areas in Brazil. *Scientific Reports, 7*(1), 1–7. doi:10.1038rep45254 PMID:28382937

Devi, T. S., Das, D., Ansari, R. A., Rizvi, R., Sumbul, A., & Mahmood, I. (2020). Role of organic additives in the sustainable management of phytoparasitic nematodes. *Management of Phytonematodes: Recent Advances and Future Challenges*, 279-295.

Elkelany, U., Hammam, M., El-Nagdi, W., & Abd-El-Khair, H. (2021). Field application of Trichoderma spp. for controlling the root-knot nematode, Meloidogyne javanica in peanut plants. *Egyptian Journal of Agronematology, 20*(2), 85–100. doi:10.21608/ejaj.2021.183249

Elsadr, H. (2016). *A Genome Wide Association Study of Flowering and Fruit Quality Traits in Peach [(Prunus persica (L.) Batsch]* [Doctoral dissertation]. University of Guelph.

Ferreira, F. V., & Musumeci, M. A. (2021). Trichoderma as biological control agent: Scope and prospects to improve efficacy. *World Journal of Microbiology and Biotechnology, 37*(5), 90.

Hajihassani, A., Lawrence, K. S., & Jagdale, G. B. (2018). Plant parasitic nematodes in Georgia and Alabama. Plant Parasitic Nematodes in Sustainable Agriculture of North America: Vol. 2-Northeastern, Midwestern and Southern USA, 357-391.

Hamouda, R. A., & El-Ansary, M. S. M. (2017). Potential of Plant-Parasitic Nematode Control in Banana Plants by Microalgae as a New Approach Towards Resistance. *Egyptian Journal of Biological Pest Control, 27*(2).

Handoo, Z. A., Nyczepir, A. P., Esmenjaud, D., Van Der Beek, J. G., Castagnone-Sereno, P., Carta, L. K., Skantar, A. M., & Higgins, J. A. (2004). Morphological, molecular, and differential-host characterization of Meloidogyne floridensis n. sp.(Nematoda: Meloidogynidae), a root-knot nematode parasitizing peach in Florida. *Journal of Nematology*, *36*(1), 20. doi:10.21307/jofnem-2017-005 PMID:19262784

Hewezi, T., & Baum, T. J. (2017). Communication of sedentary plant-parasitic nematodes with their host plants. *Advances in Botanical Research*, *82*, 305–324. doi:10.1016/bs.abr.2016.11.004

Hooks, C. R., Wang, K. H., Ploeg, A., & McSorley, R. (2010). Using marigold (Tagetes spp.) as a cover crop to protect crops from plant-parasitic nematodes. *Applied Soil Ecology*, *46*(3), 307–320. doi:10.1016/j.apsoil.2010.09.005

Kamira, M., Hauser, S., Van Asten, P., Coyne, D., & Talwana, H. L. (2013). Plant Parasitic Nematodes Associated with Banana and Plantain in Eastern and Western Democratic Republic of Congo [Nematodos Fitoparásitos Asociados A Los Cultivos De Banano Y Plátano En El Este Y Oeste De La República Democrática Del Congo]. *Nematropica*, *43*(2), 216–225.

Khan, M. R. (2015). Nematode diseases of crops in India. *Recent advances in the diagnosis and management of plant diseases*, 183-224.

Khoja, S., Eltayef, K. M., Baxter, I., Myrta, A., Bull, J. C., & Butt, T. (2021). Volatiles of the entomopathogenic fungus, Metarhizium brunneum, attract and kill plant parasitic nematodes. *Biological Control*, *152*, 104472. doi:10.1016/j.biocontrol.2020.104472 PMID:33390683

Kumar, Y., & Yadav, B. C. (2020). Plant-parasitic nematodes: Nature's most successful plant parasite. *International Journal of Research and Review, 7*(3), 379-386.

Lazarova, S., Coyne, D., Rodriguez, M. G., Peteira, B., & Ciancio, A. (2021). Functional diversity of soil nematodes in relation to the impact of agriculture—A review. *Diversity (Basel)*, *13*(2), 64. doi:10.3390/d13020064

Lesmes-Vesga, R. A., Cano, L. M., Ritenour, M. A., Sarkhosh, A., Chaparro, J. X., & Rossi, L. (2022). Rootstocks for Commercial Peach Production in the Southeastern United States: Current Research, Challenges, and Opportunities. *Horticulturae*, *8*(7), 602. doi:10.3390/horticulturae8070602

Liu, X., Zhang, D., Li, H., Qi, X., Gao, Y., Zhang, Y., Han, Y., Jiang, Y., & Li, H. (2020). Soil nematode community and crop productivity in response to 5-year biochar and manure addition to yellow cinnamon soil. *BMC Ecology*, *20*(1), 1–13. doi:10.118612898-020-00304-8 PMID:32677934

Loos, C. A. (1962). Studies on the life-history and habits of the burrowing nematode, Radopholus similis, the cause of black-head disease of banana. *Proceedings of the Helminthological Society of Washington*, *29*(1), 43–52.

Luambano, N. D., Kashando, B. E., Masunga, M. M., Mwenisongole, A. E., Mziray, M. F., Mbaga, J. E., Polini, R. M., & Mgonja, D. M. (2019). Status of Pratylenchus coffeae in banana-growing areas of Tanzania. *Physiological and Molecular Plant Pathology*, *105*, 102–109. doi:10.1016/j.pmpp.2018.08.002 PMID:31007378

Luo, C. X., Schnabel, G., Hu, M., & De Cal, A. (2022). Global distribution and management of peach diseases. *Phytopathology Research*, *4*(1), 1–15. doi:10.118642483-022-00134-0

Malik, I. M., Tak, H., Lone, G. M., & Dass, W. M. (2022). Phytoparasitic nematodes as the major threat to viticulture. *Environmental and Experimental Biology*, *20*(1), 1–10. doi:10.22364/eeb.20.01

Mgonja, D. M., Temu, G. E., Lyantagaye, S. L., Makaranga, A., Ndunguru, J. C., & Luambano, N. D. (2020). Plant parasitic nematodes occurrence and genetic diversity of banana cultivars grown in Tanzania. *Plant Omics*, *13*(1), 21–29. doi:10.21475/POJ.13.01.20.p2085

Miller, H., Fletcher, M., Primitivo, M., Leonard, A., Sutphin, G. L., Rintala, N., Kaeberlein, M., & Leiser, S. F. (2017). Genetic interaction with temperature is an important determinant of nematode longevity. *Aging Cell*, *16*(6), 1425–1429. doi:10.1111/acel.12658 PMID:28940623

Nyczepir, A. P. (2011). Host suitability of an endophyte-friendly tall fescue grass to Mesocriconema xenoplax and Pratylenchus vulnus.. *Nematoropica*, *41*(1), 45–51.

Poveda, J., Abril-Urias, P., & Escobar, C. (2020). Biological control of plant-parasitic nematodes by filamentous fungi inducers of resistance: Trichoderma, mycorrhizal and endophytic fungi. *Frontiers in Microbiology*, *11*, 992. doi:10.3389/fmicb.2020.00992 PMID:32523567

Ramclam, W., & Araya, M. (2006). Frequency of occurrence and abundance of root nematodes on banana (Musa AAA) in Belize. *International Journal of Pest Management*, *52*(1), 71–77. doi:10.1080/09670870600581009

Ravichandra, N. G., & Ravichandra, N. G. (2014). Nematode diseases of horticultural crops. *Horticultural Nematology*, 127-205.

Rech, C., Ribeiro, L. P., Bento, J. M. S., Pott, C. A., & Nardi, C. (2022). Monocrotaline presence in the Crotalaria (Fabaceae) plant genus and its influence on arthropods in agroecosystems. *Brazilian Journal of Biology*, 84. PMID:35043839

Reddy, P. P., & Reddy, P. P. (2021). Nematode diseases of crop plants: an overview. *Nematode diseases of crops and their management*, 3-32.

Sasanelli, N., Konrat, A., Migunova, V., Toderas, I., Iurcu-Straistaru, E., Rusu, S., Bivol, A., Andoni, A., & Veronico, P. (2021). Review on control methods against plant parasitic nematodes applied in southern member states (C Zone) of the European Union. *Agriculture*, *11*(7), 602. doi:10.3390/agriculture11070602

Shirley, A. M., Noe, J. P., Nyczepir, A. P., Brannen, P. M., Shirley, B. J., & Jagdale, G. B. (2019). Effect of spirotetramat and fluensulfone on population densities of and on peach. *Journal of Nematology*, *51*(1), 1–10. doi:10.21307/jofnem-2019-012 PMID:31088024

Sikora, R. A., Coyne, D., Hallmann, J., & Timper, P. (2018). Reflections and challenges: nematology in subtropical and tropical agriculture. In *Plant parasitic nematodes in subtropical and tropical agriculture* (pp. 1–19). CAB International. doi:10.1079/9781786391247.0001

Smant, G., Helder, J., & Goverse, A. (2018). Parallel adaptations and common host cell responses enabling feeding of obligate and facultative plant parasitic nematodes. *The Plant Journal*, *93*(4), 686–702. doi:10.1111/tpj.13811 PMID:29277939

Smiley, R. W. (2021). Root-lesion nematodes affecting dryland cereals in the semiarid pacific Northwest USA. *Plant Disease*, *105*(11), 3324–3343. doi:10.1094/PDIS-04-21-0883-FE PMID:34236212

Song, D., Tariq, A., Pan, K., Chen, W., Zhang, A., Sun, X., Ran, Y., & Zeng, F. (2020). Effects of straw mulching practices on soil nematode communities under walnut plantation. *Scientific Reports*, *10*(1), 1–10. doi:10.103841598-020-72530-5 PMID:32948816

Sumner, D. R. (2018). Crop rotation and plant productivity. In *CRC handbook of agricultural productivity* (pp. 273–314). CRC Press.

Teketel, M. F. (2019). *Field evaluation of two tomato (Lycopersicon esculentum L.) varieties for resistance against root knot nematode (Meloidogyne incognita) and bacterial wilt (Ralstonia solanecearum) co-inoculated with bacterial antagonists in Ziway* [Doctoral dissertation]. Hawasssa University.

Tranier, M. S., Pognant-Gros, J., Quiroz, R. D. L. C., González, C. N. A., Mateille, T., & Roussos, S. (2014). Commercial biological control agents targeted against plant-parasitic root-knot nematodes. *Brazilian Archives of Biology and Technology*, *57*(6), 831–841. doi:10.1590/S1516-8913201402540

Xu, D. L., Long, H., Liang, J. J., Zhang, J., Chen, X., Li, J. L., Pan, Z. F., Deng, G. B., & Yu, M. Q. (2012). De novo assembly and characterization of the root transcriptome of Aegilops variabilis during an interaction with the cereal cyst nematode. *BMC Genomics*, *13*(1), 1–9. doi:10.1186/1471-2164-13-133 PMID:22494814

Zasada, I. A., Halbrendt, J. M., Kokalis-Burelle, N., La Mondia, J., Mc Kenry, M. V., & Noling, J. W. (2010). Managing nematodes without methyl bromide. *Annual Review of Phytopathology*, *48*(1), 311–328. doi:10.1146/annurev-phyto-073009-114425 PMID:20455696

Zotti, M., Dos Santos, E. A., Cagliari, D., Christiaens, O., Taning, C. N. T., & Smagghe, G. (2018). RNA interference technology in crop protection against arthropod pests, pathogens and nematodes. *Pest Management Science*, *74*(6), 1239–1250. doi:10.1002/ps.4813 PMID:29194942

Chapter 8
Pathogenicity of Citrus Nematode (*Tylenchulus semipenetrans*)

Eman Samir Hagag
Plant Protection and Biomolecular Diagnosis Department, Arid Lands Cultivation Research Institute, City of Scientific Research and Technological Applications, Egypt

ABSTRACT

Citrus slow decline nematode, Tylenchulus semipenetrans, causes significant damage to all citrus-producing regions of the world and limits the production of citrus fruit under a wide range of environmental and edaphic conditions. T. semipenetrans is a semiendoparasite with a fairly narrow host rang that includes olive, grape, and citrus. No species in the genus citrus has been found to have resistance to T. semipenetrans but the only source of resistance incorporated into citrus is a dominate trait derived from Poncirus trifoliate. This chapter covers host preference among the citrus species and other hosts, biotypes, or races of T. semipenetrans and the relationship between different initial densities of T. semipenetrans and nematode infestation. Consequently, it is necessary to understand the environmental factors affecting parasitism to T. semipenetrans and management practices.

INTRODUCTION

Citrus is considered the largest fruit crop, which is widely grown in more than 140 countries around the world, with over 4000 years of cultivation history (Xu et al., 2018). The major citrus producing countries include China, Brazil and the United

DOI: 10.4018/978-1-6684-8083-0.ch008

States of America (FAO, 2022). The expansive trend is constantly growing and we can affirm with certainty that Citrus is the most extensively produced tree fruit crop in the world, with total global production reported to be 143.755 million tons in 2019 (FAO 2020). Nowadays, citrus fruits are grown in tropical, subtropical, and Mediterranean climates, in the "citrus belt" between approximately 40° N and 40° S latitude, where the winter temperatures are suitable for tree survival and escaping freeze destruction, and where there are sufficient water and appropriate soils to sustain tree growth and fruit production.

TYLENCHULUS SEMIPENETRANS

Tylenchulus semipenetrans is a semiendoparasite, which causes the disease "slow decline" of citrus. It is a parasite of many plant species since more than 75 rutaceous species (especially citrus and their close relatives) are suitable hosts (Duncan 2009). Moreover, a few non-rutaceous hosts such as olive, grape, lilac, and persimmon are also infected. Economically, the most important nematode pest associated with citrus in the world is *T. semipenetrans*. Most studies estimate yield losses due to *T. semipenetrans* to be in the range of 10% to 30% depending on the level of infection (Verdejo-Lucas and McKenry 2004). It was first detected on citrus roots in California in 1912, named and described during the next two years (Cobb 1913, 1914).

Symptoms

The symptoms of *T. semipenetrans* infection on citrus trees develop slowly, and because of this, the disease is commonly called "slow decline". If susceptible trees are planted in lightly infested soil, they may grow for many years without apparent problems and only decline slowly over years. Typical above ground signs consist of reduced vigor, the death of terminal buds, chlorosis and dying of leaves, early wilting under moisture stress, and twig dieback (Figure 1). Fruit is reduced in size, quantity, and quality. Symptom severity and speed of the tree decline depend on tree age and vigor, nematode biotype's parasitic ability, nematode population density, and the susceptibility of the rootstock. The leaves of infected trees are smaller than normal. Wilting is more pronounced in infected trees than in healthy trees during conditions of water stress. The fibrous roots of host trees are more often infected and appear thicker than healthy roots because soil particles adhere to the gelatinous egg masses of the nematode and are retained on the root surface (Figure 2). Infected fibrous roots decay because of the lesions and because secondary organisms infect them at the sites of nematode penetration and feeding. Heavy root infections result in root lesions and cortical sloughing (Cohn, 1965).

Figure 1. Citrus trees exhibiting symptoms of citrus decline caused by the citrus nematode, Tylenchulus semipenetrans (Cobb 1913)

Figure 2. Citrus roots infected by Tylenchulus semipenetrans. (A) Dirty appearance due to the adhesion of soil particles to the gelatinous egg masses deposited by females on the root surface. (B) Non-infected (right) and citrus nematode infected roots (left) (Verdejo-Lucas, 2022).

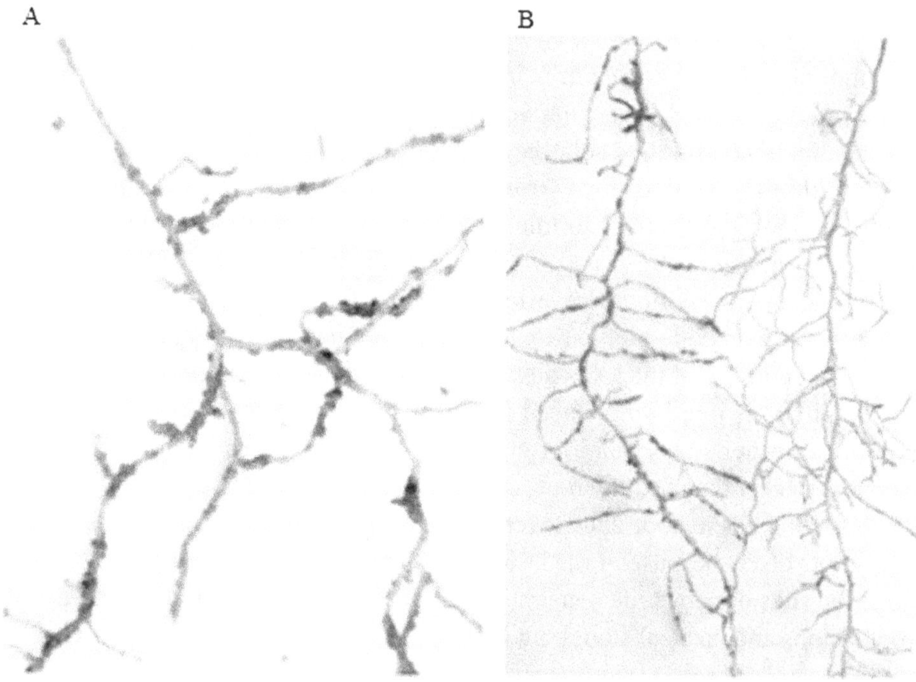

Life Cycle

The life cycle from egg to egg takes between 6 and 8 weeks (Figure 3). Eggs are laid in a gelatinous matrix deposited by the female nematode onto the root surface (Figure 4 a,b). Juveniles (Figure 5) penetrate the root 2–3 weeks after hatching, burrowing their anterior end deep inside the root cortex while the posterior end remains outside in the soil. Feeding occurs on six to ten so-called "nurse cells," which are cortical parenchyma cells around the nematode anterior regions Van Gundy (1958).

Biotypes

Biotypes or physiological races of *T. semipenetrans* exist based on host suitability (Baines et al., 1969). Information on biotypes of *T. semipenetrans* is necessary because of its relevance to breeding programs, the choice of rootstock selection, and management practices. No species in the genus citrus has been found to have resistance to *T. semipenetrans* but the only source of resistance incorporated into citrus is a dominate trait derived from *Poncirus trifoliata* (Verdejo – Lucas and Kaplan, 2002). Biotypes with different reproduction potentials on different plant hosts

Figure 3. The life cycle of the citrus nematode (Hannes Visagie, North-West University, Potchefstroom, South Africa)

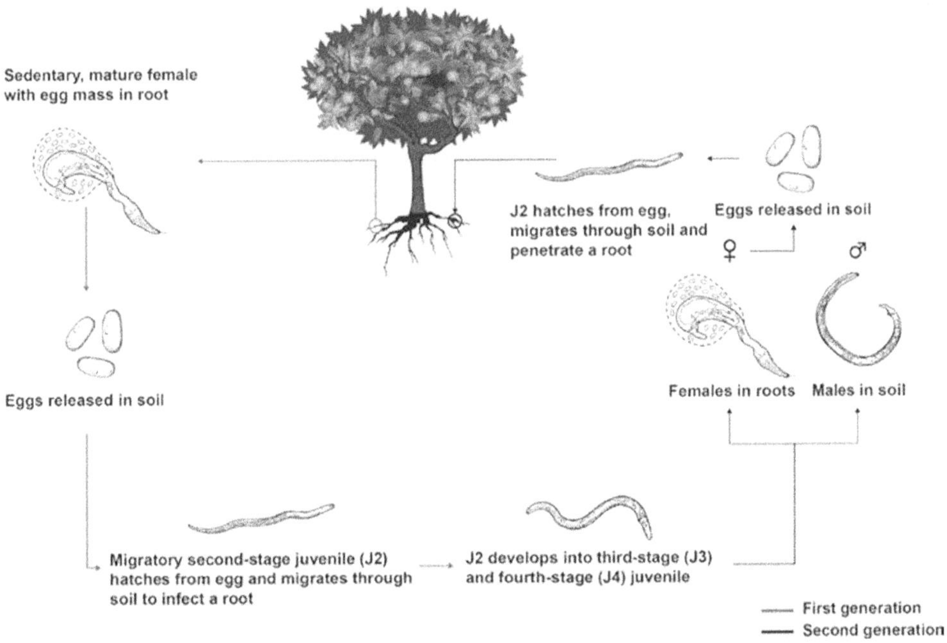

Figure 4. Female body outside root swells, and eggs are produced in a gelatinous matrix under scanning electron microscope (a) and a dissection microscope (b)

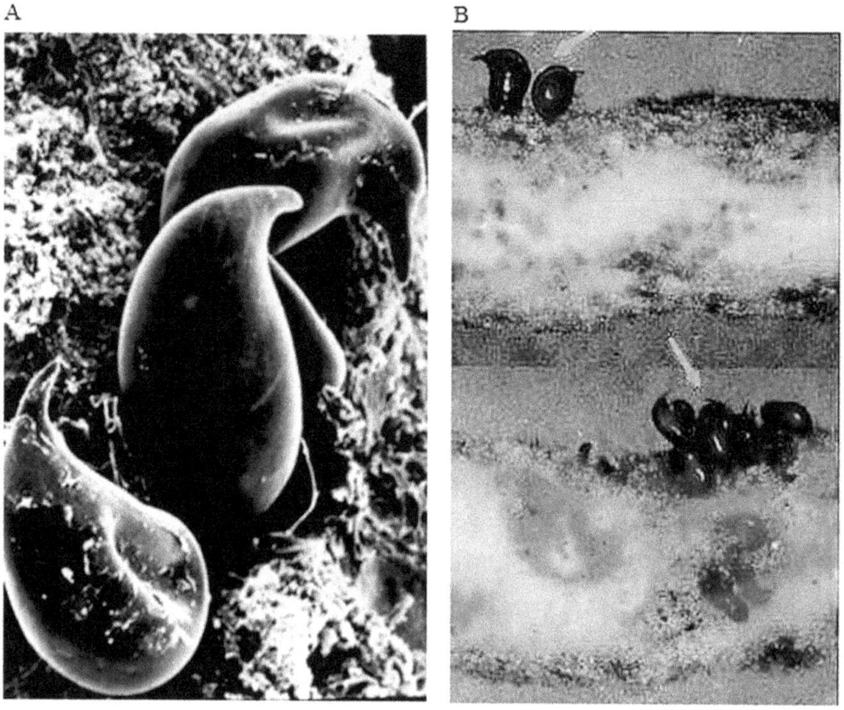

Figure 5. Vermiform second stage juvenile (J2) of the citrus nematode, Tylenchulus semipenetrans (Cobb 1913)

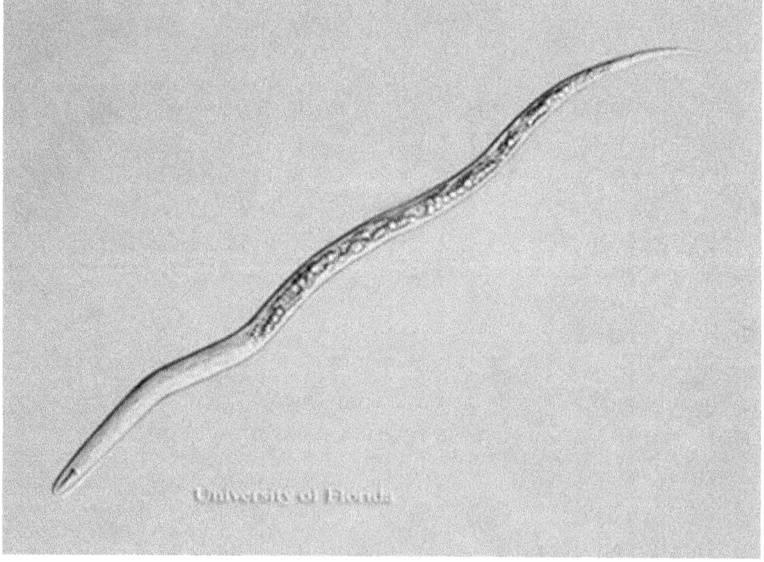

have been recognized with in the species. At present, three biotypes are commonly recognized citrus, Mediterranean and poncirus. A "Citrus" biotype was described from populations found throughout the United States citrus-growing regions and Italy. It reproduces poorly on *P. trifoliata* but will reproduce on Citrus spp. and on the hybrids "Carrizo" and "Troyer" Citrange as well as on olive (*Olea europeae*) grape (*Vitis vinifera*) and persimmon (*Diospyros* spp.). The "Poncirus" biotype, found in California, reproduces on most citrus including *P. trifoliata*, and on grape but not olive. A "Mediterranean" biotype is similar to the "Citrus" biotype, except that it does not reproduce on olive. It has been identified in rnediterranean countries and South Africa (Inserra et al., 1980; Lo Giudice & Inserrra, 1980; Verdejo-Lucas, 1992). Also, the Mediterranean biotype was identified in Spain (Verdejo - Lucas, 1992 and Verdejo-Lucas et al., 1997). The citrus populations that occur in Argentina, Australia, Brazil, India, and Venezuela may be either the citrus or the rnediterranean biotype because they reproduce poorly on *P. trifoliata*, but they have not been tested on olive (Inserra et al., 1994).

A "Grass" biotype was described from Florida, infecting *Andropogon rhizomatus*, but not citrus. "Grass" biotypes were eventually shown to be new species, *Tylenchulus graminis* and *T. palustris* (Inserra et al., 1988).

Other Hosts

Citrus species are the preferred hosts of *T. semipenetrans*, although related species, such as *Poncirus trifoliata* and its hybrids, can be parasitized as well. Citrus nematodes also infect non-rutaceous woody plants such as grape (Raski et al., 1956), olive (Baines and Thorne, 1961), and persimmon (Raski et al., 1956). Other non-rutaceous plants such as *Clerodendron inerme*, *Garcinia mangostana* and *Jasminum sambac* are reported as hosts of *T. semipenetrans* in India (Chawla et al., 1980; Bajaj et al., 1988; Nand et al., 1992). Also, *Musa textilis* and *Cunninghamia lanceolata*, have been reported in Borneo and China, respectively (Macaron, 1972; Song et al., 1994). However, the identity of these populations of *T. semipenetrans* needs to be confirmed (CABI, 2022). *T. semipenetrans* was identified morphologically from root samples collected from a pomegranate plant in a garden in the Shahdad region of southeastern Iran by Rashidifard et al., (2015) as the first record of *T. semipenetrans* on pomegranate in Iran and worldwide.

TYLENCHULUS SEMIPENETRANS ON CITRUS

Tylenchulus semipenetrans is the most nematode species capable of damaging mature citrus tend to be regional or local problems, due either to edaphic conditions or to the natural distribution of a particular nematode. *T. semipenetrans* reported in every citrus growing areas of the world (Duncan, 2005). The intensity of *T. semipenetrans* on citrus species may differ, it is an important phytopathogenic nematode and responsible for serious damage in citrus and citrus relatives growing areas in the world (CABI, 2022).

In the United States, field infestations involve 50-90% of the citrus orchards in Arizona, California, Florida, and Texas, and localized vineyards in California (O'Bannon and Esser 1985, Duncan 2005). Thorne (1961) examined more than 100 samples from citrus orchards in various parts of California and found all but one to be infested with *T. semipenetrans*). This is the only major nematode pathogen in California citrus and the disease it causes is called "slow decline" (Chitambar et al., 2018). In Turkey, *T. semipenetrans* was firstly reported from mandarin growing areas in Mersin and Adana provinces (Istiranca, 1940). Later, it was noted from other areas and became as a damaging pathogen for citrus and citrus relatives producing areas in Adana, Çanakkale, and Rize provinces (Bodenheimer, 1958; Goffart, 1951). Currently, it is an important phytopathogenic nematode and responsible for serious damage in citrus and citrus relatives growing areas in Turkey (Bozbuga et al. 2022). *T. semipenetrans*, known to be capable of damaging citrus are very limited in distribution in Egypt (Abd-Elgawad, 1992a, b; Korayem and Hassabo 2005; Abd-Elgawad, 2020, Hammam et al., 2021). In Egypt, Various surveys estimated that the citrus nematode, *T. semipentrans* was found to be very common (frequency of occurrence (FO) =51-100% in all the surveys targeting areas famous for citriculture in Egypt (Abd-Elgawad, 1992, Abd-Elgawad et al. 2016 Orf et al., 2019 and Abu Habib et al., 2020 Sweelam et al., 2022). *T. semipentrans* is the most predominate plant parasitic nematode species (88%) on citrus in Morocco (Zoubi et al., 2022) and Kenya in all the surveyed localities (Kiptoo et al., 2022). In India and Spain, about 80% of the citrus trees were reported to be infected with the nematode (Verdejo-Lucas, 2022 & Keshari and Mallikarjun 2022). Disease in most of the citrus trees in India is major problem that is estimated at 8.7-12.2% and prevalent in all citrus growing areas of Nepal (Karki, 1997 and Pandey 2022). *T. semipentrans* a major threat to the citrus industry in Pakistan where disease prevalence (20-100%) was recorded in citrus growing areas (Ali et al., 2018 and Asghar et al., 2021).

ENVIRONMENTAL FACTORS AFFECTING PARASITISM OF *TYLENCHULUS SEMIPENETRANS*

Temperature

The life cycle of *T. Semipenetrans* is regulated by host penology interacting with geographic variation and temporal changes in the soil environment. When conditions are otherwise favorable, populations will increase temperatures of 20 and 31°C, with maximum development at 25°C and prolonged development at the extremes (O'Bannon *et al.*, 1966 & Saeed *et al.*, 2019). Low winter temperatures frequently regulate the population growth of the nematode (Duncan *et al.*, 1993 & Saeed *et al.*, 2019). Summer soil temperatures in places such as Egypt, Texas, Oman, and Spain approach the upper limit of this range and often correspond to the population decline of the nematode (Sorribas *et al.*, 2000 & Saeed *et al* ., 2019). Similarly, in Arizona and Florida, population growth was slow on young trees until canopies developed sufficiently to shade the soil and provide optimum soil temperatures (Duncan, 2005).

Soil Type and Texture

Citrus nematode occurs over a wide range of soil conditions throughout the world. Certain soil factors readily influence infection and reproduction. Van Gundy *et al.*, (1964) &Duncan, (2005) reported that the development and reproduction of citrus nematodes occurred on citrus seedlings in soils containing 50% clay. The rate of reproduction, however, was notably lower in soils of 50% clay than in soils containing 5, 15, or 30% clay. In general, citrus nematode invasion and reproduction are slower in very sandy, coarse-textured soils than in other soil types (Ahmad *et al.*, 2007, Iqbal *et al.*, 2007 & Saeed et al., 2019). Organic matter in soils up to 9% encouraged nematode growth and infection, which led to early damage, that organic debris created a thin protective cover over citrus roots that enhanced nematode infectivity (Saeed *et al* ., 2019). On the other hand, Soil texture greatly influences the nematode's ability to infect tree roots. Trees can infect almost any soil, but early infection in high numbers and subsequent population increase occurs more rapidly in fine-textured organic soils than in coarse-textured sands (Heald, 1994). Maximum and minimum abundances of nematode population, in either roots or soil, were seen in soils of loam and clay textures respectively. In soils containing high levels of silt and sand, the nematode population was increased, but a high clay content resulted in low numbers of nematodes in soil and in citrus roots. A high nematode population level was recorded, when percentages of sand, silt, and clay in the soil were 48, 35, and 17 respectively (Ardakani *et al.*, 2014).

Soil Moisture

Soil moisture is often inversely related to population growth of *T. semipenetrans* (Sorribas *et al.*, 2000; Galeano, 2002& Saeed *et al* ., 2019) even though, compared with many plant parasitic nematodes, *T. semipenetrans* has little capacity for anhydrobiotic survival and nematode numbers decline quickly when trees become drought stressed (Tsai and Van Gundy, 1988 &Duncan, 2005). Nevertheless, populations in extremely dry parts of the rhizosphere can either grow rapidly or decline precipitously, depending on whether part or all of the root system is affected by drought. Hydraulic lift of water deep in the soil drier surface soil horizons via the root xylem (Caldwell *et al.*, 1991&Duncan, 2005) creates an environment highly favorable for population growth of *T. semipenetrans* (Duncan and El-Morshedy, 1996). It is not known whether this is due to increased oxygen, passive movement of nematodes deeper in soil with precipitation, increased activity of natural enemies or other factors (Sorribas *et al.*, 2000& Ardakani *et al.*, 2014). *T. semipenetrans* may have experienced less selection pressure for anhydrobiotic survival through co-evolution exclusively with deep-rooted woody perennials. The potential importance of hydraulic lift for *T. semipenetrans* is consistent with reports that peak population densities of this nematode in subtropical regions tend to be bimodal, occurring in the dry months that precede and follow the summer rainy season (O'Bannon *et al.*, 1972& Duncan, 2005) and the observation in Mediterranean climates that higher population densities tended to occur under drip compared with flood irrigation (Sorribas *et al.*, 2000& Ardakani *et al.*, 2014).

Soil pH

It was observed that fluctuation of *T. semipenetrans* populations at different soil pH levels. The maximum total population of the nematode was observed at pH 7.4-7.8; optimum pH for the maximum population (male, female, larva, and total) was 7.6 (Pandey *et al.*, 2004 & Saeed *et al.*, 2019). The maximum population of males was recorded at pH 7.4-7.6, female population was the maximum at pH 7.8, and the larval population increased at pH 7.6.

Soil Nutrients

Many studies have revealed that standard critical (minimum) levels of N, P, and K for the growth of citrus were also appropriate for nematode activities. An increase in the level of these elements in soil resulted in a similar increase in the nematode population. This result is in line with other reports, except for the case of nitrogen. Sorribas *et al.*, (2008) reported citrus nematode population densities in the soil as negative related to N concentration, and positively related to K content in the soil. Other studies have also shown ammonia to have a nematicidal effect (Rodriguez-Kábana, 1986) that is enhanced when ammonia-releasing fertilizers are combined with alkaline amendments (Oka *et al.*, 2006). Furthermore, the application of urea at the rate of 160 kg ha-1 prevented *T. Semipenetrans* infection and development on rough lemon in a pot test (Mangat and Sharman, 1981). The potassium content of the soil was positively related to densities of J_2 and males in the soil, a relationship known to occur in other crops (Pettigrew *et al.*, 2005; Sharma *et al.*, 2005 Ardakani *et al.*, 2014).There was a low nematode population was recorded in the soil and in citrus roots when the level of organic carbon was either below 0.5% or higher than 1.7%. While it was noted that the maximum population of the nematode was seen in soils with 1.5% of organic carbon (Ardakani *et al.*, 2014).

INOCULUM DENSITIES OF *TYLENCHULUS SEMIPENETRANS*

The severity of plant disease depends on the level of abundance of the pathogen, which itself is dependent upon the soil conditions for growth and survival of the organism (Ardakani *et al.*, 2014). The population density of citrus nematode has been reported to fluctuate throughout the year (Singh, 1999; Sorribas *et al.*, 2000& Saeed *et al.*, 2019) and often exhibits two distinct periods of growth (Al-Qasem and Abu-Gharbieh, 1995& Saeed *et al.*, 2019). While, the peak population developed during corresponding periods of increased root growth that occurred in April-May and November- December. Infection and subsequent population cycles are restricted to primary roots because citrus nematodes feed only in the cortex of primary roots (Iqbal *et al.*, 2004 and Al-Azzeh and Gharbieh, 2005) .Several other investigators observed some seasonal variations (Yokoo, 1964; Prasad and Chawla, 1965). Duncan and Cohn (1990) & Saeed *et al.*, (2019) found that population densities of citrus nematode generally increase in autumn and spring, presumably in response to flushes in the growth of new fibrous roots on which the nematode feeds however, found that root availability did not adequately explain population declines in nematode levels in summer, since root mass density increased by 75% during that period.

DAMAGE THRESHOLD LEVELS OF *T. SEMIPENETRANS*

Most studies estimate yield losses due to *T. Semipenetrans* to be in the range of 0 to 30%, depending on the infestation level and orchard management. Damage thresholds, the minimum nematode population densities that reduce tree growth and yield, are influenced by nematode reproduction rate, susceptibility of the rootstock, climatic conditions, soil properties, the presence of other pathogens or microbial antagonists, and orchard management practices (Sorribas *et al.*, 2008). It is well known that young trees appeared to poorly grow when planted into nematode–infested soil (Verrdejo-Lucas and McKenry, 2004 & Saeed *et al.*, 2019). Damage thresholds must be based on the factors that affect the nematode in a given region. For example, nematicide treatments are recommended when the population exceeds 400 females per gram of root in February to April or 700 in May to June in California (Sorribas *et al.*, 2008), 100 females per gram of root in South Africa (Le Roux *et al.,* 2000), or 20 juveniles/cm3 of soil in Cyprus (Philis, 1989). In Egypt, the economic threshold figures per 1 kg soil and 5 g roots were 13,000 and 15,000 nematodes during periods of low populations in February and August but 36000 nematodes during the peak period in March for Navel orange trees grafted on sour orange rootstock (Abd-Elgawad, 2016 & Korayem and Hassabo, 2005). Likewise, the economic thresholds (the population density of *T. semipenetrans* at which the value of the damage caused is equal to the cost of both sampling and control) for lemon trees were 1810 and 141 second-stage juveniles (150 cm3 soil)−1.Verdejo-Lucas and Mckenry, (2004) and Ardakani *et al.*, (2014) decided to nematicide treatments are recommended for various nematode populations, ranging from 100-400 females root and 2,000-4,000 juveniles/100 ml of soil while, Bozbuga *et al.*, (2022) revealed to the economic-injury level and economic damage threshold are defined as the level at which the number of nematodes [2nd stage juveniles (J_2s) + male nematodes] in 100 g of soil exceeds 3600 and 1600, respectively, for this nematode species.

MANAGEMENT OF *T. SEMIPENETRANS*

Resistant Rootstocks

Resistance is generally the most useful, environmentally sound solution to increasing yields by suppression of nematode population densities. A rootstock is classified resistant when it greatly inhibits nematode reproduction relative to a known susceptible standard. The only source of genetic resistance identified against *T. semipenetrans* is derived from *Poncirus trifoliata*. Some selections of *P. trifoliata* are highly resistant (Pf/Pi < 1) to *T. semipenetrans*, whereas others are only moderately resistant (Pf/Pi > 1) relative to the reference rootstock (Verdejo-Lucas and Kaplan, 2002). The hybrid rootstock Swingle citrumelo (*Citrus paradisi* × *P. trifoliata*) is highly resistant to the citrus nematode (Kaplan and O'Bannon, 1981). However, both *P. trifoliata* and Swingle citrumelo grow poorly in alkaline soils. Troyer and Carrizo citranges (*Citrus sinensis* × *P. trifoliata*) are considered moderately resistant in some regions. Continuous cultivation of resistant rootstocks may lead to the development of new biotypes or select virulent populations of the nematode. Baines et al. (1974) reported that citrus nematode biotypes could develop on all the resistant rootstocks available at that time. Populations of *T. semipenetrans* capable of reproducing well on Swingle citrumelo have been reported in Florida (Duncan et al., 1994) and South Africa (Le Roux et al., 2000), but these populations appeared to be confined to the sites where they were detected. Also, a progressive adaptation of the nematode to reproduce on rootstocks previously described as moderately resistant (i.e., Troyer and Carrizo citranges) can occur as a result of their continuous cultivation (Verdejo-Lucas et al., 1997). The presence of high nematode levels at the time of replanting or when interplanting moderately resistant and susceptible rootstocks can reduce their relative resistance level to *T. semipenetrans* (Verdejo-Lucas et al., 2003).

Nematicides

Various nematicides have successfully been used to decrease densities of *T. semipenetrans* on citrus in many regions. However, repeated applications are needed to maintain reduced densities and consistent yield increases. Little effect of treatment on yield and fruit quality may be obtained the first year after their application, but efficacy can be demonstrated in the second year (Verdejo-Lucas, and McKenry, 2004). Two main groups of nematicides, oxime-carbamates (aldicarb, oxamyl, carbofuran) and organophosphates (fenamiphos, ethoprophos, and cadusaphos), are available. Of these, granular Cadusafos has shown superior efficacy against the citrus nematode (Walker and Morey, 1999). Additionally, the broad-spectrum product Enzone, which liberates low concentrations of carbon bisulfide within soil, can reduce citrus nematode populations if properly applied. All of these products are directly lethal to the nematodes and have activity against insects. Their primary action is a result of direct contact. Once these nematicides reach more than 8 cm into soil, their action is mostly due to sub lethal effects including modification of nematode behavior. For example, stimulation of egg hatches, inability of males to find females, or inability of females to find roots are known effects of aldicarb (Verdejo-Lucas, and McKenry, 2004). Most of the early formulations of the nematicides developed in the 1960s were as granules that were spread on the soil surface and incorporated mechanically into the top 10 cm of soil. Currently, emulsifiable concentrates are available for application through drip irrigation systems. Nematicides break down in the soil or plants by hydrolysis and oxidation, and their properties relative to movement in soil and persistence should be considered. They include the solubility of the nematicides in water. For example, oxamyl is highly soluble (280,000 ppm at 25 °C), followed by aldicarb (6,000 ppm at 25 °C), ethoprophos (750 ppm), fenamiphos (700 ppm at 20 °C), carbofuran (700 ppm at 25 °C), and cadusafos, which has very low solubility (248 ppm) (Hague and Gowen, 1987). Generally, the higher the solubility, the greater the potential movement of the active ingredient of the target. The amount of soil moisture is another important factor influencing the distribution of the nematicide in soil. Products of low solubility also can be highly adsorptive to dry soil particles, rendering them unavailable for deeper movement. Hence, irrigation is often recommended before nematicide application. In these systems, application technology becomes a critical issue for effective control.

Biological Control

A diversity of antagonists of the citrus nematode, including trapping and parasitic fungi, bacteria, and predacious nematodes and mites, occur naturally in citrus orchards (Roccuzzo et al., 1992). As with nematicides, soil moisture, and temperature affect the persistence of biological control agents (BCA) once applied to the soil. Biological control may be helpful in regions where chemical control is not available or affordable, and in orchards under organic farming. Biological control may be more successful at moderate rather than at high pest pressure due to the inverse relationship between nematode densities and the level of control of the BCA (Bourne and Kerry, 1999). Parasitism of *T. semipenetrans* by *Pasteuria spp.* has been reported to occur in many countries (Sorribas et al., 2000). The bacterial parasite produces mature spores inside the vermiform stages of the nematode. The *Pasteuria* spp. associated with nematode pests represent a valuable and widespread natural resource, whose benefits and application potential are not yet completely explored (Ciancio, 2018). The egg-parasitic fungus, *Paecilomyces lilacinus*, reduced *T. semipenetrans* densities in pot experiments (Parvatha Reddy et al., 1991), and the results were best when combined with organic amendments (i.e., oil cakes). Similar results were observed in an 8-year-old Valencia orchard on rough lemon rootstock (Le Roux et al., 2000). This fungus now produces commercially in several countries, including Australia, Colombia, Germany, and South Africa. A product of biological origin from the hyphomycete *Myrothecium*, DiTera, has been developed as an environmentally compatible biological alternative to chemical control. It is noteworthy that metabolites, rather than live organisms, are added to soil, killing nematodes in the soil contact. At present, DiTera has registrations in the United States, Chile, and Panama (Warrior et al., 1999). The application of certain soil bacteria as seed or root inoculants to improve plant growth and productivity is known as plant-growth-promoting rhizobacteria PGPR (Khan et al., 2012). Well-known PGPRs include Bacillus, Azospirillium, Azotobacter, Burkholderia, Pseudomonas, Rhizobium, Serratia, and several novel PGPRs like *Methylobacterium, Azoarcus, Exiguobacterium, Paenibacillus* and *Pantoea* etc., (Hemlata et al., 2015). According to their mode of action, they can be classified as biofertilizers and biopesticides (Martinez et al., 2010). Thus, PGPR has grabbed the scientists' attention to be used as an alternative eco-friendly biological control agent. PGPR operates through either direct or indirect mechanisms or a combination of both (Orf et al., 2019).

Cultural Practices

Many properly established orchards generate high yields in the presence of nematodes and that when following recommended agronomic and cultural practices as rotations with annual crops for 1 to 3 years before replanting citrus help to reduce citrus nematode populations, but the economics of such rotations limit their use. Controlling weeds will reduce competition for water and nutrients, flowing for 4 months to 1 year before replanting an orchard is a common practice, but may not be sufficient due to nematode survival within old roots (Sorribas et al., 2003). Mulching is a cultural practice that can help reduce water loss from the soil as it reduces evaporation, moderates extreme daily soil temperatures, and helps suppress weed competition. As a consequence, the crop's environment is modified, promoting tree vigor and increased yield. (Verdejo-Lucas, and McKenry, 2004).

REFERENCES

Abd-Elgawad, M. M. (2016). Use of Taylor's power law parameters in nematode sampling. *International Journal of Pharm Tech Research*, *9*(12), 999–1004.

Abd-Elgawad, M. M. (2020). Managing nematodes in Egyptian citrus orchards. *Bulletin of the National Research Center*, *44*(1), 1–15.

Abd-Elgawad, M. M., Koura, F. F. H., Montasser, S. A., & Hammam, M. M. A. (2016). Distribution and losses of *Tylenchulus semipenetrans* in citrus orchards on reclaimed land in Egypt. *Nematology*, *18*(10), 1141–1150. doi:10.1163/15685411-00003020

Abu Habib, A. H. A., Younes, H. A., Ibrahim, I. K. A., & Khalil, A. E. (2020). Plant parasitic nematodes associated with citrus trees and reaction of two citrus cultivars to *Tylenchulus semipenetrans* in northern Egypt. *Journal of the Advances in Agricultural Researches*, *25*(2), 166–175. doi:10.21608/jalexu.2020.161764

Ahmad, R., Khan, M. A., Mukhtar, T., & Javed, N. (2007). Infestation of citrus orchards by citrus nematode (*Tylenchulus semipenetrans*) and screening of some rootstocks for resistance or susceptibility. *Proceedings of the International Symposium on Prospects of Horticultural Industry in Pakistan*, 165-169.

Al-Azzeh, T. K., & Gharbieh, W. A. (2005). Factors related to seasonal changes in numbers of *Tylenchulus semipenetrans* in the Jordan valley. *Pakistan Journal of Nematology*, *23*, 241–250.

Al-Qasem, M. S., & Abu-Gharbieh, W. I. (1995). Occurrence and distribution of citrus nematode, *Tylenchulus semipenetrans* in Jordan. *Nematologia Mediterranea*, *23*, 335–339.

Ali, Y., Abbas, G., Sajid, M., & Imran, M. (2018). Response of Citrus Slow Decline Causing Nematode (*T. semipenetrans*) against Different Cultivars (Kinnow, Musambi, Shakri, Feutrel Early, Red Blood) of Citrus. *Advances in Zoology and Botany.*, *6*(1), 1–18. doi:10.13189/azb.2018.060101

Ardakani, S. A., Mafi, Z. T., Mokaram, A. H., & Goltappeh, E. M. (2014). Relationship between soil properties and abundance of *Tylenchulus semipenetrans* in citrus orchards, Kohgilouyeh va Boyerahmad Province. *Journal of Agricultural Science and Technology*, *16*(7), 1699–1710.

Asghar, M. N., Khan, S. A., Javed, N., & Arshad, M. (2021). Exploring Citrus Nematodes and its Antagonists in Citrus Growing Areas of Punjab, *Pakistan. J. Plant Environ*, *03*(02), 91–101. doi:10.33687/jpe.003.02.3714

Bahadur, A. (2022). Nematodes Diseases of Fruits and Vegetables Crops in India. In C. Belle & T. Kaspary (Eds.), *Nematodes - Recent Advances, Management and New Perspectives* (pp. 1–18). IntechOpen. doi:10.5772/intechopen.98850

Baines, R. c., Miyakawa, T., Cameron, J. W., & Small, R. H. (1969). Biotypes of the citrus nematode. *Proceedings of the lst International Citrus Symposium*, 955-956.

Baines, R. C., Cameron, J. W., & Soost, R. K. (1974). Four biotypes of *Tylenchulus semipenetrans* in California identified, and their importance in the development of resistant citrus rootstocks. *Journal of Nematology*, *6*, 63–66. PMID:19319369

Baines, R. C., & Thorne, G. (1961). The olive tree as a host of the citrus-root nematode. *Phytopathology*, *42*, 77–78.

Bajaj, H.K., Kanwar, R.S., Baghel, P.P.S. & Bhatti, D.S. (1988). Motia (*Jasminum sambac*) - a new host record of citrus nematode, *Tylenchulus semipenetrans* Cobb, 1913. *Haryana Agricultural University Journal of Research*, *18*(2), 146-147.

Bodenheimer, F. S. (1958). *The history of biology: an introduction*. Dawson.

Bourne, J. M., & Kerry, B. R. (1999). Effect of the host plant on the efficacy of *Verticillium chlamydosporium* as a biological control agent of root-knot nematodes at different nematode densities and fungal application rates. *Soil Biology & Biochemistry*, *31*(1), 75–84. doi:10.1016/S0038-0717(98)00107-2

BozbugaS.YildizS.YukselE.ÖzerG.DababatA. A.İmrenM. (2022). Nematode-citrus plant interactions: host preference, damage rate and molecular characterization of Citrus root nematode *Tylenchulus semipenetrans*. Research square. doi:10.21203/rs.3.rs-1502604/v1

CABI. (2022). https://www.cabi.org/isc/datasheet/61059

Caldwell, M. M., Richards, J. H., & Beyschlag, W. (1991). Hydraulic lift: ecological implications of water efflux from roots. In D. Atkinson (Ed.), *Plant Root Growth: an Ecological Perspective* (pp. 423–436). Blackwell Scientific.

Chawla, M. L., Samathanam, G. J., & Sharma, S. B. (1980). Occurrence of citrus nematode (*Tylenchulus semipenetrans* Cobb, 1913) on roots of mangos teen (*Garcinia mangostana*). *Indian Journal of Nematology*, *10*(2), 240–242.

Chitambar, J. J., Westerdahl, B. B., & Subbotin, S. A. (2018). Plant Parasitic Nematodes in California Agriculture. In Plant Parasitic Nematodes in Sustainable Agriculture of North America. Sustainability in Plant and Crop Protection. Springer. doi:10.1007/978-3-319-99585-4_6

Ciancio, A. (2018). Biocontrol potential of *Pasteuria* spp. for the management of plant parasitic nematodes. *Perspectives in Agriculture, Veterinary Science, Nutrition and Natural Resources*, *13*(013), 1–13. doi:10.1079/PAVSNNR201813013

Cobb, N. A. (1913). Notes on Mononchus and Tylenchulus. *Journal of the Washington Academy of Sciences*, *3*, 287–288.

Cobb, N. A. (1914). Citrus-root nematode. *Journal of Agricultural Research*, *2*, 217–230.

Cohn, E. (1965). The development of the citrus nematode on some of its hosts. *Nematologica*, *11*(4), 593–600. doi:10.1163/187529265X00744

Cohn, E. (1972). Nematode diseases of citrus. In J. M. Webster (Ed.), *Economic Nematology* (pp. 215–244). Academic Press.

Duncan, L. W. (1999). Nematode diseases of citrus. In L. W. Timmer & L. W. Duncan (Eds.), *Citrus Health Management* (pp. 136–148). American Phytopathological Society Press.

Duncan, L. W. (2005). Nematode parasites of citrus. In M. Luc, R. A. Sikora, & J. Bridge (Eds.), *Plant Parasitic Nematodes in Subtropical and Tropical Agriculture* (pp. 437–466). CAB International. doi:10.1079/9780851997278.0437

Duncan, L. W., & Cohn, E. (1990). Nematode parasites of citrus. In M. Luc, R. A. Sikora, & J. Bridge (Eds.), *Plant Parasitic Nematodes in Subtropical and Tropical Agriculture* (pp. 321–346). Commonwealth Agricultural Bureau International.

Duncan, L. W., & Eissenstat, D. M. (1993). Response to *Tylenchulus semipenetrans* to citrus fruit removal: Implications for carbohydrates composition. *Journal of Nematology*, *25*, 7–14. PMID:19279735

Duncan, L. W., & El-Morshedy, M. M. (1996). Population changes of *Tylenchulus semipenetrans* under localized versus uniform drought in the citrus root zone. *Journal of Nematology*, *28*(3), 360. PMID:19277154

Duncan, L. W., El-Morshedy, M. M., & McSorley, R. (1994). Sampling citrus fibrous roots and *Tylenchulus semipenetrans*. *Journal of Nematology*, *26*, 442–451. PMID:19279914

Duncan, L. W., Graham, J. H., & Timmer, L. W. (1993). Seasonal pattern associated with *Tylenchulus semipenetrans* and *Phtophthora parasitic* in the citrus rhizosphere. *Phytopathology*, *83*(5), 573–581. doi:10.1094/Phyto-83-573

FAOSTAT. (2020). *Citrus fruit fresh and processed – Statistical Bulletin*. Author.

FAOSTAT. (2022). *Markets and Trade*. Citrus.

Galeano, M. (2002). Dinamica di popolazione di *Tylenchulus semipenetrans* e della nematofauna di vita libera nella rizosfera di agrumi in Spagna. *Supplemento Nematologia Mediterranea*, *30*, 49–53.

Goffart, H. (1951). *Nematoden der Kulturpflanzen Europas*. Paul Parey.

Hague, N. G. M., & Gowen, S. (1987). Chemical control of nematodes. In R. H. Brown & B. R. Kerry (Eds.), *Principles and practice of nematode control in crops* (pp. 131–178). Academic Press.

Hammam, M. M. A., Abd-Elgawad, M. M. M., Ruan, W., & El-bahrawy, A. (2021). Management of pests and pathogens affecting citrus yield in Egypt with special emphasis on nematodes. *Egyptian Journal of Agronematology*, *20*(2), 64–84. doi:10.21608/ejaj.2021.183231

Heald, C. M. (1994). *Citrus Declines Caused by Nematodes in Florida V. Slow Decline. University of Florida Cooperative Extension Service, Institute of Food and Agriculture Sciences*. EDIS.

Hemlata, C., Bagyaraj, G., Selvakumar, G., & Sundaramc, S. P. (2015). Novel plant growth promoting rhizobacterial. App. *Soil Ecology*, *95*, 38–53. doi:10.1016/j.apsoil.2015.05.011

Inserra, R. N., Duncan, L. W., O'Bannon, I. H., & Fuller, S. A. (1994). Citrus nematode biotypes and resistance of citrus rootstocks in Florida. Florida. Dept. Agric. & Consumer Serv. Div. Pl. Ind.

Inserra, R. N., & Vovlas, N. (1978). *Tylenchulus semipenetrans* on olive trees in Northern Italy. *3rd International Congress of Plant Pathology, München, GFR, 16-23 Aug., 1978. Abstracts of Papers: l'51.*

Inserra, R. N., Vovlas, N., & O'Bannon, J. H. (1980). A classification of *Tylenchulus semipenetrans* biotypes. *Journal of Nematology*, *12*, 283–287. PMID:19300704

Inserra, R. N., Vovlas, N., O'Bannon, J. H., & Esser, R. P. (1988). Tylenchulus graminis n. sp. and T. palustris n. sp. (Tylenchulidae) from native flora of Florida, with notes on *T. semipenetrans* and *T. furcus*. *Journal of Nematology*, *20*, 266–287. PMID:19290211

Iqbal, M. A., Ahmad, R., Mukhtar, T., & Khan, H. U. (2004). Interrelationship of citrus nematode (*Tylenchulus semipenetrans* Cobb) population and seasonal fluctuations in Sargodha district. *Pakistan Journal of Phytopathology*, *16*, 60–62.

Iqbal, M. A., Mukhtar, T., & Ahmad, R. (2007). Effect of soil types on the movement and infection of citrus nematode, *Tylenchulus semipenetrans*. *Pakistan Journal of Nematology*, *25*, 69–72.

Istiranca, R. (1940). *Narenciye Nematodu*. Seyhan Ziraat Mücadele İstasyonu.

Jaffee, B. A. (1993). Density-dependent parasitism in biological control of soil borne insects, nematodes, fungi and bacteria. *Biocontrol Science and Technology*, *3*(3), 235–246. doi:10.1080/09583159309355279

Kaplan, D. T., & O'Bannon, J. H. (1981). Evaluation and nature of citrus nematode resistance in Swingle citrumelo. *Proceedings of the Annual Meeting of the Florida State Horticultural Society*, *94*, 33–36.

Karki, P. B. (1997). Status of nematode problems and research in Nepal. Diagnosis of key nematode pests of chickpea and pigeonpea and their management. *Proceedings of a Regional Training Course*, 112.

Keshari, N., & Mallikarjun, G. (2022). Plant Parasitic Nematodes: A Major Constraint in Fruit Production. In C. Belle & T. Kaspary (Eds.), *Nematodes - Recent Advances, Management and New Perspectives* (pp. 1–33). IntechOpen. doi:10.5772/intechopen.101696

Khan, M. R., Khan, M. M., Anwer, M. A., & Haque, Z. (2012). Laboratory and field performance of some soil bacteria used as seed treatments on *Meloidogyne incognita* in chickpea. *Nematologia Mediterranea*, *40*, 143–151.

Kiptoo, J., Mutisya, D., Ndegwa, P., Amata, R., Irungu, L., & Godfrey, R. (2022). Influence of soil types on citrus nematode species diversity and abundance on varied ecological conditions in Kenya. *Pakistan Journal of Nematology*, *40*(1), 12–21. doi:10.17582/journal.pjn/2022/40.1.12.21

Korayem, A. M., & Hassabo, S. A. A. (2005). Citrus yield in relation to Tylenchulus semipanetrans in silty loam soil. *International Journal of Nematology*, *15*(2), 179–182.

Le Roux, H. F., Pretorius, M. C., & Huisman, L. (2000). Citrus nematode IPM in Southern Africa. *Proc. Int. Soc. Citricult.*, *2*, 823–827.

Macaron, J. (1972). *Contribution to the study of the plant-parasitic nematode* Tylenchulus semipenetrans *Cobb 1913 (Nematoda-Tylenchida)* [PhD Thesis]. University of Science and Technology of Languedoc.

Mangat, B. P. S., & Sharma, N. K. (1981). Influence of Host Nutrition on Multiplication and Development of Citrus Nematode. *Indian Phytopathology*, *34*, 90–91.

Martinez, O. V., Jorquera, M. A., Crowley, D. E., Gajardo, G., & Mora, M. L. (2010). Mechanisms and practical considerations involved in plant growth promotion by rhizobacterial. *Journal of Soil Science and Plant Nutrition*, *3*, 293–319.

Nand, S., Meher, H. C., & Singh, G. (1992). Clerodendron inerme (L.) Gaertn: A new host of *Meloidogyne incognita* and Tylenchulus semipenetrans. *Indian Journal of Nematology*, *22*(2), 149.

Navas, A., Nombela, G., & Bello, A. (1992). Caracterización de la modalidad de dis-tribución de *Tylenchulus semipenetrans* en el levante español. *Nematropica*, *22*, 205–216.

O'Bannon, J. H., & Esser, R. P. (1985). *Citrus declines caused by nematodes in Florida. II. Physiological races. Nematology Circular No. 116*. Florida Department of Agriculture and Consumer Services.

O'Bannon, J. H., Radewald, J. D., & Tomerlin, A. T. (1972). Population fluctuation of three parasitic nematodes in Florida citrus. *Journal of Nematology*, *4*, 194–199. PMID:19319266

O'Bannon, J. H., Reynolds, H. W., & Leathers, C. R. (1966). Effects of temperature on penetration, development, and reproduction of *Tylenchulus semipenetrans*. *Nematologica*, *12*(4), 483–487. doi:10.1163/187529266X00301

Oka, Y., Tkachi, N., Shuker, S., Rosenberg, R., Suriano, S., & Fine, P. (2006). Laboratory Studies on the Enhancement of Nematicidal Activity of Ammonia-releasing Fertilizers by Alkaline Amendments. *Nematology*, *8*(3), 335–346. doi:10.1163/156854106778493466

Orf, O. M. H., Soad, Y. S. E.-S., & Sahar, H. A.-B. (2019). Impact Of Plant Growth Promoting Rhizobacteria As Bio-Control Agents On Citrus Nematode, *Tylenchulus Semipenetrans* Infecting Balady Orange (*Citrus Sinensis* L.) And Improving Its Productivity. *The Future of Agriculture*, (4), 73–87.

Pandey, P. (2022). A Review on Major Diseases of Citrus in Nepal and their Management. *Journal of Tikapur Multiple Campus*, *5*(June), 152–167. doi:10.3126/jotmc.v5i1.46522

Pandey, R. K., Goswami, B. K., & Singh, S. (2004). Influence of soil pH on population dynamics of *Tylenchulus semipenetrans* infesting citrus and its bio-management using VAM fungi. *International Journal of Nematology*, *14*, 174–176.

Parvatha Reddy, P., Khan, R. M., & Rao, M. S. (1991). Integrated management of the citrus nematode *Tylenchulus semipenetrans* using oil-cakes and *Paecilomyces lilacinus*. *Afro-Asian Journal of Nematology*, *1*, 221–222.

Pettigrew, W. T., Meredith, W. R. Jr, & Young, L. D. (2005). Potassium Fertilization Effects on Cotton Lint Yield, Yield Components, and Reniform Nematode Populations. *Agronomy Journal*, *7*(4), 1245–1251. doi:10.2134/agronj2004.0321

Philis, J. (1989). Yield loss assessment caused by the citrus nematode *Tylenchulus semipenetrans* on Valencia oranges in Cyprus. *Nematologia Mediterranea*, *17*, 5–6.

Prasad, S. K., & Chawla, M. L. (1965). Observation on the population fluctuation of citrus nematode, *Tylenchulus semipenetrans* Cobb, 1913. *Indian Journal of Entomology*, *27*, 450–454.

Rashidifard, M., Shokoohi, E., Hoseinipour, A., & Jamali, S. (2015). *Tylenchulus semipenetrans* (Nematoda: Tylenchulidae) on pomegranate in Iran. *Australasian Plant Disease Notes, Australasian Plant Pathology Society, 10*(1), 2. doi:10.100713314-014-0157-7

Roccuzzo, G., Ciancio, A., & Lo Giudice, V. (1992). Some observations on the ecology of the citrus nematode *Tylenchulus semipenetrans* Cobb in Southern Italy. *Proceedings of the VII International Citrus Congress, 3*, 950–952.

Rodriguez-Kábana, R. (1986). Organic and Inorganic Nitrogen Amendments to Soil as Nematode Suppressants. *Journal of Nematology, 18*, 129–135. PMID:19294153

Saeed, M., Mukhtar, T., & Rehman, M. A. (2019). Temporal fluctuations in the population of citrus nematode (*Tylenchulus semipenetrans*) in the Pothowar region of Pakistan. *Pakistan Journal of Zoology, 51*(6), 2257–2263. doi:10.17582/journal.pjz/2019.51.6.2257.2263

Sharma, G. C., Thakur, B. S., & Kashyap, A. S. (2005). Impact of NPK on the Nematode Populations and Yield of Plum (*Prunus salicina*). *Acta Horticulturae*, (696), 433–436. doi:10.17660/ActaHortic.2005.696.77

Shokoohi, E., & Duncan, L. W. (2018). Nematode parasites of citrus. In R. Sikora, P. Timper, & D. Coyne (Eds.), *Plant-parasitic nematodes in tropical & subtropical agriculture* (3rd ed., pp. 446–476). CAB International. doi:10.1079/9781786391247.0446

Siddiqi, M. R. (1996). Occurrence of the citrus nematode, *Tylenchulus semipenetrans* Cobb, 1913, and the reniform nematode, *Rotylenchulus reniformis* in India (Abstr.). *Proceedings of 48th Indian Science Congress,* 504.

Singh, B. (1999). Occurrence and distribution of *Tylenchulus semipenetrans* in Nagpur mandarin orchards of Vidarbha, Maharashtra. *Indian J. Nematol, 29*, 149–153.

Song Z., Chen, X. Z. & Huang T. Z. (1994). Studies on soil micro-organisms and biochemical activities of Chinese fir woodland infected by *Tylenchulus semipenetrans* II. Soil enzyme activity and soil chemical properties. *Journal of Fujian College of Forestry, 14*(1), 81-85.

Sorribas, F. J., Verdejo-Lucas, S., Forner, J. B., Alcaide, A., Pons, J., & Ornat, C. (2000). Seasonality of *Tylenchulus semipenetrans* Cobb and *Pasteuria* sp., in citrus orchards of spain. *Journal of Nematology, 32*, 622–632. PMID:19271018

Sorribas, F. J., Verdejo-Lucas, S., Galeano, M., Pastor, J., & Ornat, C. (2003). Effect of 1,3-dichloropropene and rootstocks alone and in combination on *Tylenchulus semipenetrans* and citrus tree growth in a replant management program. *Nematropica*, *34*, 149–158.

Sorribas, F. J., Verdejo-Lucas, S., Pastor, J., Ornat, C., Pons, J., & Valero, J. (2008). Population densities of *Tylenchulus semipenetrans* related to physicochemical properties of soil and yield of Clementine Mandarin in Spain. *Plant Disease*, *92*(3), 445–450. doi:10.1094/PDIS-92-3-0445 PMID:30769676

Sweelam, M. E., Abokorah, M. S., & Elsebaay, W. A. (2022). Plant parasitic nematodes associated with fruit trees in the newly reclaimed soils at El-Sadat province, Menoufia governorate. *Egyptian Journal of Agronematology*, *21*(1), 1–13. doi:10.21608/ejaj.2022.218676

Thorne, G. (1961). *Principles of nematology*. McGraw-Hill Book Company.

Tsai, B. Y., & Van Gundy, S. D. (1988). Comparison of anhydrobiotic ability of the citrus nematode with other plant parasitic nematodes. *Proceedings of the International Society of Citriculture*, *2*, 983–992.

Van Gundy, S. D. (1958). The life history of the citrus nematode Tylenchulus semipenetrans Cobb. *Nematologica*, *3*(4), 283–294. doi:10.1163/187529258X00049

Van Gundy, S. D., & Kirkpatrick, J. D. (1964). Nature of resistance in certain citrus rootstocks to citrus nematode. *Phytopathology*, *54*, 419–427.

Verdejo-Lucas, S. (1992). On the occurrence of the Mediterranean biotype of *Tylenehulus semipenetrans* in Spain. Fundam. appl. *NemaLOI.*, *15*, 475–477.

Verdejo-Lucas, S. (2022). *Pre-planting solutions for the slow decline of citrus caused by Tylenchulus semipenetrans*. https://cabidigitallibrary.org

Verdejo-Lucas, S., Galeano, M., Sorribas, F. J., Forner, F. B., & Alcaide, A. (2003). Effect on resistance to *Tylenchulus semipenetrans* of hybrid citrus rootstocks subjected to continuous exposure to high population densities of the nematode. *European Journal of Plant Pathology*, *109*(5), 427–433. doi:10.1023/A:1024220417737

Verdejo-Lucas, S., & Kaplan, D. T. (2002). The citrus nematode: *Tylenchulus semipenetrans*. In J. L. Starr, R. Cook, & J. Bridge (Eds.), *Plant resistance to parasitic nematodes* (pp. 207–219). CABI. doi:10.1079/9780851994666.0207

Verdejo-Lucas, S., & Mckenry, M. V. (2004). Management of the Citrus Nematode, *Tylenchulus semipenetrans*. *Journal of Nematology*, *36*, 424–432. PMID:19262822

Verdejo-Lucas, S., Sorribas, F. J., Pons, J., Forner, J. B., & Alcaide, A. (1997). The mediterranean biotypes of *Tylenchulus semipenetrans* in Spanish citrus orchards. *Fundamental and Applied Nematology, 20*(4), 399–404.

Walker, G. E., & Morey, B. G. (1999). Effects of chemicals and microbial antagonists on nematodes and fungal pathogens of citrus roots. *Australian Journal of Experimental Agriculture, 39*(5), 629–637. doi:10.1071/EA99003

Warrior, P., Linda, A., Beach, M., Grau, P. A., Kirfman, G. W., & Conley, J. M. (1999). Commercial development and introduction of DiTera™, a new nematicide. *Pesticide Science, 55*(3), 343–389. doi:10.1002/(SICI)1096-9063(199903)55:3<376::AID-PS918>3.0.CO;2-S

Xu, J., Zhang, Y., Zhang, P., Trivedi, P., Riera, N., Wang, Y., Liu, X., Fan, G., Tang, J., Coletta-Filho, H. D., Cubero, J., Deng, X., Ancona, V., Lu, Z., Zhong, B., Roper, M. C., Capote, N., Catara, V., Pietersen, G., ... Wang, N. (2018). The structure and function of the global citrus rhizosphere microbiome. *Nature Communications, 9*(1), 4894. doi:10.103841467-018-07343-2 PMID:30459421

Yokoo, T. (1964). Studies on the citrus nematode *Tylenchulus semipenetrans* Cobb, 1913) in Japan. *Agric. Bull. Saga Univ., 20*, 71–109.

Zoubi, B., Mokrini, F., Dababat, A. A., Amer, M., Ghoulam, C., Lahlali, R., Laasli, S.-E., Khfif, K., Imren, M., Akachoud, O., Benkebboura, A., Housseini, A. I., & Qaddoury, A. (2022). Occurrence and Geographic Distribution of Plant-Parasitic Nematodes Associated with Citrus in Morocco and Their Interaction with Soil Patterns. *Life (Chicago, Ill.), 12*(5), 637. doi:10.3390/life12050637 PMID:35629306

Chapter 9
Biological Control of Nematodes in Citrus Orchards

Fouad Abobatta
https://orcid.org/0000-0002-1063-3165
Agriculture Research Center, Egypt

Chidanandamurthy Thippeswamy Swamy
https://orcid.org/0000-0003-0989-4381
Captain Srinivasa Murthy Central Ayurveda Research Institute, India

ABSTRACT

There are various challenges facing citrus orchards, including abiotic and biotic threats. In citrus orchards, nematodes are one of the significant biotic stressors that cause severe threats on growth and fruiting of citrus trees. Managing nematodes in citrus orchards reduces tree decline, decreasing various hazards of chemical nemacides on environment and humanity. Using biological control enhances tree growth, improves productivity, and protects the environment. There are different methods to manage the nematode problem in different crops, including physical techniques, crop rotation, resistant varieties, in addition to biological control. Using multiple strategies to control nematodes is beneficial and safer through reducing the use of nematicides and utilizing environmentally friendly substances, which are safer for human health and protect the environment. Presently, there is more interest in using organic nematicides to control plant-parasitic nematodes because of their positive impacts on crop productivity, soil, and the environment.

DOI: 10.4018/978-1-6684-8083-0.ch009

1. INTRODUCTION

Citrus is one of the most famous and extensively cultivated fruit crops worldwide. Citrus fruits are a rich source of vitamins, minerals, and dietary fiber that are crucial for a person's general nutritional health (Abobatta, 2019a).

The most frequently traded citrus fruit and one of the most generally farmed varieties, oranges account for more than half of global citrus output. Tangerines, lemons, and grapefruits are the next most traded citrus fruits. Over the past three decades, global citrus output and export have increased gradually, but more slowly than rival products like mangoes, avocados, and melons (FAO, 2022).

There are various challenges facing citrus groves that including abiotic and biotic threats (Abobatta, 2020). Nematode worms constitute one of the most common creatures on the planet. They comprise free-living organisms, humans, animals, plants, and insect parasites.

Nematode plant-parasitic (NPPs) are one of Ecdysozoa family, they are multicellular, microscopic, and bilaterally worms. Nematodes vary in size, they could reach up to 12 mm in a length with a width reach to 35 μ (Khan, 2015). Nematodes have a body cavity and are rounded, they possess one-way intestines with a mouth and anus at either end. They have a pseudocoelom that has been mesoderm-lined on one side and endoderm-lined on the other. The cuticle appears to be an intricate and plastic characteristic that has evolved through time, serving crucial roles in protection, body mobility, and shape maintenance. Nematodes only have longitudinal muscles, which accounts for their apparent whirling motion. Nematodes do not possess distinct circulatory or respiratory systems, although they have digestive, reproductive, nervous, and excretory systems. Nematodes can orient themselves and react to various environmental stimuli thanks to chemosensory and mechanosensory neurons located in the cuticle. About 1000 somatic cells make up an adult; hundreds of these cells are usually connected to the reproductive systems (Basyoni & Rizk, 2016).

The environmental infective larval stage of many parasitic nematodes involves host searching, a mechanism in which the infective larvae employ sensory signals to seek hosts. Olfaction, gustation, thermosensation, and humidity sensation are only a few sensory modalities used in a complicated host-seeking activity. Host seeking might be a potent target for prophylactic intervention since it is the first stage of the parasite-host interaction. Host-seeking behavior is still not well understood, though (Gang & Hallem, 2016).

Infection with nematodes interacting with soilborne disease leads to more hazards for host plants and increases deterioration symptoms. The magnitude of the damage was clearly correlated with the level of soil infection of nematodes that reduced agricultural plant output, particularly the beet cyst nematode, the precise nature of this correlation remained unclear to most researchers. Nevertheless, attempts to advise farmers on crop rotation to protect themselves against cyst nematode assault had already begun in the first half of the last century, most likely because the ability to quantify cysts in soil samples led researchers to interpret the numbers discovered (Seinhorst, 1986).

Nematodes are one of the major obstacles facing farmers around the world. They affect the various metabolic activities of plants, negatively affecting the foliage and producing organs, thus, causing significant economic losses in crops and reducing food production, there are plentiful of nematodes found in the soil, while, still some more need to explore.

Nematodes inhibit the growth and development of plants infecting the vascular system by reducing the nutrient supply, which causes high economic loss in fruit crop production. The root damaged by nematodes is very vulnerable to bacterial and fungal infection, further deteriorating the plant roots and stunting the plant growth.

The plant parasitic nematodes feed from plants (roots, stems, leaves, flowers, and seeds) with the help of stylet a specialized spear. The scars are made on plants after the nematode infection due to the withdrawal of cell contents and making them die. They affect the various metabolic activities of plants, negatively affecting the foliage and producing organs, thus, causing significant economic losses in crops and reducing food production (Khanna, *et al*, 2019).

Nematodes are considered the main factor in limiting citrus cultivation and cause economic loss for citrus growers in various regions. The first infection with citrus nematode *Tylenchulus semipenetrans*, was recorded in California (1913), and it was described as a causative agent of slow deterioration in citrus by Nathan Cobb (1913). It was found in the majority of citrus groves in the United States, where the infection ranged from 50-90% in citrus groves in Arizona, California, Florida, and Texas (Keshari & Mallikarjun, 2022).

In citrus groves, the nematode is one of the significant biotic stress that causes severe threats to growth and fruiting of citrus trees. The root-knot nematodes like *Meloidogyne incognita*, citrus nematode (*Tylenchulus semipenetrans), and* the soil burrowing nematode (*Radopholus similis)* are the major nematode pests that infect these fruit crops (Keshari & Mallikarjun, 2022). The citrus nematode (*Tylenchulus semipenetrans*) is responsible for thc “slow decline” of citrus, the burrowing nematode (*Radopholus similis*), which is responsible for the “spreading decline” of citrus, the coffee lesion nematode (*Pratylenchus coffeae*), these three are endoparasites. Further, the sting nematode (*Belonolaimus longicaudatus*) and the dagger (*Xiphinema vulgare*) are ectoparasites that are present at the root tip and stop the root growth (Duncan, *et al*, 2022).

Tylenchulus semipenetrans of the parasitic nematodes, and considered the most significant nematode species that parasitizes citrus globally, is commonly called a citrus nematode. While, the citrus slow decline is the term used to describe the progressive hunger, damage, and subsequent demise of the tree because of citrus nematode infection. This nematode is of particular significance accounting for 10-30% economic loss of citrus worldwide (Forner-Giner, *et al*, 2020).

There are different methods to manage the nematode problem in different crops, including physical techniques, crop rotation, resistant varieties, and organic nematicides, in addition to biological control. Using multiple strategies to control nematodes is beneficial and safer, through reducing the use of nematicides and utilizing environmentally friendly substances, which are safer for human health and protect the environment (Ahirwar, *et al,* 2020). Since no single strategy provides appropriate control of the worm, managing the citrus nematode is challenging. Therefore, site selection, nursery stock that isn't infected, using at least one post-plant nematode control method, and careful management of other environmental factors that could stress the trees are all necessary for an overall management strategy. The treatment of this pest still heavily relies on nematicides, furthermore, application methods need to be carefully considered for the best outcomes (Verdejo-Lucas & McKenry, 2004).

There are other nematode pests of citrus include *Meloidogyne indica, Pratylenchus coffeae, Radopholus spp., Longidorus spp., Hoplolaimus spp., Helicotylenchus spp., Hemicycliophora spp., Trichodorus spp., Xiphinema spp., Pratylenchus spp. Belonolaimus longicaudatus* is also responsible for the loss of various fruit yields (Zoubi, *et al*, 2022).

Growing resistant plant varieties, rotating crops, chemical and biological control methods are some of the effective ways to suppress nematodes (Davide, 1988). As of now chemical control strategies are highly effective against nematodes in citrus, however, it is toxic to humans as well as the environment. Managing nematodes in citrus orchards reduce tree decline, decreasing various hazards of chemical nematicides on the environment and humanity. Unlike chemical control methods, biological control exhibits lower toxicity and improved safety against non-targets (Alabouvette, *et al,* 2006; S. Kumar & Singh, 2015; Lahlali, *et al,* 2022). Thus, using the biological control method is one of the most effective and safest control strategies against plant pests.

Presently, there is more interest in using organic and bio-nematicides to control plant-parasitic nematodes (NPPs), in different crops, because of their positive impacts on crop productivity, soil, and the environment (Khan, *et al,* 2022).

Using biological control enhances tree growth, improves productivity, protects the environment and is effective in the control, reduction, or regulation of nematode population by using living organisms (Poveda, *et al,* 2020). Furthermore, various fungi, bacteria, and the bio-actives produced by them are used as active biological control agents against nematodes affecting citrus plants (Kumar & Arthurs, 2021; Abd-Elgawad, 2020).

The efficiency of biological control depends on reducing the number of nematodes in the soil and breaking their life cycle, by attacking nematodes during the vulnerable periods of the life cycle, infecting, and parasitizing the egg and juvenile stages, as well as competing with nematodes for food (Mulusa, 2021).

There are bacterial species that could use as biocontrol for nematodes, they act through reduce the activity oflarva and decrease egg hatching of nematodes. Labiadh, *et al,* (2023) reported that *Bacillus spp. species* that could use as biocontrol for nematodes, they act through reduce the activity of citrus nematode *(T. semipenetrans)* and improving the growth of citrus seedlings. Particularly *B. weihenstephanensis, B. subtilis, B. thuringiensis, B. pumilus* and *B. cereus,* which repressed egg hatching and decreasing larva survival to less than 15% and improve root system of sour orange seedlings.

Hence, this chapter focuses on the nematodes, their life cycle, infection in the citrus groves, controlling infection, and primary focus on the biological control of nematodes through the practical use of some organic and bio substances, to manage nematodes in various farming systems, to achieve sustainable production of citrus.

2. OVERVIEW FOR NEMATODE PEST MANAGEMENT

Control of nematode infestation is more difficult than that of other threats because nematodes mostly inhabit the soil and usually attack the root system. Chemical nematicides cause serious threats to the ecosystem and human health. Presently, there is increasing interest of alter methods that include bionematicides, organic substances, plant derivatives, microorganisms, etc., which are effective approaches for controlling nematodes and the accompanied diseases in citrus groves.

There are several techniques, including chemical nematicides, organic substances, and biological agents are used for controlling nematode infection in various plants. While, chemical nematicides are considered the effective way to fight nematode, but with harmful effects on the environment and causes health hazards.

There are various strategies and substances used to control nematode infection (Figure 1), which include:

1. Chemical nematicides.
2. Agriculture management practices such as (crop rotations, cover crops, and soil amendments).
3. Physical method of soil solarization.

Figure 1. Schematic for the main controlling techniques for nematodes

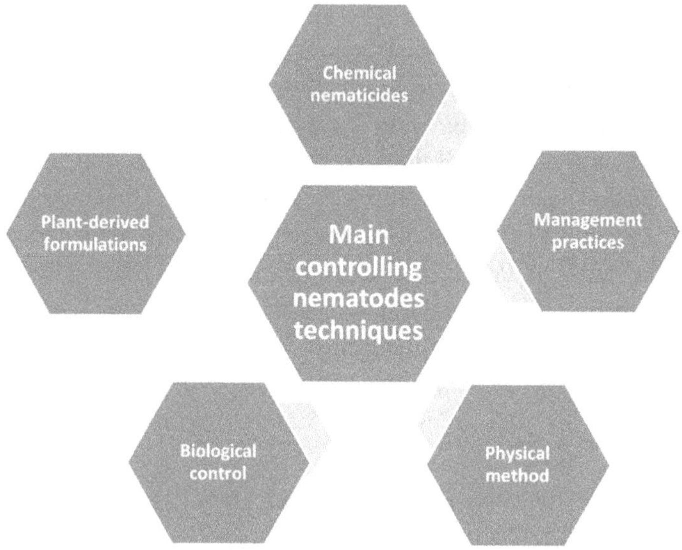

4. Application of biological control such as fungi, bacteria and their derivatives, and biopesticides).
5. Plant-derived formulations.

Currently, due to the harmful effects of chemical nematicides, thus, it is required to manage nematode infection in citrus groves in an efficient manner that protects the environment and reduces the hazards of synthetic nematicides (Abd-Elgawad, 2021), therefore, using various techniques to combat nematodes is required to efficient control nematodes (Stirling, 2018). Bio-control of nematodes acts reduces the nematode population density in the rhizosphere, improves plant growth, and sustains the environment.

Management practices could play an effective role in fighting nematodes that include crop rotation, using non-host plants, or resistant cultivars.

Developing an integrated strategy to combat nematodes requires identifying the various factors that affect their activity, such as soil texture, weeds, host phenological growth stage, and the nutritional status of plants.

Furthermore, biocontrol agents, fungicides, and plant materials derivatives could be effective components of an integrated strategy for managing nematodes and as alternatives to synthetic nematicides.

Previous researches evaluate various microorganisms as bio-control agents for citrus nematodes, also, predatory mites (predatory mites) use to control citrus nematodes instead of chemical control (Kumar & Arthurs, 2021).

3. THE MAIN HARMFUL GENERA OF PLANT PARASITIC NEMATODES

Various parasitic nematodes infested citrus orchards include *Tylenchulus semipenetrans, Radopholus similis* and *R. citri, Pratylenchus, Belonolaimus longicaudatus, Meloidogyne, Trichodorus, Paratrichodorus,* and *Xiphinema*, while, Tylenchulus semipenetrans, Radopholus similis, Pratylenchus coffee, and Meloidogyne spp. are considered main nematode that causes serious threats of citrus groves and economic losses in several regions worldwide (Shokoohi & Duncan, 2018).

Citrus growers facing a serious problem in managing nematode infection, until now, there is no single technique could completely control infection.

Therefore, nematode infection required integration strategy to combat nematodes in citrus groves, including farm choosing, using non-infected seedlings, treating groves with control methods, and using proper practice management.

4. IMPACTS OF NEMATODES IN CITRUS ORCHARDS

Plant-parasitic nematodes cause negative impacts on various crop productivity, which tremendously affects the economic main crops.

The harmful effects of citrus nematodes are due to their rapid reproduction and high density in a short time, consequently infested in more areas in the grove and destroying more and more tissues and root decay (Figure 2), which ultimately led to complete destruction of the whole tree within a short period.

Nematodes infection increases fungal disease in citrus groves, which reduces productivity and accelerates tree decline. Furthermore, Nematodes cause decreasing growth, production, and deterioration of citrus trees (Figure 3).

5. LIFE CYCLE OF *T. SEMIPENETRANS*

T. semipenetrans is sexually dimorphic at the juvenile as well as the adult stages. The life cycle of a female lasts for 4-8 weeks. Nematode begins as an egg, which is the first stage of juvenile (J1). Molting occurs inside the egg where it develops into the second stage juvenile (J2). The egg hatches and the motile J2 searches for the root

Figure 2. Destructive citrus trees by nematodes
Source: Dr. Abobatta (2021)

Figure 3. Infected citrus trees with nematodes
Source: Dr. Abobatta (2021)

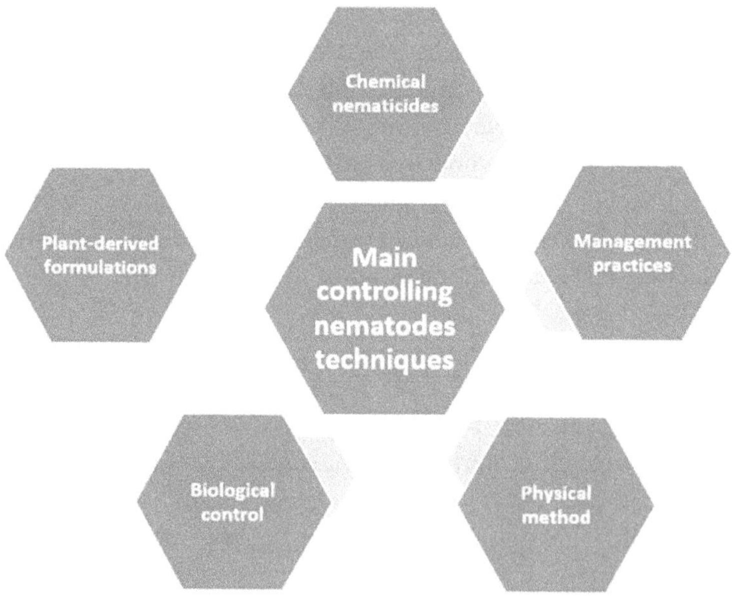

of the host plant. The immature juvenile then penetrates into the deep cortical layers where it becomes sedentary and establishes a feeding site. The immature juvenile feeds on the nurse cells and is matured to J3 and eventually to J4 and adult form. When maturation occurs, the posterior body of the nematode swells and protrudes outside the root while the head remains inside the cortex. The completely mature adult females produce eggs covered in gelatinous matrix that is secreted via the excretory pore {Cobb., 1913} (Hunt, *et al,* 2018).

5.1 Managing Nematodes in Citrus Orchards

There are different parasitic nematodes that infect citrus groves worldwide, while, *Tylenchulus semipenetrans, Radopholus similis, Pratylenchus coffeae*, and *Meloidogyne spp.*, are considered the main species that cause severe hazards on citrus trees (Kumar & Arthurs, 2021). Managing nematode populations required integrated control techniques such as applying biological agents, organic formulations, farm yard manure, practice management, etc…

Due to serious threats caused by chemical nematicides to the ecosystem, managing nematodes in citrus groves required integrated approaches of all techniques to reduce the count of nematodes in the rhizosphere.

Integrated management of nematodes aims to reduce nematode count by various mechanisms such as manipulation of the environment, attacking vulnerable stages of nematodes like egg and juvenility, or the introduction of antagonists.

Biological control is considered an alternative approach to managing citrus nematode infection. Various fungi, bacteria, mites, and microbial formulations have been used. Previous studies reported that citrus nematode almost infected most citrus groves worldwide, and causes economic loss ranging between 10-30% approximately, depending on the infection level of the nematode and the density population (Abu Habib, *et al,* 2020).

6. ORGANIC NEMATICIDES AND BIOLOGICAL CONTROL IN CITRUS GROVES

Currently, there is an increasing interest in protecting the environment and public health from the negative harmful effects of synthetic nematicides. Therefore, the alternative methods to control parasitic nematodes got more attention, due to achieving sustainable production and reducing agrochemicals, through utilizing alter substances to control nematodes or reduce their count in the rhizosphere (El-Saadony, *et al,* 2021). Furthermore, as shown in Table (1), there has been a growing interest in the

Table 1. The biological control agents of nematodes

Crop	Biological Agents	Nematodes Spp.	Reference
Sour orange	Baciluus spp.,	Tylenchulus semipenetrans	Labiadh, *et al*, (2023)
Mandarin	*Dactylaria brochopaga and P. chlamydosporia*	*T. semipenetrans*	Kumar& Arthurs (2021)
Navel Orange	*A. conoides* spore	*T. semipenetrans*	Sweelam, *et al.*, (2019)
Oranges	*A. robusta, A. eudermata, A. oligospora, A. musiformis and Dactylella leptospora*	*P. jaehni*	Martinelli, *et al.*, (2012)
Mandarin and Rough lemon	P. lilacinum Tylenchulus	Tylenchulus semipenetrans Cobb	Le Roux, *et al.*, (2000).
Guava and Fig trees	Fungal species (Trichoderma harzianum, Verticillium chlamydosporium, and Purpureocillium lilacinum),- the bacterium Pasteuria penetrans,	Meloidogyne javanica, Tylenchorhynchus mediterraneus, Hoplolaimus seinhorsti, Longidorus latocephalus, and Xiphinema elongatum on guava and fig trees	Dawabah, et al, 2019
Tomato	Purpureocillium lilacinum	Meloidogyne spp.	Dahlin, *et al*, (2019)
Sugarcane	Pasteuria penetrans	root-knot nematodes	Bhuiyan, *et al.* (2018)
Tomato	*T. harzianum*	nematode eggs	Martínez-Medina et al., (2017)
Wheat	*Trichoderma, T. longibrachiatum*	*Heterodera avenae*	Zhang, et al., (2017)
Guava	T. harzianum (Jindapunnapat et al. 2013).	the root-knot nematode, M. enterolobii Yang & Eisenback	Jindapunnapat, et al., (2013).
Papaya	P. lilacinum and the bacteria Pseudomonas fluorescens Migula were combined to enrich the farm yard manure,	Rotylenchulus reniformis Linford & Oliveira and Meloidogyne incognita (Kofoid & White) Chitwood	Rao (2010)
Papaya	T. harzianum and the rhizobacteria P. fluorescens,	M. incognita	Rao (2007)
Grape	P. penetrans	M. incognita	Al-Rehiayani (2007)
Peach	*G. intraradices, G. mosseae, G. etunicatum*	*M. javanica* root penentration	Calvet, et al., (2001)
Tomato	Verticillium chlamydosprium	M. arenaria.	De Leij, *et al.* (1992)

biological control of nematodes in citrus groves to produce healthy fruits, reduce residues of chemical pesticides in fruits, and protect the environment (Timper, 2011).

Rhizobacteria such as Baciluus spp., enhance the growth of citrus, while, using of *B. weihenstephanensis, B. cereus, B. mycoides, B. thuringiensis, L. sphaericus,* and *Alcaligenes faecalis* (sour orange seedlings) Labiadh, *et al*, (2023).

There are various biological control agents of nematodes (Table 1) including different microorganisms like fungi, bacteria, and some acari spp., while the combination of fungi and bacteria could be more effective in controlling nematodes.

For instance, *Pasteuria penetrans* has demonstrated a great ability to adhere or infest hundreds of parasitic nematodes and control populations of plant-parasitic nematode species, including root-knot nematodes (*Meloidogyne spp.*) in different areas worldwide (Chen and Dickson, 1998).

a. Fungal Biocontrol Agents

Fungi used in bio-control of plant parasitic nematodes include *Trichoderma* spp., *Purpureocillium lilacinum*, *Pochonia chlamydosporia,* mycorrhizae (Glomus spp.).

(i) *Trichoderma* spp.

Trichoderma is a rapidly growing anamorphic filamentous fungi reproducing asexually (Poveda, *et al,* 2020). It is prevalent worldwide and is a predominant constituent of the soil ecosystem. Trichoderma acts as a symbiont and colonizes plant roots. However, colonization is limited only to the outer layer without disturbing the vascular bundle (Hermosa, *et al,* 2012; Tyśkiewicz, *et al,* 2022). Rapid growth and quick adaptation to soil and climatic conditions have made Trichoderma, a promising biocontrol agent (Poveda, 2020). Biocontrol of Trichoderma spp. against plant parasitic nematodes is by various modes of action including parasitism, antibiosis, and competing for resources thereby indirectly reducing the chance of colonization (Dababat, *et al,* 2006). The production of hydrolytic enzymes and secondary metabolites also suppresses plant parasitic nematodes.

Reduction in the hatching of nematode eggs and a decline in the disease was observed when tomato seeds were treated with *T. harzianum* (Martínez-Medina, *et al,* 2017; Sahebani & Hadavi, 2008). Cell free cultures of *T. viridae* and *T. harzianum* have been found to be effective against *M. incognita*. It acted by inhibiting egg hatching and causing J2 mortality (Annapurna, *et al,* 2018; Khan, *et al,* 2018; Rajinikanth, *et al,* 2016). Other species of *Trichoderma, T. longibrachiatum* have shown strong inhibitory activity cyst hatching of *Heterodera avenae* (Zhang, *et al,* 2017).

Other than the above mechanisms of action, *Trichoderma* acts as a systemic resistance inducer for nematode control. It has the ability to stimulate the innate defense of the plant against nematodes (Hermosa, *et al,* 2012). Plants themselves have innate defense mechanisms to protect them from various pathogens. The response against an antigen can be either local or systemic. When a beneficial microbe such as Trichoderma attacks, initially a hypersensitivity response occurs. This response will become systemic or broad-spectrum immunity when a re-infection occurs thereby inducing the host response (Hermosa, *et al,* 2012; Mendoza-Mendoza, *et al*, 2018).

(ii) Mycorrhiza

Mycorrhizae are symbionts of roots of 97% of vascular plants. Mycorrhiza is widely used as biofertilizers in agriculture (Ferlian, *et al,* 2018). Two major groups of mycorrhiza with particular significance in agronomy are Ectomycorrhial fungi and Arbuscular mycorrhizal fungi (AMF).

Arbuscular mycorrhizal fungi are potential biocontrol agents. AMF belongs to the phylum Glomeromycota. It forms symbiosis of plant roots when specialized structures called arbuscles come in contact with plant cytosol (Liao, *et al.,* 2018). AMF acts against the nematodes by different modes, which include depleting the resources and competing with the nematode, increasing plant tolerance, altering rhizosphere interactions and induced systemic resistance. However, direct mechanisms are yet to be described (Abobatta. 2019b; Schouteden, *et al,* 2015).

There are only fewer studies on activity of arbuscular mycorrhizal fungi in citrus. Among the versicular- arbuscular fungi, Glomus species have found to have good nematocidal activity. Application of *G. intraradices, G. mosseae, G. etunicatum* reduced *M. javanica* root penentration in peach trees (Calvet, *et al.,* 2001). However, few conflicting results causing increase in *Pralytenchus brachyurus* population was observed in cotton. Thus, these mechanisms depend on specific characteristics and rely on several factors (Ferreira, *et al*, 2018).

(iii) Nematophagous Fungi

Nematophagous or predatory fungi capable of capturing and digesting nematodes are also known. The trapping mechanism occurs via hyphal modification, adhesive nets, constriction rings etc. Once capture occurs, digestion is by degrading enzymes. Mixed formulations of nematophagous fungi containing *A. robusta, A. eudermata, A. oligospora, A. musiformis and Dactylella leptospora* was effective in pera sweet oranges against *P. jaehni* (Martinelli, *et al*, 2012). In Naval sweet orange trees, about 67% reduction of *T. semipenetrans* was observed when *A. conoides* spore suspension was added (Sweelam, *et al*, 2019).

Purpureocillium lilacinum formerly known as *Paecilomyces lilacinus* is a filamentous, ubiquitous saprobe colonizing the rhizosphere (Atkins, *et al*, 2003). On the other hand, *Pochonia chlamydosporia* formely known as *Verticillium chlamydosporium* is a microfungi with nematicidal activity. It is most effective and most studied against genera such as Globodera, Heterodera and Meloidozyne. *P. chlamydosporia* forms environmentally resistant chlamydospores as an adaptive method for survival in the rhizosphere (Manzanilla-López, *et al*, 2013). Both these fungi have known nematophagous activity. A mixture of *Dactylaria brochopaga and P. chlamydosporia* spore formulations reduced *T. semipenetrans* infesting mandarin in roots as well in the soil (Noweer, 2018). Apart from being nematophagous, it also acts by producing extracellular enzymes such as proteases, which degrades the proteinaceous outer vitelline layer of parasitic nematode eggs, or by producing secondary metabolites such as phomalactone (Khambay, *et al*, 2000).

b. Bacterial Control Agents

Bacteria exist freely as well as in symbiosis with plant phyllosphere in the soil. They are beneficial also known as plant growth promoting bacteria (PGPB) which promotes plant growth by increasing nitrogen, phosphorous and iron content, increasing plant biomass, enhancing seed germination, increasing resistance against stress, pathogens and overall plant nutrition enhancement (Miliute, *et al*, 2015).

Various bacterial species are reported for their activity against nematodes which includes *Bacillus spp., Serratia spp., Streptomyces spp., Pseudomonas spp., Agrobacterium spp., Arthrobacter sp., Corynebacterium spp., Azotobacter spp., Desulfovibrio spp.,Clostridium spp., Burkholderia spp., Azospirillum spp., and Chromobacterium spp.* (Jeyanthi & Kanimozhi, 2018; Ngalimat, *et al,* 2021; Ruiu, 2020). Among the various bacterial species with known nematicidal activity, *Bacillus spp., Pseudomonas spp.* and *Streptomycin spp.* are of particular significance.

(i) *Bacillus spp.*

Bacillus spp. are Gram positive, rod shaped, endospore forming bacteria. Several reports have been published on *Bacillus spp.* acting as a bio control agent against parasitic nematodes of citrus plants. Application of *Bacillus spp.* with compost to the soil has shown to substantially reduce the *T. semipenetrans* to 65-70% (Kumar & Arthurs, 2021). Bacterial filtrates of *B. subtilis, B. thurengenesis, B. marcescens* have shown excellent nematicidal activity in citrus groves by interfering with the nematode reproductive cycle or forming endospores thereby causing immobilization of juveniles of *T. semipenetrans and M. incognita.*(Migunova & Sasanelli, 2021; Topalović, *et al,* 2020). Commercially two bacillus products have been registered for use against citrus nematodes namely BioArc® and Bionemagon®. BioArc containing *B. megaterium* strain has shown 91% mortality of juvenile in both *in-vitro* studies as well as greenhouse trials. Bionemagon® contains the strain, *B. firmus* exhibiting good activity against *T. semipenetrans.*

(ii) *Pseudomonas spp.*

Pseudomonas fluorescens is used as a biocontrol agent against *T. semipenetrans* and *Meloidogyne* spp. *P. fluorescens* acts against nematodes in citrus plants by producing secondary metabolites such as 2,4- diacetylphloroglucinol (2,4-DAPG), pyoluteorin, pyrrolnitrin,phenazines and hydrogen cyanide (HCN) etc. The other modes of action are by inducing systemic resistance in host and inhibiting nematode egg hatching (Das, *et al*, 2021). *P. fluorescens* have exhibited high nematocidal activity against *T. semipenetrans* in citrus plants. Plants repotted with *P. fluorescens* and neem cake has shown substantial reduction in *T. semipenetrans.*

(iii) Streptomycin spp.

Various antibiotics and pesticides are known to be produced from *Streptomyces spp.* Abamectin is one such natural fermentation product produced by *S. avermitilis* which has the potential to kill nematodes from citrus groves (Vurukonda, *et al*, 2018). The mode of action is by causing immobilization leading to death of J2 stage of plant parasitic nematodes. Several Streptomycin spp. including *S. avermitilis* are known to reduce egg hatching and induce mortality of J2 nematodes (El-Saedy, *et al*, 2019).

7. ORGANIC NEMATICIDES

7.1. Nemastop

Nemastop is a commercially produced organic nematicide used for killing nematodes. It is a combination of plant extract and fatty acids. Nemastop is an inexpensive liquid formulation used on fruit crops, ornamental plants and turf grass. It is most effective against root node nematodes.

7.2. Volatile Organic Substances (VOSs)

VOCs, which are extracted from plant materials and microorganisms, are promising substances for the development of organic nematicides. There is more interest in developing these substances, which could play a significant role in controlling nematodes (Deng, *et al*, 2022)

8. CONCLUSION

Citrus crops face many challenges, both abiotic and biotic threats. Parasitic nematodes are one of the most dangerous biotic threats to citrus cultivation worldwide, there are various nematodes species infesting citrus orchards including Tylenchulus semipenetrans, Radopholus similis, Pratylenchus coffee, and Meloidogyne spp. which causes serious threats of citrus growers and economic losses in several regions worldwide. Tylenchulus semipenetrans, which is known as citrus nematode in particular, is considered the most dangerous species that weakens trees slowly and destroys citrus trees.

There are various techniques used to control nematodes, while, but there is no one method that can control nematode infection completely. Therefore, managing nematode infection required integrated control methods that include using biological agents, organic formulations, and practice management.

Currently, there is great interest in the use of organic and biocontrol methods to control nematode infection and reduce the harmful effects of chemicals nematicides on human health and the environment. Using biological control stimulates tree growth and improves fruit quality, furthermore, it is safer for humanity and protects the environment.

There are various fungi and bacteria and their bioactive products, which act as effective biological control agents against citrus nematodes such as Trichoderma spp., Purpureocillium lilacinum, Pochonia chlamydosporia, mycorrhizae fungi. While, bacteria species that are used in the biological control of citrus nematodes include Bacillus spp., Streptomyces spp., Pseudomonas spp.

REFERENCES

Abd-Elgawad, M. M. (2020). Managing nematodes in Egyptian citrus orchards. *Bulletin of the National Research Center*, *44*, 1–15.

Abd-Elgawad, M. M. (2021). Optimizing safe approaches to manage plant-parasitic nematodes. *Plants*, *10*(9), 1911. doi:10.3390/plants10091911 PMID:34579442

Abobatta, W. F. (2019a). Nutritional Benefits of Citrus Fruits. *Am J Biomed Sci & Res.* *3*(4), 303-306. doi:10.34297/AJBSR.2019.03.000681

Abobatta, W. F. (2019b). *Arbuscular Mycorrhizal* and Citrus Growth: Overview. *Acta Scientific Microbiology*, *2*(6), 14–17.

Abobatta, W. F. (2020). Citriculture and Climate Change. *Modern Concepts & Developments in Agronomy*, *6*(3), 649–650. doi:10.31031/MCDA.2020.06.000639

Abu Habib, A. H. A., Younes, H. A., Ibrahim, I. K. A., & Khalil, A. E. (2020). Plant parasitic nematodes associated with citrus trees and reaction of two citrus cultivars to Tylenchulus semipenetrans in northern Egypt. *Journal of the Advances in Agricultural Researches*, *25*(2), 166–175. doi:10.21608/jalexu.2020.161764

Ahirwar, N. K., Singh, R., Chaurasia, S., Chandra, R., & Ramana, S. (2020). Effective role of beneficial microbes in achieving the sustainable agriculture and eco-friendly environment development goals: A review. *Frontiers in Microbiology*, *5*(6), 111–123. doi:10.11648/j.fem.20190506.12

Al-Rehiayani, S., & Motawei, M. I. (2014) Detection and molecular characterization of Pasteuria penetrans isolates parasitizing root-knot nematodes in AL-Qassim area, Saudi Arabia. *Biotechnology*. http://www.omicsgroup.com/conferences/ ACS/ conference/pdfs/2113-Poster-Pdf.pdf

Alabouvette, C., Olivain, C., & Steinberg, C. (2006). Biological Control of Plant Diseases: The European Situation. *European Journal of Plant Pathology*, *114*(3), 329–341. doi:10.100710658-005-0233-0

Annapurna, M., Bhagawati, B., & Uday, K. (2018). In-vitro Efficacy of Native Fungal Bioagents against Meloidogyne incognita. *International Journal of Current Microbiology and Applied Sciences*, *7*(11), 396–410. doi:10.20546/ ijcmas.2018.711.048

Atkins, S. D., Hidalgo-Diaz, L., Kalisz, H., Mauchline, T. H., Hirsch, P. R., & Kerry, B. R. (2003). Development of a new management strategy for the control of root-knot nematodes (Meloidogyne spp) in organic vegetable production. *Pest Management Science*, *59*(2), 183–189. doi:10.1002/ps.603 PMID:12587872

Basyoni, M. M. A., & Rizk, E. M. A. (2016). Nematodes ultrastructure: Complex systems and processes. *Journal of Parasitic Diseases : Official Organ of the Indian Society for Parasitology*, *40*(4), 1130–1140. doi:10.100712639-015-0707-8 PMID:27876901

Bhuiyan, S. A., Garlick, K., Anderson, J. M., Wickramasinghe, P., & Stirling, G. R. (2018). Biological control of root-knot nematode on sugarcane in soil naturally or artificially infested with Pasteuria penetrans. *Australasian Plant Pathology*, *47*(1), 45–52. doi:10.100713313-017-0530-z

Calvet, C., Pinochet, J., Hernández-Dorrego, A., Estaún, V., & Camprubí, A. (2001). Field microplot performance of the peach-almond hybrid GF-677 after inoculation with arbuscular mycorrhizal fungi in a replant soil infested with root-knot nematodes. *Mycorrhiza*, *10*(6), 295–300. doi:10.1007/PL00009998

Dababat, A. A., Sikora, R. A., & Hauschild, R. (2006). Use of Trichoderma harzianum and Trichoderma viride for the biological control of Meloidogyne incognita on tomato. *Communications in Agricultural and Applied Biological Sciences*, *71*(3 Pt B), 953–961. PMID:17390844

Dahlin, P., Eder, R., Consoli, E., Krauss, J., & Kiewnick, S. (2019). Integrated control of Meloidogyne incognita in tomatoes using fluopyram and Purpureocillium lilacinum strain 251. *Crop Protection (Guildford, Surrey)*, *124*, 104874. doi:10.1016/j.cropro.2019.104874

Das, S., Abdul Wadud, M., & Atiqur Rahman Khokon, M. (2021). Functional evaluation of culture filtrates of Bacillus subtilis and Pseudomonas fluorescens on the mortality and hatching of Meloidogyne javanica. *Saudi Journal of Biological Sciences*, *28*(2), 1318–1323. doi:10.1016/j.sjbs.2020.11.055 PMID:33613061

Davide, R. G. (1988). Nematode problems affecting agriculture in the Philippines. *Journal of Nematology*, *20*(2), 214–218. PMID:19290204

Dawabah, A. A., Al-Yahya, F. A., & Lafi, H. A. (2019). Integrated management of plant-parasitic nematodes on guava and fig trees under tropical field conditions. *Egyptian Journal of Biological Pest Control*, *29*(1), 1–9. doi:10.118641938-019-0133-9

Deng, X., Wang, X., & Li, G. (2022). Nematicidal effects of volatile organic compounds from microorganisms and plants on Plant-parasitic nematodes. *Microorganisms*, *10*(6), 1201. doi:10.3390/microorganisms10061201 PMID:35744719

Duncan, L. W., Noling, J. W., & Inserra, R. N. (2022). *2022–2023 Florida Citrus Production Guide: Nematodes*. EDIS. doi:10.32473/edis-cg010-2022

El-Saadony, M. T., Abuljadayel, D. A., Shafi, M. E., Albaqami, N. M., Desoky, E. S. M., El-Tahan, A. M., Mesiha, P. K., Elnahal, A. S. M., Almakas, A., Taha, A. E., Abd El-Mageed, T. A., Hassanin, A. A., Elrys, A. M., & Saad, A. M. (2021). Control of foliar phytoparasitic nematodes through sustainable natural materials: Current progress and challenges. *Saudi Journal of Biological Sciences*, *28*(12), 7314–7326. doi:10.1016/j.sjbs.2021.08.035 PMID:34867034

El-Saedy, M. A. M., Sandy, E. H., & Sherin, F. A. (2019). Nematicidal effect of abamectin, boron, chitosan, hydrogen peroxide and Bacillus thuringiensis against citrus nematode on Valencia orange trees. *Journal of Plant Science and Phytopathology*, *3*(3), 111–117. doi:10.29328/journal.jpsp.1001041

Ferlian, O., Biere, A., Bonfante, P., Buscot, F., Eisenhauer, N., Fernandez, I., Hause, B., Herrmann, S., Krajinski-Barth, F., Meier, I. C., Pozo, M. J., Rasmann, S., Rillig, M. C., Tarkka, M. T., van Dam, N. M., Wagg, C., & Martinez-Medina, A. (2018). Growing Research Networks on Mycorrhizae for Mutual Benefits. *Trends in Plant Science*, *23*(11), 975–984. doi:10.1016/j.tplants.2018.08.008 PMID:30241736

Ferreira, B. S., Santana, M., Macedo, R. S., Silva, J. O., Carneiro, M. A. C., & Rocha, M. R. (2018). Co-occurrence patterns between plant-parasitic nematodes and arbuscular mycorrhizal fungi are driven by environmental factors. *Agriculture, Ecosystems & Environment*, *265*, 54–61. doi:10.1016/j.agee.2018.05.020

Food and Agriculture Organization. (2022, December 23). *Citrus*. Https://Www.Fao.Org/Markets-and-Trade/Commodities/Citrus/En/

. Forner-Giner, M. A., Continella, A., & Grosser, J. W. (2020). Citrus Rootstock Breeding and Selection. *The Citrus Genome*, 49-74.

Gang, S. S., & Hallem, E. A. (2016). Mechanisms of host seeking by parasitic nematodes. *Molecular and Biochemical Parasitology*, *208*(1), 23–32. doi:10.1016/j.molbiopara.2016.05.007 PMID:27211240

Hermosa, R., Viterbo, A., Chet, I., & Monte, E. (2012). Plant-beneficial effects of Trichoderma and of its genes. *Microbiology*, *158*(1), 17–25. doi:10.1099/mic.0.052274-0 PMID:21998166

Hunt, D. J., Palomares-Rius, J. E., & Manzanilla-López, R. H. (2018). Identification, morphology and biology of plant parasitic nematodes. In *Plant parasitic nematodes in subtropical and tropical agriculture* (pp. 20–61). CAB International. doi:10.1079/9781786391247.0020

Jeyanthi, V., & Kanimozhi, S. (2018). Plant Growth Promoting Rhizobacteria (PGPR) - Prospective and Mechanisms: A Review. *Journal of Pure & Applied Microbiology*, *12*(2), 733–749. doi:10.22207/JPAM.12.2.34

Keshari, N., & Mallikarjun, G. (2022). Plant Parasitic Nematodes: A Major Constraint in Fruit Production. In Nematodes - Recent Advances, Management and New Perspectives. IntechOpen. doi:10.5772/intechopen.101696

Khambay, B. P., Bourne, J. M., Cameron, S., Kerry, B. R., & Zaki, M. J. (2000). A nematicidal metabolite fromVerticillium chlamydosporium. *Pest Management Science*, *56*(12), 1098–1099. doi:10.1002/1526-4998(200012)56:12<1098::AID-PS244>3.0.CO;2-H

Khan, F., Shariq, M., Asif, M., Ansari, T., Fatima, S., Khan, A., Ikram, M., & Siddiqui, M. A. (2022). Organic Nematicides: A Green Technique and Its Overview for Nematode Pest Management. In Sustainable Management of Nematodes in Agriculture, Vol. 1: Organic Management (pp. 39-66). Springer.doi:10.1007/978-3-031-09943-4_2

Khan, M. R. (2015). Nematode diseases of crops in India. *Recent advances in the diagnosis and management of plant diseases*, 183-224.

Khan, M. R., Ahmad, I., & Ahamad, F. (2018). Effect of pure culture and culture filtrates of Trichoderma species on root-knot nematode, Meloidogyne incognita infesting tomato. *Indian Phytopathology*, *71*(2), 265–274. doi:10.100742360-018-0031-1

Khanna, K., Jamwal, V. L., Sharma, A., Gandhi, S. G., Ohri, P., Bhardwaj, R., Al-Huqail, A. A., Siddiqui, M. H., Marraiki, N., & Ahmad, P. (2019). Evaluation of the role of rhizobacteria in controlling root-knot nematode infection in *Lycopersicon esculentum* plants by modulation in the secondary metabolite profiles. *AoB Plants*, *11*(6), plz069. doi:10.1093/aobpla/plz069

Kumar, K. K., & Arthurs, S. (2021). Recent advances in the biological control of citrus nematodes: A review. *Biological Control*, *157*, 104593. doi:10.1016/j.biocontrol.2021.104593

Kumar, S., & Singh, A. (2015). Biopesticides: Present Status and the Future Prospects. *Journal of Fertilizers & Pesticides*, *06*(02). Advance online publication. doi:10.4172/2471-2728.1000e129

Labiadh, M., Mhamdi, B., Loulou, A., & Sadreddine, K. (2023). Impact of rhizobacteria community of citrus root on Tylenchulus semipenetrans and on Citrus plant growth. *Biocontrol Science and Technology*, *33*(3), 1–17. doi:10.1080/09583157.2023.2175785

Lahlali, R., Ezrari, S., Radouane, N., Kenfaoui, J., Esmaeel, Q., el Hamss, H., Belabess, Z., & Barka, E. A. (2022). Biological Control of Plant Pathogens: A Global Perspective. *Microorganisms*, *10*(3), 596. doi:10.3390/microorganisms10030596 PMID:35336171

Liao, D., Wang, S., Cui, M., Liu, J., Chen, A., & Xu, G. (2018). Phytohormones Regulate the Development of Arbuscular Mycorrhizal Symbiosis. *International Journal of Molecular Sciences*, *19*(10), 3146. doi:10.3390/ijms19103146 PMID:30322086

Martinelli, P. R. P., Santos, J. M., & Barbosa, J. C. (2012). Efficacy of formulations containing five nematophagous fungi for the management of Pratylenchus jaehni in Citrus. *Nematologia Brasileira, 36*, 1–8.

Martínez-Medina, A., Fernandez, I., Lok, G. B., Pozo, M. J., Pieterse, C. M. J., & van Wees, S. C. M. (2017). Shifting from priming of salicylic acid- to jasmonic acid-regulated defences by *Trichoderma* protects tomato against the root knot nematode *Meloidogyne incognita. The New Phytologist, 213*(3), 1363–1377. doi:10.1111/nph.14251 PMID:27801946

Mendoza-Mendoza, A., Zaid, R., Lawry, R., Hermosa, R., Monte, E., Horwitz, B. A., & Mukherjee, P. K. (2018). Molecular dialogues between Trichoderma and roots: Role of the fungal secretome. *Fungal Biology Reviews, 32*(2), 62–85. doi:10.1016/j.fbr.2017.12.001

Migunova, V. D., & Sasanelli, N. (2021). Bacteria as Biocontrol Tool against Phytoparasitic Nematodes. *Plants, 10*(2), 389. doi:10.3390/plants10020389 PMID:33670522

Miliute, I., Buzaite, O., Baniulis, D., & Stanys, V. (2015). Bacterial endophytes in agricultural crops and their role in stress tolerance: A review. *Zemdirbyste, 102*(4), 465–478. doi:10.13080/z-a.2015.102.060

Mulusa, L. (2021). Biological Control of Root-Knot Nematodes Using Trichoderma Spp. *Nematodes-Recent Advances, Management and New Perspectives*, 139.

Ngalimat, M. S., Mohd Hata, E., Zulperi, D., Ismail, S. I., Ismail, M. R., Mohd Zainudin, N. A. I., Saidi, N. B., & Yusof, M. T. (2021). Plant Growth-Promoting Bacteria as an Emerging Tool to Manage Bacterial Rice Pathogens. *Microorganisms, 9*(4), 682. doi:10.3390/microorganisms9040682 PMID:33810209

Noweer, E.M.A. (2018). Effect of the Nematode-Trapping Fungus Dactylaria Brochopaga and the Nematode Egg Parasitic Fungus Verticilium Chlamydosporium in Controlling Citrus Nematode Infesting Mandarin, And Interrelationship with the Co Inhabitant Fungi. *International Journal of Engineering & Technology, 7*(3.32), 19. doi:10.14419/ijet.v7i3.32.18383

Poveda, J. (2020). Trichoderma parareesei Favors the Tolerance of Rapeseed (Brassica napus L.) to Salinity and Drought Due to a Chorismate Mutase. *Agronomy (Basel), 10*(1), 118. doi:10.3390/agronomy10010118

Poveda, J., Abril-Urias, P., & Escobar, C. (2020). Biological Control of Plant-Parasitic Nematodes by Filamentous Fungi Inducers of Resistance: Trichoderma, Mycorrhizal and Endophytic Fungi. *Frontiers in Microbiology*, *11*, 992. Advance online publication. doi:10.3389/fmicb.2020.00992 PMID:32523567

Rajinikanth, R., Sreenivasa, R. M., & Pavani, K. V. (2016). Evaluation of potential bio-control agents on root-knot nematode Meloidogyne incognita and wilt causing fungus Fusarium oxysporum f.sp. conglutinans in vitro. *African Journal of Biotechnology*, *15*(19), 798–805. doi:10.5897/AJB2015.15235

Ruiu, L. (2020). Plant-Growth-Promoting Bacteria (PGPB) against Insects and Other Agricultural Pests. *Agronomy (Basel)*, *10*(6), 861. doi:10.3390/agronomy10060861

Sahebani, N., & Hadavi, N. (2008). Biological control of the root-knot nematode Meloidogyne javanica by Trichoderma harzianum. *Soil Biology & Biochemistry*, *40*(8), 2016–2020. doi:10.1016/j.soilbio.2008.03.011

Schouteden, N., de Waele, D., Panis, B., & Vos, C. M. (2015). Arbuscular Mycorrhizal Fungi for the Biocontrol of Plant-Parasitic Nematodes: A Review of the Mechanisms Involved. *Frontiers in Microbiology*, *6*. Advance online publication. doi:10.3389/fmicb.2015.01280 PMID:26635750

Seinhorst, J. W. (1986). Effects of Nematode Attack on the Growth and Yield of Crop Plants. In *Cyst Nematodes* (pp. 191–209). Springer US. doi:10.1007/978-1-4613-2251-1_11

Shokoohi, E., & Duncan, L. W. (2018). Nematode parasites of citrus. In *Plant parasitic nematodes in subtropical and tropical agriculture* (pp. 446–476). CAB International. doi:10.1079/9781786391247.0446

Stirling, G. R. (2018). Biological control of plant-parasitic nematodes. In *Diseases of nematodes* (pp. 103–150). CRC Press.

Sweelam, M. E., Abokorah, M. S., Taka, S. M. A., & Ellatief, S. M. A. (2019). Biological control of the citrus nematode, Tylenchulus semipenetrans attacking Citrus sinensis Navel orange trees. *Menoufia Journal of Plant Protection*, *4*(5), 179–188. doi:10.21608/mjapam.2019.123021

Timper, P. (2011). Utilization of biological control for managing plant-parasitic nematodes. *Biological control of plant-parasitic nematodes: building coherence between microbial ecology and molecular mechanisms*, 259-289.

Topalović, O., Hussain, M., & Heuer, H. (2020). Plants and Associated Soil Microbiota Cooperatively Suppress Plant-Parasitic Nematodes. *Frontiers in Microbiology, 11*, 313. Advance online publication. doi:10.3389/fmicb.2020.00313 PMID:32184773

Tyśkiewicz, R., Nowak, A., Ozimek, E., & Jaroszuk-Ściseł, J. (2022). Trichoderma: The Current Status of Its Application in Agriculture for the Biocontrol of Fungal Phytopathogens and Stimulation of Plant Growth. *International Journal of Molecular Sciences, 23*(4), 2329. doi:10.3390/ijms23042329 PMID:35216444

Verdejo-Lucas, S., & McKenry, M. v. (2004). Management of the Citrus Nematode, Tylenchulus semipenetrans. *Journal of Nematology, 36*(4), 424–432. PMID:19262822

Vurukonda, S. S. K. P., Giovanardi, D., & Stefani, E. (2018). Plant growth promoting and biocontrol activity of Streptomyces spp. as endophytes. *International Journal of Molecular Sciences, 19*(4), 952. doi:10.3390/ijms19040952 PMID:29565834

Zhang, S., Gan, Y., Ji, W., Xu, B., Hou, B., & Liu, J. (2017). Mechanisms and Characterization of Trichoderma longibrachiatum T6 in Suppressing Nematodes (Heterodera avenae) in Wheat. *Frontiers in Plant Science, 8*, 1491. Advance online publication. doi:10.3389/fpls.2017.01491 PMID:28966623

Zoubi, B., Mokrini, F., Dababat, A. A., Amer, M., Ghoulam, C., Lahlali, R., Laasli, S.-E., Khfif, K., Imren, M., Akachoud, O., Benkebboura, A., Housseini, A. I., & Qaddoury, A. (2022). Occurrence and Geographic Distribution of Plant-Parasitic Nematodes Associated with Citrus in Morocco and Their Interaction with Soil Patterns. *Life (Chicago, Ill.), 12*(5), 637. doi:10.3390/life12050637 PMID:35629306

Chapter 10
Eco-Friendly Nematodes:
Bioindicators for Promoting Sustainable Soil Health

El-Sayed Ewis Omran
Independent Researcher, Egypt

Mahmoud Farag Mahmoud
Independent Researcher, Egypt

Heba M. Abd El-Nabi
Independent Researcher, Egypt

ABSTRACT

Sustainable agriculture relies on meticulous soil health assessment, including nematode indicators. They serve as vital bio-indicators, yet the study of beneficial nematodes remains unexplored. The goal of this chapter is to explore nematodes' significance in agriculture, covering their role in soil health, responses to climate change affecting crop production, and contributions to SDGs. Nematodes are valuable indicators due to their roles in nutrient cycling. Nematodes enhance soil health through various strategies, contributing to plant growth. Resistant pests are driving a shift towards eco-friendly pesticides, promoting biological control for plant protection and environmental preservation. EPNs rely on understanding adaptation and establishment in agricultural ecosystems. EPNs, known for their virulence and adaptable cultivation, combat insect pests effectively across various agricultural systems. The connections between nematodes, soil health, and pest management promise to advance sustainable agriculture, bolster environmental stability, and ensure food security.

DOI: 10.4018/978-1-6684-8083-0.ch010

1. INTRODUCTION

Research on soil biodiversity is being used to promote local (water quality, food security), regional (land degradation), and global (climate change) solutions to problems like biodiversity loss. Resources, particularly food, are in greater demand as the human population rises. Modern agriculture must be made sustainable in order to fulfill the rising demand for food. Managing soils as the dynamic living systems that they are opens up new possibilities for coordinated actions and fixes. Improper soil management, such as the heavy use of chemical fertilizers and pesticides, endangers the health of the soil. Sustainable agricultural development depends on healthy soils, and using indicators to measure soil health is becoming more and more common. In accordance with ecosystem and land-use boundaries, soil health is defined as "the persistence of soil functioning as a vibrant living system, to sustain biological productivity, preserve the quality of air and water environments, and promote plant, animal, and human health" (Doran et al., 1996). The soil quality is "the ability of a particular kind of soil to function, within natural or managed ecosystem boundaries, to sustain plant and animal productivity, maintain or improve water and air quality, and support human health and habitation" (Karlen, 1997). While soil health is essentially an ecological trait and is described as "a living, dynamic system whose activities are mediated by a range of living creatures that require management and conservation" (Doran, 2000), soil quality is related to soil function. Established soil quality determination takes into account a number of factors. Soil health assessment necessitates a thorough synthesis of chemical, physical, and biological indicators, from which trends and natural features for soil health can be identified (Mann, 2019).

The Wisconsin Soil Health Scorecard (Romig, 1995) and the Cornell Soil Health Assessment (Moebius-Clune, 2016) are only two of the many techniques used to evaluate soil health. Since living organisms mediate the majority of soil processes, soil health and quality monitoring relies heavily on biological indicators. Protozoa, metazoa, and microorganisms are examples of biological indicators. All ecosystems are supported by soil organisms, both microscopic and macroscopic. The most prevalent type of metazoan, nematodes, are found in different types of soils (Bongers, 1998). Nematode communities are well recognized as straightforward indicators of soil quality and soil health (Griffiths, 2016). Nematodes represent many trophic components of the food web (Yeates et al., 1993) and play essential roles in ecosystem functioning (Ferris, 2006). Nematode communities are sensitive to change in environmental conditions due to natural or anthropogenic causes and are therefore considered useful bioindicators for soil health assessment (Neher, 1999; Yeates and Bongers, 1999). Nematodes differ in terms of their susceptibility to contaminants and environmental disturbances. The amount of nitrogen cycling and breakdown is associated with soil nematode abundance (both free-living and plant parasitic) (Neher, 2001).

On the one hand, the building blocks for creating sustainable smallholder agro-ecosystems include integrated pest management (IPM) and enhancing soil health. Cyst, root-knot, root-lesion, and spiral plant-parasitic nematodes reduce smallholder farmers' yields. Free-living nematodes are a good indicator of the health and richness of soil ecosystems (Urzelai et al., 2000). Numerous soil nematode indices, including the Shannon index, the maturity index (MI) for free-living nematodes, and the plant-parasitic index (PPI) for parasitic nematodes, are used to assess the health of agricultural and natural soils. The indexes are used to track shifts in land use, alterations to the environment, and the results of management practices. Forests, grasslands, and agricultural systems are frequently assessed using soil nematodes (Liu et al., 2019). Currently, morphological and molecular techniques are both widely utilized for identification (Seesao et al., 2017). Caenorhabditis elegans has been the most often utilized soil nematode in laboratory tests to determine soil toxicity (Nigon and Félix, 2017). This nematode has a lot of benefits. Recent advances in analytical and identification techniques have made it more common to evaluate soil health using soil nematodes.

On the other hand, soil degradation and plant-parasitic nematodes on smallholder farms lower yields and restrict smallholder food security. Due to the prevalence of nematodes in biologically and nutritionally deteriorated soils, current smallholder farming techniques have the potential to worsen soil degradation and thus increase losses from worms. The three most dangerous and serious nematodes in the world are root-knot, cyst, and root-lesion nematodes. Because nematodes have wide host ranges, populations frequently mix different species, few crops have broad-spectrum resistance, and chemical nematicides are either unfavorable, prohibitively expensive, or both, management is difficult.

The most effective way to control plant-parasitic nematodes sustainably is through a soil health management approach, taking into account the mechanisms by which nematodes reduce yields and the fact that smallholders frequently cultivate damaged soils. The most sensible approaches for smallholder farmers are those that enhance soil health and lower population densities of harmful nematodes within a context of conservation agriculture practices and sustainable intensification. By incorporating soil health information into smallholder agriculture, nematodes and other pests have less of an impact on food, nutrition, and income.

Because it is challenging, sustainable soil health remains a goal despite a substantial body of knowledge. The balance of biological, physicochemical, nutritional, structural, and water-holding elements required to support plant productivity is known as soil health. However, this balance is affected by climate change in unsustainable way. Climate change affects the distribution of insects and fungal pests, and thus agricultural chemical use patterns, with the consequent risk of chemical residues in food in areas where they were not previously present. Soil health has become an urgent necessity to cope with climate change. In line with Egypt's Vision 2030 and the National Climate Change Strategy 2050, we belief that COP27 hosted by Egypt, is a pivotal opportunity for the world to join forces to accelerate climate action plans to continue green projects in order to achieve sustainable development goals. So, the goals of this chapter are to (a) identify the nematodes role in agricultural practices (APs) that contribute to soil health development, (b) explain how the nematode community can be exploited to promote soil health development, c) assess the impact of climate change on nematode and its consequences to crop production, and d) describe the role of soil nematodes in achieving SDGs.

2. WHAT IS THE SOIL NEMATODES ROLE FOR SUSTAINABLE SOIL HEALTH?

Nematodes are indeed used as green biological indicators for assessing soil health and sustainability. These microscopic, worm-like organisms play a crucial role in soil ecosystems, participating in nutrient cycling, organic matter decomposition, and influencing the balance of other soil organisms. Their populations and diversity can provide valuable insights into the overall health and quality of the soil environment.

Although some nematodes are pests that injure plants, others serve as plant protectors and have advantageous effects on the soil environment. Nematodes can be divided into three functional groups according to how they feed: bacterial-feeders, fungal-feeders, and plant parasites. Nematodes are tiny, worm-like animals that dwell in soil pore pores and water-filled aqueous films. They are typically more prevalent in the upper soil layers where there is more organic matter, plant roots, and other supplies. The number of nematodes per square meter of soil, both managed and uncontrolled, ranges from 1 to 10 million individuals/m^2 (Lavelle and Spain, 2001; Peterson and Luxton, 1982). Due to their diverse range of prey, including predators who feed on other carnivores as well as bacteria, fungus, and plant roots, beneficial nematodes play a vital role in the soil food web. Nematode indices (Ferris et al., 2001) are derived from nematode identification and can provide important inferences to nutrient cycling, ecosystem disturbance, decomposition pathways, and community complexity. Despite the valuable data that can be obtained, nematode indices have mainly been left out of the soil health framework due of the specialized training needed for microscope identification. Understanding the role of beneficial nematodes in soil health is a key step toward optimizing crop production and sustaining your farm's future. Nematodes support a number of activities in the soil.

1) Nematodes can improve nutrient mineralization in agricultural systems, serve as biological control agents, and maintain adequate levels of nitrogen that is available to plants by utilizing organic fertility sources. Like protozoa, nematodes are important in mineralizing, or releasing, nutrients (Nutrient cycling) in plant-available forms. Nematodes are important nutrient mineralizers. When nematodes consume bacteria or fungi, they release excess ammonium (NH_4^+). Bacteria and fungi both have more ammonium than what the nematode needs so the extra is released in a plant available form. Nematodes contribute to the nitrogen cycle, the ability of organic matter to decompose, and the management of pests in soil systems, among other ecosystem services. The effects on the nitrogen cycle have increased interest in the usefulness of nematodes in raising soil nitrogen levels. The availability of nitrogen is specifically increased as well as the quantity of bacteria that compete for nitrogen after soil disruption caused by practices like ploughing. Bacterial feeding nematodes multiply in response to an increase in bacteria, which stabilizes nitrogen availability and boosts nitrogen mineralization in the soil, which is essential for crop growth and soil fertility.
2) Nematodes contribute to the breakdown of organic matter in soil. Free-living nematodes are very important because they recycle nutrients in the soil, and their indirect contribution is balanced by that of nematodes that feed on the bacteria and fungi that break down organic materials. Together, these worms' feeding activity speeds up the breakdown process and makes nutrients and minerals available to roots of growing plants.
3) Nematodes are beneficial biological pest control agents. As they consume organisms like protozoa, insects, and other nematode species while feeding, predatory nematodes control the populations of other organisms in soil. As a result, nematode regulate the population of other nematodes, and research has demonstrated that nematodes such as the *Steinernematidae* and *Heterorhabditidae* are beneficial to the environment and can kill soil-based insect pests within 48 hours.
4) Nematodes are now important biological indicators of soil health. Nematode communities' diversity and richness are particularly instructive because they serve as proxies for revealing data on the concentrations of nutrients, the availability of fungi, and the presence of bacteria in soil. Nematode communities offer a wealth of data on soil health and have a variety of positive effects on food production systems. Maintaining healthy soil is key to the success of your farm, as it helps sustain a productive environment for your crops, both today and in the future. Understanding and managing soil health is vital for growers. Doran describes soil health as "The capacity of soil to function as a vital living system, within ecosystem and land-use boundaries, to sustain plant and animal

productivity, maintain or enhance water and air quality, and promote plant and animal health" (Doran et al., 1996). No matter the soil type, sound agricultural practices can help improve soil health. Some important farming fundamentals include conservation tillage, crop rotation, nutrient management and the use of cover crops. It is important to understand that soil is a living and active part of the farm which lies beneath your feet. It contains a variety of beneficial organisms including bacteria, fungi, protozoa, macro- and micro-arthropods and nematodes. They contribute to nutrient cycling, natural disease and pest suppression, and improvement of your soils structure.

5) Nematodes may be useful indicators of soil quality because of their tremendous diversity and their participation in many functions at different levels of the soil food web. Several researchers have proposed approaches to assessing the status of soil quality by counting the number of nematodes in different families or trophic groups. In addition to their diversity, nematodes may be useful indicators because their populations are relatively stable in response to changes in moisture and temperature (in contrast to bacteria), yet nematode populations respond to land management changes in predictable ways. Because they are quite small and live in water films, changes in nematode populations reflect changes in soil microenvironments.

6) Beneficial nematodes contribute in various ways to the soil and are involved in nutrient cycles that boost plant growth. Different types of nematodes feed on bacteria, fungi, insects and even harmful nematodes. These helpful organisms play an important role in keeping destructive plant pests and diseases at bay, thus promoting both soil biodiversity and health. Healthy soils are rich in beneficial soil organisms that help plants improve water and nutrient usage, and suppress pests and diseases. They enhance the tolerance of crops to other environmental stresses such as weather events. This ultimately results in better long-term productivity of your land and the profitability of your farm.

7) Finally, nematodes are considered grazers. They move through the soil profile devouring smaller organisms as well as distributing any bacteria or fungi that are on them as well as any that are in their digestive system. If the nematode population is low, they will stimulate the growth rate of prey populations. If the nematode population is high, they have the potential to have negative impact on soil health by devouring too much of their prey especially micorrhizal fungi. There are also predatory nematodes that balance the population of other nematodes. At low nematode densities, feeding by nematodes stimulates the growth rate of prey populations. That is, bacterial-feeders stimulate bacterial growth, plant-feeders stimulate plant growth, and so on. At higher densities, nematodes will reduce the population of their prey. This may decrease plant productivity, may negatively impact mycorrhizal fungi, and can reduce

decomposition and immobilization rates by bacteria and fungi. Predatory nematodes may regulate populations of bacterial-and fungal-feeding nematodes, thus preventing over-grazing by those groups. Nematode grazing may control the balance between bacteria and fungi, and the species composition of the microbial community. Nematodes also help distribute bacteria and fungi through the soil and along roots by carrying live and dormant microbes on their surfaces and in their digestive systems.

3. WHAT DO THE ENTOMOPATHOGENIC NEMATODES (EPNS) DO IN INSECT PESTS' BIOCONTROL?

Steinernema and Heterorhabditis nematodes are intriguing model systems for use in biological control because of two distinct characteristics. They begin by developing a sophisticated nematode-bacterium mutualistic relationship. The nematodes' bodies contain the bacteria, which are then discharged into their hosts (Poinar, 1990). The bulk of insect orders are included in their wide host spectrum, which makes them insect pathogens. Second, they can be mass produced in both in vitro and in vivo cultures. The two types of nematodes have distinct life cycles, despite the fact that they both have the ability to infect, kill, and produce new generations from insects.

The so-called larvae phases are a subset of the life cycle of the Entomopathogenic Nematodes (EPNs) Steinernema and Heterorhabditis. The only stage of the nematode outside of its insect victim is the infective juvenile (IJ) or (dauer). The nematode is currently a non-feeding, soil-dwelling larval with a closed mouth and anus, protected by a double cuticle, and capable of long-term survival in the soil. The so-called cruiser approach is used by IJs of the Heterorhabditiae family to aggressively scour the soil for suitable insect larvae. The Steinernematidae family of nematodes adapted the ambusher strategy, passively waiting near the soil surface for prey to pass by.

The nematode sheds its outer cuticle when it detects an insect, exposing its mouth and anus. It then penetrates the insect through its natural openings, including its anus, mouth, and spiracles, and moves to the insect blood cavity (Griffin et al., 2005). Heterorhabditis, as opposed to Steinernema, can use a dorsal tooth to pierce straight through the delicate intersegmental areas of the insect integument (Griffin et al., 2005).

It has been demonstrated that other nematode species parasitize insect hosts by using pathogenic bacteria. By using insect larvae as bait to attract nematodes, a method used to identify EPNs, two Oscheius (=Heterorhabditoides) species, *O. chongmingensis* and *O. carolinensis*, and *Caenorhabditis briggsae*, have been discovered to be potential insect pathogens (Nguyen and Hunt, 2007). Insect pathogenic bacteria of the genus Serratia have been discovered to associate with all of these, and *O. carolinensis* may have additional associates (Torres-Barragan et al., 2011). When an insect is killed, *O. chongmingensis* and *C. briggsae* need their bacterial partners to kill the host, develop, and reproduce there. Emerging dauer juveniles are also connected to the vectored pathogen (Ye et al., 2010).

Classical, augmentative, and conservation control are the three types of biological control (Bale et al., 2008). The traditional method of controlling an exotic pest that has spread to a new area entails importing and unleashing the parasitoid or predator of the pest. The term "conservation biocontrol" refers to a broad range of actions that can be done to encourage the growth of local natural enemies of (native) pests. (e.g., creating refuges and providing alternative food for natural enemies).

The majority of practical study has concentrated on their potential for augmentation of biological control agents through inundative application (Grewal et al., 2005). These have a number of qualities that make them suitable for integrated pest control and sustainable agriculture. Some species can also recycle and persist in the environment, affect plant pathogen and parasitic nematode populations directly or indirectly, indirectly improve soil quality, and are compatible with a variety of chemical and biological pesticides used in IPM programmes. Table 1 depicts the current use of Steinernema and Heterorhabditis nematodes as biological control organisms (Shapiro-Ilan and Gaugler, 2010), which have been modified by Mahmoud (Mahmoud, 2016). His paper concentrate on some chosen reviews on the effective use of EPNs in biocontrol (Mahmoud, 2014b; Mahmoud and Osman, 2014). Growers have been using commercially produced EPN in their groves for many years to control root weevils because it is one of the most efficient methods to reduce populations of the Diaprepes root weevil, *Diaprepes abbreviatus* (L.), in Florida citrus groves (Shapiro-Ilan et al., 2005).

Neoaplectana (=Steinernema) carpocapsae fully eradicates the carpenter worm, *Prionoxytus robiniae*, in commercial fig orchards (Lindegren and Barnette, 1982). Similar to that, this nematode eliminated 85–90% of *Zeuzera pyrina* larvae, an Italian pest of fruit plants (Deseo and Docci, 1985). In the field trial in Egypt, *Steinernema riobrave* also resulted in 92.8 to 94.7% mortalities in *Zeuzera pyrina* larvae when used at concentrations of 3000 IJs and 5000 IJs (Shamseldean et al., 2009). To control wood borers in the Sessidae family, Steinernema species have been created for widespread use. With the treatment of *S. bibionis* to currants, *Synanthedon tipuliformis* borer mortality has increased by more than 90%.

In the United States, Canada, and Western Europe, the black vine weevil, *Otiorhynchus sulcatus* (Curculionidae), is one of the most significant pest species of strawberries, cranberries, and seedling ornamentals. Field tests with EPNs revealed that Heterorhabditis species outperform Steinernema species in terms of managing the larvae (van Tol and Raupp, 2006). But not all Heterorhabditid species and strains work as well in spring as they do in autumn applications (van Tol and Raupp, 2006).

Table 1 . Current use of entomopathogenic nematodes Steinernema and Heterorhabditis nematodes as biocontrol agents of insect pests

Crops/Targets	Insect Pests		Nematodes*
	Common Name	Scientific Name	
Artichokes	Artichoke plume moth	*Platyptilia carduidactyla*	Sc
Vegetables	Armyworm	Lepidoptera: Noctuidae	Sc, Sf, Sr
Ornamentals	Banana moth	*Opogona sachari*	Hb, Sc
Bananas	Banana root borer	*Cosmopolites sordidus*	Sc, Sf, Sg
Turf	Billbug	*Sphenophorus* spp. (Coleoptera: Curculionidae)	Hb,Sc
Turf, vegetables	Black cutworm	*Agrotis ipsilon*	Sc
Canola	Black cutworm	*Agrotis ipsilon*	Sc, Hb
Berries, ornamentals	Black vine weevil	*Otiorhynchus sulcatus*	Hb, Hd, Hm, Hmeg, Sc, Sg
Fruit trees, ornamentals	Borer	*Synanthedon* spp. and other sesiids	Hb, Sc, Sf
Home yard, turf	Cat flea	*Ctenocephalides felis*	Sc
Citrus, ornamentals	Citrus root weevil	*Pachnaeus* spp. (Coleoptera: Curculionidae)	Sr, Hb
Pome fruit	Codling moth	*Cydia pomonella*	Sc, Sf
Canola	Diamondback moth	*Plutella xylostella*	Sc, Hb
Vegetables	Corn earworm	*Helicoverpa zea*	Sc, Sf, Sr
Vegetables	Corn rootworm	*Diabrotica* spp.	Hb, Sc
Cranberries	Cranberry girdler	*Chrysoteuchia* topiaria	Sc
Turf	Crane fly	Diptera: Tipulidae	Sc
Citrus, ornamentals	Diaprepes root weevil	*Diaprepes abbreviatus*	Hb, Sr
Mushrooms	Fungus gnat	Diptera: Sciaridae	Sf, Hb
Grapes	Grape root borer	*Vitacea polistiformis*	Hz, Hb
Iris	Iris borer	*Macronoctua onusta*	Hb, Sc
Forest plantings	Large pine weevil	*Hylobius albietis*	Hd, Sc
Vegetables, ornamentals	Leafminer	*Liriomyza* spp. (Diptera: Agromyzidae)	Sc, Sf
Turf	Mole cricket	*Scapteriscus* spp.	Sc, Sr, Sscap
Nut and fruit trees	Navel orangeworm	*Amyelois transitella*	Sc
Fruit trees	Plum curculio	*Conotrachelus nenuphar*	Sr
Stone fruit orchards	Flat-headed root borer	*Capnodis tenebrionis*	Sc, Sf
Date palm	Red palm weevil	*Rhynchophorus ferrugineus*	Sc
Turf, ornamentals	Scarab grub[3]	Coleoptera: Scarabaeidae	Hb, Sc, Sg, Ss, Hz
Ornamentals	Shore fly	*Scatella* spp.	Sc, Sf
Berries	Strawberry root weevil	*Otiorhynchus ovatus*	Hm
Bee hives	Small hive beetle	*Aethina tumida*	Hi, Sr
Sweet potato	Sweetpotato weevil	*Cylas formicarius*	Hb, Sc, Sf
Termites hills	Subterranean termites	*Psammotermes hypostoma*	Sc, Hb

* Nematode species are abbreviated as follows: Hb = *Heterorhabditis bacteriophora*, Hd = *H. downesi*, Hi = *H. indica*, Hm = *H. marelata*, Hmeg = *H. megidis*, Hz = *H. zealandica*, Sc = *Steinernema carpocapsae*, Sf = *S. feltiae*, Sg = *S. glaseri*, Sk = *S. kushidai*, Sr = *S. riobrave*, Sscap = *S. scapterisci*, Ss = *S. scarabaei*

In the Netherlands, England, and Germany, *Steinernema feltiae* is an effective and affordable alternative to chemical pesticides for controlling fungus gnats (Diptera: Sciaridae). (Jagdale et al., 2004). *S. feltiae*, a cold-adapted nematode, has been used to effectively eradicate fungus gnats in environments with temperatures between 12 and 25 $^{o}_{C}$ (Jagdale et al., 2004). *Lycoriella* spp. fungus gnats can be controlled at rates of 1.0 to 1.5 x 10^6 *S. feltiae* IJs/m^2 that are as good as or better than those achieved with pesticides frequently used in the mushroom industry (Jagdale et al., 2004). In Egypt, *S. feltiae* was highly virulent towards second- and third-instar larvae of *Musca domestica*, *Stomoxys calcitans*, *Lucilia sericata*, and *Calliphora vicina*; however, it was less virulent towards fly pupae, especially in dung (Mahmoud and Osman, 2007). Additionally, it works well against *Bactrocera zonata*, the peach fruit fly. In Petri plates lined with moist filter paper, *S. feltiae* Cross N 33 in Egypt demonstrated effectiveness against second- and third-instar larvae of *B. zonata* (Mahmoud and Osman, 2007).

Nematodes may be used as foliar treatments to combat the larval stages of various leafminers, according to greenhouse experiments (Head and Walters, 2003; Williams and Walters, 2000). In order to ensure dependable control, it is generally necessary to use the right amount of spray so that the nematodes can reach the larval stages. Successful control of nematodes depends on maintaining high relative humidity (above 90%) in the greenhouse and/or wetness on the plants for at least 6 to 8 hours after nematode applications (Williams and Walters, 2000). Applications of *S. carpocapsae* or *S. feltiae* at 1 × 10^6 IJs/m^2 against the second and third instar larvae provided the best control of *Liriomyza trifolii* Burgress (Williams and Walters, 2000).

Black cutworm *Agrotis ipsilon* (Hufnagel), Silver Y moth *Phytometra gamma* L. (Lepidoptera: Noctuidae), Diamondback moth *Plutella xylostella* L. (Lepidoptera: Plutellidae) and Beet armyworm *Spodoptera exigua* (Hubner) are significant insect pests on canola in Egypt (Mahmoud and Shebl, 2014). In greenhouse tests, EPN, *Steinernema carpocapsae* was highly effective when applied in an aqueous suspension at 6000 IJs per 25 ml to larvae on canola leaves, killing the larvae of *A. ipsilon*, *P. xylostella*, and *S. exigua*, respectively, by 74%, 72%, and 68% (Mahmoud, 2014a).

The most frequently used species for the management of foliar and other above-ground pests is *Steinernema carpocapsae*. They are the best prospects for pest insects found on the surface soil when they descend from foliage because of their ambusher host-finding strategy. According to Belair et al. (Belair et al., 2003), under the environmental circumstances in Québec, foliar applications of *S. carpocapsae* did not offer an adequate degree of control of the imported cabbage worm *Artogeia rapae* (Lepidoptera: Pieridae). On the other hand, studies on *S. carpocapsae* and *S. feltiae* showed that they had the potential to control the leafminers (Diptera: Agromyzidae), including *Liriomyza trifolii* (Tomalak et al., 2005), *Liriomyza huidobrensis* (Williams and Walters, 2000), *Tuta absoluta* (Batalla-Carrera et al., 2010).

A great example of how to use EPNs effectively in cryptic habitats is the codling moth, *Cydia pomonella* (Lepidoptera: Tortricidae), a pest of apple and other pome fruit that is present throughout the globe. They make up the entire community of codling moths after harvest. If these larvae were controlled, fewer adult moths would appear the following spring. *S. carpocapsae*, *S. feltiae*, *H. bacteriophora*, and *H. zealandica* are the species that have received the most evaluation for controlling codling moths. Temperature, moisture, and habitat type are the abiotic variables that have the biggest an impact on their larvicidal activity against *C. pomonella* (de Waal et al., 2011). Applications of EPNs to apple tree trunks for control of codling moth, were improved when the treatments included the sprayable fire-gel or wood flour foam as a protective agent.

Scapteriscus vicinus, an invasive mole cricket from South America (Orthoptera: Gryllotalpidae), is a significant pest of grass and lawn in the Southern United States. *Steinernema scapterisci*, an EPN found in the alleged cricket genesis region of Uruguay, effectively controls the cricket using traditional biological methods (Parkman and Smart, 1996). After the introduction of *S. scapterisci*-infested cadavers and applications in small plots at a rate equal to 2 x 10^9 IJs/ha, the nematode was effectively established (Parkman and Smart, 1996). Infected mole crickets also self-dispersed *S. scapterisci* to produce new sites of infection (Parkman et al., 1993). *S. vicinus* is highly territorial, so created a novel treatment strategy that used sound devices to draw in and infect crickets. There was a 98% reduction in mole cricket populations at release locations three years after *S. scapterisci* was first introduced. Commercial nematode production has made it easier to apply *S. scapterisci* to untreated areas and to make additional applications (Grewal et al., 2001).

Cutworms, or Noctuidae (Lepidoptera: Noctuidae), include the species Agrotis, Amathes, Noctua, Peridroma, and Prodenia. They eat leaves, buds, and stems, as well as some kinds of roots. They come into contact with the soil during some or all of their feeding phases. In the soil or beneath dead leaves and other detritus at the soil surface, many species overwinter as larvae or pupae in their penultimate or last instar. When the soil moisture is adequate for IJ survival and infectivity, they make excellent targets for ambusher EPNs while feeding or resting on the soil's surface. Although several studies (Ebssa and Koppenhöfer, 2012; Mahmoud, 2014a) have shown effective cutworm management in crops and turf, they have not yet been widely applied.

Every two weeks in Spain and other Mediterranean nations, EPNs are administered through drip irrigation to sweet peppers to combat the soil-dwelling stages of thrips (*Frankliniella occidentalis*), which have about 100.000 infectious juveniles (IJs) m^{-1} (Ehlers, 2011). As EPN penetrate the mines and destroy the larval stages, this foliar application also controls the tomato leaf miner, *Tuta absoluta*. This pest has lately been introduced to the Mediterranean region, including Egypt. Susceptibility of *T. absoluta* larvae and pupae to the entomopathogenic nematode (EPNs) was determined under both laboratory and field condition (Shamseldean et al., 2014). Field efficacy of *H. bacteriophora* varied along 3 consecutive years from around 60% in years 2011 and 2012 to 80% mortality in 2013. Whereas, *S. monticolum* efficacy mounted to 58 to 61% in years 2011 and 2012, respectively and reached a maximum of 67% in 2013 (Shamseldean et al., 2014). In Mediterranean stone fruit plantations, the Flat-headed Root Borer *Capnodis tenebrionis* (Linné) (Coleoptera: Buprestidae) is a significant pest. (peaches, plums, apricots, cherries etc). The roots and trunk are attacked. More than 3.000 ha of Prunus orchards in Spain have had the larvae and adults of *C. tenebrionis* effectively controlled by applying 1 million IJ m^{-1} of *Steinernema carpocapsae* or *S. feltiae* in the spring and the fall (Ehlers, 2011). Nemastar® can be applied to all life phases, including adults after pupation. During the high soil humidity months of April/May and September/October, it is administered via drip irrigation, drench, or soil injection. A minimum of 30 litres of water and 1-3 million nematodes per tree are used as the dosage rate. Irrigate both before and after application for best outcomes. Nemastar has been used effectively on thousands of hectares of land in Spain. Independent of the application technique, the efficacy in field trials in apricot trees varied from 75 to 90%. Applying three years in a row is essential for long-term pest suppression (Martinez de Altube et al., 2008). Greece and Italy are presently introducing this application. Chitosan can significantly boost nematode efficacy when added to the IJ suspension, according to studies conducted in Spain against the newly introduced invasive Red Palm Weevil *Rhynchophorus ferrugineus* (Coleoptera, Curculionidae). IJs of *S. carpocapsae* can live for about a month inside the trunk of a palm tree. For each tree, it takes a monthly application of 1–5 million IJs to destroy the larvae and defend the tree from further invasions. Application of the therapy can be made simpler and more affordable by installing a tube system in the tree canopy. After EPN + Chitosan was successfully used to combat the Red Palm Weevil in Spain, the technique is now being used in Greece and Italy. The Spanish firm Idebio has a patent for the use of EPN and Chitosan together in Europe (Ehlers, 2011).

4. WHAT IS THE ECONOMIC IMPORTANCE OF PLANT PARASITIC NEMATODES (PPNs)?

Plant parasitic nematodes (PPNs) have emerged as a serious biotic stress and significantly impacted the yield potentials of agricultural and horticultural crops. Nematodes constitute one of the most important groups of organisms which inhabit the soil around the roots of plants and which frequently play a vital role in their growth and production. Rarely any crop is free from their attacks, yet we usually are unaware of their presence because of their microscopic size and protected position Within the soil. Plant parasitic nematodes (PPNs) have emerged as a serious biotic stress and significantly impacted the yield potentials of agricultural and horticultural crops. Nematode-induced mechanical injuries and physiological alterations in the plant systems also facilitate the infection from other pathogens.

The destructive plant parasitic nematodes are one of the major limiting factors in the production of agricultural and horticultural crops throughout the world. Some plant parasitic nematodes (PPNs) are capable of causing disease on many economically important crops grown throughout the world and attained the status of pests for substantial reduction of crop yield. For centuries, humans have been plagued by these microscopic organisms feeding on the roots of crop plants essential to their survival and well-being. Roots damaged by the nematodes are not efficient in the utilization of available moisture and nutrients in the soil resulting in reduced functional metabolism. Stunting of individual plants (reduced growth), yellowing of leaves (varying degrees of chlorosis), wilting of the foliage, and sometimes death of plants are some of the visible symptoms of nematode attack. Further, nematodes interact with soil-borne pathogens in inducing disease complexes. Reduced yields and poor quality of crops are some of the deleterious effects on plant growth. In view of the high cost of modern crop production, nematode management is therefore important for high yields and quality of the produce.

On a worldwide basis, the 10 most important genera of plant parasitic nematodes were reported, as listed in Table 2. The estimated overall average annual yield loss of the world's major crops due to damage by plant parasitic nematodes is 12.3% (Table 3). For the 20 crops that stand between man and starvation, an estimated annual yield loss of 10.7% is reported. For the 20 crops that represent an important for food or export value, an estimated annual yield loss of 14% is reported (Sasser and Freckman, 1987). Based on 1984 production figures and prices, the nematodes were responsible for monetary crop losses to the extent of US$ 77 billion annually on 21 crops, 15 of which are life sustaining. These figures are staggering, and the real figure, when all crops are considered, probably exceeds US$ 100 billion annually. The losses are 5.8% greater in developing countries than in developed countries (Sasser and Freckman, 1987). Abad et al. (Abad et al., 2008) reported that the crop losses caused by phytonematodes in economic terms were estimated to be US$ 157 billion annually to the world agriculture.

The changing cropping patterns, introduction of new crops, crop diversification, agronomic practices, etc. also change the spectrum of pests and pathogens, including plant parasitic nematodes. In the near future, the management strategies like use of regulatory (seed certification), physical (soil solarization and hot water treatment of planting material), cultural (crop rotation, organic amendments), chemical (naturally

Table 2. The 10 most important genera of plant parasitic nematodes on a worldwide basis (Sasser & Freckman, 1987)

Serial Number	Important Genera
1	*Meloidogyne*
2	*Pratylenchus*
3	*Heterodera*
4	*Ditylenchus*
5	*Globodera*
6	*Tylenchulus*
7	*Xiphinema*
8	*Radopholus*
9	*Rotylenchulus*
10	*Helicotylenchus*

Table 3. Estimated annual yield losses due to damage by plant parasitic nematodes on a worldwide basis (Sasser & Freckman, 1987)

Life-Sustaining Crops	Crops Loss (%)	Economically Important Crops	Crops Loss (%)
Banana	19.7	Cocoa	10.5
Barley	6.3	Citrus	14.2
Cassava	8.4	Coffee	15.1
Chickpea	13.7	Cotton	10.7
Coconut	17.1	Cowpea	15.1
Corn	10.2	Eggplant	16.9
Field bean	10.9	Forages	8.2
Millet	11.8	Grapes	12.5
Oat	4.2	Guava	10.8
Peanut	12.0	Melons	13.8
Pigeon pea	13.2	Misc. other	17.3
Potato	12.2	Okra	20.4
Rice 10.0	10.0	Ornamentals	11.1
Rye 3.3	3.3	Papaya	15.1
Sorghum	6.9	Pepper	12.2
Soybean	10.6	Pineapple	14.9
Sugar beet	10.9	Tea	8.2
Sugarcane	15.3	Tobacco	14.7
Sweet potato	10.2	Tomato	20.6
Wheat	7.0	Yam	17.7
Average	**10.7%**		**14.0%**

occurring nematicides), and biological (natural enemies) methods and host resistance (induced resistance, interruption in recognition of host, and genetically modified crops) will form components of integrated nematode management (Reddy, 2021).

5. WHAT IS THE BENEFICIAL VS. THE HARMFUL SIDE OF SOIL NEMATODES?

Nematodes are one of the most diverse and abundant groups of animals in the soil, and plants almost always interact with nematodes during their lifecycles. The majority of soil nematodes are beneficial to soil health and the environment. Nematodes are often talked about in a quiet fearful voice. The image of the small microscopic worms can bring grown men to their knees. Unfortunately, like many things in our world, a few "bad" apples have ruined the entire bushel. Attention has been given extensively to a small segment of the nematode population that negatively impacts crops but those nematodes are a very small percent of the nematode population. Some nematodes cause disease and used for disease suppression and development. Others consume disease-causing organisms, such as root-feeding nematodes, or prevent their access to roots. These may be potential biocontrol agents. A major function of soil nematodes is that they are biocontrol agents, meaning they can be used to eliminate disease causing nematodes and other organisms. This trait causes predatory nematodes to be a great resource in the battle against soil borne diseases.

Nematodes are food source for higher level predators, including predatory nematodes, soil microarthropods, and soil insects. They are also parasitized by bacteria and fungi. Nematodes are not the highest organism in the soil food web. Soil microarthropds and insects as well as bacteria and fungi feed on nematodes. As stated earlier there are also predatory nematodes in the soil that consume nematodes.

Conserving the beneficial nematodes and controlling the plant-parasitic ones can make a significant impact on your crops. "You could spend a lot of money putting nutrients back into the soil with blended fertilizers, but it could be for nothing if nematode feeding has damaged those roots and the plant can't access the nutrients." Using a nematicide that's gentle on the beneficial nematodes while stopping the plant-parasitic ones before they can find and infect the plant roots helps sustain the natural productivity of soil every agricultural cycle year after year. Doing all you can to maintain the health of your soil supports your profitability today and the legacy of your farm for future generations.

Soil nematodes, as integral components of soil ecosystems, have both beneficial and potentially harmful aspects. Here's a breakdown of their positive and negative contributions:

5.1 Beneficial Side of Soil Nematodes

- Nutrient Cycling: Many soil nematodes are involved in nutrient cycling by feeding on bacteria, fungi, and organic matter, thus aiding in the decomposition process and releasing nutrients back into the soil, which supports plant growth.
- Microbial Regulation: Some nematodes are predators of bacteria and fungi, helping to control the populations of these microorganisms and maintain a balanced microbial community in the soil.
- Soil Aeration: Burrowing nematodes create small channels in the soil as they move, promoting better air circulation and water movement within the soil profile.
- Indicator of Soil Health: The presence of diverse nematode species in the soil can indicate a healthy and balanced soil ecosystem. Monitoring their populations can provide insights into overall soil health.
- Biocontrol Agents: Entomopathogenic nematodes (EPNs) are used as biocontrol agents to combat insect pests. They infect and kill insects, making them a natural and environmentally friendly alternative to chemical pesticides.

5.2 Harmful Side of Soil Nematodes

- Plant Parasitism: Some nematode species are plant parasites and can cause significant damage to crops by feeding on plant roots, leading to reduced nutrient uptake and stunted growth. These nematodes are known as plant-parasitic nematodes.
- Crop Losses: Plant-parasitic nematodes can lead to reduced crop yields, economic losses, and decreased food production. They are considered a major threat to agricultural productivity.
- Vector of Plant Diseases: Certain nematodes can transmit plant viruses and other pathogens, contributing to the spread of diseases among crops.
- Root Damage: Nematodes that feed on plant roots can cause physical damage to the root system, creating entry points for other pathogens and reducing the plant's ability to absorb water and nutrients.
- Influence on Soil Structure: Large populations of some nematodes can affect soil structure by creating channels and voids in the soil, potentially leading to soil compaction and decreased water infiltration.
- Loss of Biodiversity: An imbalance in nematode populations can lead to a reduction in overall soil biodiversity, impacting the functioning of the soil ecosystem.

In summary, soil nematodes play essential roles in nutrient cycling, microbial regulation, and soil health maintenance. However, certain plant-parasitic nematodes can negatively impact agricultural productivity by damaging crops and spreading diseases. The overall impact of nematodes in soil ecosystems depends on their diversity, abundance, and ecological interactions. Proper soil management practices can help maintain a healthy balance and mitigate the potentially harmful effects of nematodes.

6. HOW THE SUSTAINABLE DEVELOPMENT GOALS (SDGS) CONNECTED TO NEMATODE AND SOIL HEALTH?

More than 7.5 billion people depend on soils for their livelihood, and they form the basis for many ecosystem processes that both directly and indirectly promote human health and well-being. Land and soil are addressed both directly and indirectly by all international policy frameworks, including the Sustainable Development Goals (SDGs) of the United Nations. Without sound soils and sustainable land use, several of these SGDs are impossible to accomplish. This is especially clear when we consider that practically all food crops and animal fodder, including pastures, are closely related to the achievement of Sustainable Development Goals (SDGs) 2 and 3. The processes essential to plant growth are driven by the living communities found in soil, or soil biodiversity, which has a direct impact on human health and wellbeing through the production of crops and livestock fodder (Bender and van der Heijden, 2015; Thiele-Bruhn et al., 2012).

Soils are also home to disease-causing organisms that directly impact humans, such as helminth parasites (nematode Strongyloides), and brain encephalitis (Naegleria fowleri) (Wall et al., 2015), as well as Coccidioides immitis, the fungus that causes Valley fever (Nguyen et al., 2013; Wall et al., 2015) (SDG 3). Diverse soil communities are typically good for human health and keep pathogens in control by competing with and predating on disease-causing agents for people, livestock, and crops, even if disease-causing organisms can be problematic and even dangerous (Nielsen et al., 2015; Wall et al., 2015). Both directly as a source of food and indirectly through procedures like decomposition and nutrient cycling, soil organisms support life aboveground (SDG 15) (Wardle, 2002).

SDG 13 and Aichi Target 15 both place a strong emphasis on the importance of soil in tackling global climate change (protect and increase biodiversity contributions to C stocks). Some climate change remedies already include soil.

Several Sustainable Development Goals (SDGs) outlined by the United Nations are closely connected to nematode and soil health. Nematodes play a significant role in soil ecosystems, influencing soil health, biodiversity, and agricultural productivity. As a result, their interactions with soil align with various SDGs related to environmental sustainability, food security, and well-being. Here's how nematode and soil health are linked to specific SDGs:

SDG 2: Zero Hunger: Nematodes impact crop growth and agricultural productivity by influencing nutrient cycling and plant health. Healthy soil nematode communities can contribute to increased crop yields and food security, supporting the goal of eliminating hunger.

SDG 3: Good Health and Well-Being: Nematodes can influence human health indirectly through their impact on food crops. Healthy soils with balanced nematode populations can lead to nutritious and safe food production, contributing to improved health and well-being.

SDG 6: Clean Water and Sanitation: Nematodes affect soil structure and water movement within the soil. Sustainable soil management practices that promote nematode diversity can contribute to better water retention, reduced erosion, and improved water quality.

SDG 12: Responsible Consumption and Production: Sustainable agriculture practices that promote soil health and reduce the need for synthetic pesticides align with this goal. Using nematodes as biocontrol agents instead of chemical pesticides can reduce environmental impact and promote responsible production.

SDG 13: Climate Action: Soil health, including nematode populations, influences carbon sequestration and greenhouse gas emissions. Practices that enhance soil health can contribute to climate change mitigation and adaptation efforts.

SDG 14: Life Below Water: Soil health, which includes nematode diversity, can impact terrestrial ecosystems and runoff into aquatic systems. Improved soil management can minimize soil erosion and nutrient runoff, which can have positive effects on aquatic ecosystems.

SDG 15: Life on Land: This goal is directly linked to nematode and soil health. Healthy soils support diverse ecosystems, contribute to biodiversity conservation, and provide habitats for various organisms, including nematodes.

SDG 17: Partnerships for the Goals: Collaboration between researchers, policymakers, farmers, and other stakeholders is essential for promoting sustainable soil management practices that consider nematode interactions and their impact on soil health.

In essence, nematode and soil health are interconnected with several SDGs, particularly those related to food security, sustainable agriculture, environmental conservation, and well-being. Recognizing and managing the role of nematodes in soil ecosystems is crucial for achieving these sustainable development objectives.

7. WHAT IS THE IMPACT OF CLIMATE CHANGE ON SOIL NEMATODES?

Cutting emission is done by reducing agriculture's carbon footprints through improving soil health. Carbon footprint, a measurement of the greenhouse gases produced by a person, organization, action, or product, is a key indicator of climate care. It is critical to reduce the overall carbon footprint of human activity. In order to prevent greenhouse gases (GHGs) from being released into the atmosphere, soil serves as a carbon "sink." The reduction of agriculture's carbon footprint is supported by sustainable agricultural practices. Growers can be more conscious of protecting the environment by having a better understanding of the environmental effects of various production techniques. Here are five suggestions for reducing the carbon footprint of agriculture.

The most significant resource for farmers is their soil, and adding more organic matter to the soil enhances its fertility, structure, and water-holding ability, which can have a favorable effect on yields and its capacity to sequester carbon. The carbon component of soil organic matter is known as soil organic carbon (SOC). As soil organic matter oxidizes during cultivation, carbon dioxide is released into the atmosphere. This flow of carbon from the soil can be reduced by reducing soil disturbance and switching to lower-tillage systems, although this approach won't work in all circumstances.

Where it occurs, other advantages could include lower startup costs, less soil erosion, and higher biodiversity. The most obvious technique to improve soil organic matter is to regularly add organic fertilizers like farmyard manure and compost to the soil. Increasing the use of cover crops, grass leys, and generally introducing a wider variety of crops would also aid in increasing organic matter and strengthening the structure of the soil.

After enhancing the soil's health and fertility, optimizing fertilizer applications to reduce the amount needed without lowering yields can have a major positive influence on greenhouse gas emissions from arable agricultural systems. The use of precision farming methods that permit the application of fertilizers at variable rates will also help to increase effectiveness and decrease wastage. Another option is to switch from urea to fertilizers with a lower carbon footprint. In any case, the government is currently conducting consultations on limiting the use of urea fertilizers. It's equally crucial to administer nutrients at the proper time and in the proper quantity. The amount of nitrous oxide emissions can dramatically rise when fertilizers are applied in warm or wet circumstances.

Overall, climate change can have significant effects on soil nematode populations and communities due to alterations in temperature, precipitation patterns, and overall environmental conditions. Some potential impacts include:

- **Shifts in Distribution:** As temperatures change, the geographic distribution of different nematode species may shift. Species that are adapted to specific temperature ranges might become more or less prevalent in certain regions.
- **Changes in Community Composition:** Climate change can lead to shifts in the relative abundance and composition of nematode communities. Some species may thrive under new conditions, while others could decline.
- **Altered Trophic Interactions:** Nematodes occupy various trophic levels in soil food webs. Changes in their abundance or distribution can disrupt trophic interactions and impact nutrient cycling dynamics.
- **Impact on Plant Health:** Some nematodes are plant parasites and can affect crop health. Climate change-induced shifts in nematode populations could influence crop damage levels and agricultural productivity.
- **Water Availability Effects:** Changes in precipitation patterns can impact soil moisture levels. Soil nematodes are sensitive to water availability, and shifts in moisture conditions could influence their populations.
- **Microbial Interactions:** Nematodes interact with soil microorganisms. Changes in temperature and moisture can alter microbial communities, which could indirectly affect nematode populations.
- **Invasion of Exotic Species:** Climate change can create conditions that are more favorable for the establishment of invasive nematode species, potentially disrupting local ecosystems.
- **Carbon Cycling:** Soil nematodes play a role in organic matter decomposition and nutrient cycling. Changes in nematode populations could impact soil carbon dynamics and nutrient availability.

It's important to note that the specific impacts of climate change on soil nematodes can vary depending on factors like local conditions, nematode species, and the overall ecological context. More research is needed to fully understand these interactions and their implications for soil ecosystems.

8. CONCLUSIONS AND RECOMMENDATIONS

The arrangement of nematode communities, encompassing both those that parasitize plants and those that exist independently, has the potential to function as an indicative measure of soil health and condition. This association arises from the strong correlation between community composition and critical soil ecological processes, such as nitrogen cycling and decomposition. The constructive influence of nematodes extends to several pivotal aspects of soil quality: the regulation of other soil organism populations, the conversion of nutrients into forms accessible to plants, provision of nourishment for other soil organisms, and the consumption of pathogenic entities.

Soil health evaluation involves an amalgamation of physical, chemical, and biological indicators. Among the biological indicators, encompassing microorganisms, protozoa, and metazoa, nematodes, as the most prevalent metazoans, hold particular significance. Their susceptibility to toxins and environmental disturbances differs, reflecting their varied ecological roles. Benefitting from their advantageous positioning within the food web, soil nematodes emerge as valuable bio-indicators. Both free-living and plant-parasitic nematodes furnish ecological insights due to their pivotal involvement in nutrient cycling and their roles as primary, secondary, and tertiary consumers within food webs.

Saprophytic nematodes assume the role of bio-indicators for soil health by deploying diverse strategic mechanisms that amplify soil functions. These include ecosystem management, nitrogen enrichment facilitated by nitrogen ingestion and subsequent release as NH_4, which is readily assimilated, decomposition, and the dispersion of bacteria and fungi into recently available organic remains. Nematodes act as conduits for nutrient conveyance through associated microorganisms, thus augmenting soil health and plant growth. Furthermore, alterations in soil nematode compositions reflect soil perturbations, with practices such as tillage inversion, cropping patterns, and fertilizer management potentially exerting profound effects on soil nematode communities.

Navigating the challenges posed by climate change and pursuing sustainable solutions emphasize the growing significance of comprehending the intricacies of nematode interactions. The potential of Entomopathogenic nematodes (EPNs) as biocontrol agents showcases their potential to revolutionize insect pest management while aligning with global efforts to reduce the environmental impact of synthetic pesticides.

Advancements in nematode research, from refining mass-production techniques to exploring genetic enhancements and novel insecticidal proteins, open promising avenues for practical implementation. These developments hold the promise of fostering resilient agricultural systems and safeguarding ecosystems from the pressures of climate change and chemical resistance.

Embracing the multifaceted contributions of nematodes to soil health facilitates the fostering of sustainable practices, achieving a harmonious equilibrium between agricultural productivity, environmental preservation, and human well-being. Moving forward, the imperative of ongoing collaboration among researchers, policymakers, and practitioners remains pivotal in harnessing the potential of nematodes to effectively address challenges and pave the pathway towards a more resilient and sustainable future.

Anticipated advancements are on the horizon, propelled by ongoing endeavors aimed at refining mass-production methods, reducing production costs, and devising more sophisticated carriers and formulation techniques to extend the shelf-life of infective juveniles (IJs). Additionally, genetic enhancement emerges as a promising avenue for enhancing nematode performance and effectiveness in practical applications. The exploration of insecticidal toxins derived from symbiotic bacteria like Photorhabdus and Xenorhabdus holds promise for novel insecticidal proteins, augmenting insect control capabilities. In essence, the future utility of Entomopathogenic nematodes (EPNs) is auspicious, given their intrinsic advantages and the mounting demand for potent microbial agents to alleviate environmental impacts and mitigate the resistance challenge posed by synthetic chemical insecticides.

REFERENCES

Abad, P., Gouzy, J., Aury, J., Castagnone-Sereno, P., Danchin, E. G. J., Deleury, E., Perfus-Barbeoch, L., Anthouard, V., Artiguenave, F., Blok, V. C., Caillaud, M.-C., Coutinho, P. M., Dasilva, C., De Luca, F., Deau, F., Esquibet, M., Flutre, T., Goldstone, J. V., Hamamouch, N., ... Wincker, P. (2008). Genome sequence of the metazoan plant-parasitic nematode Meloidogyne incognita. *Nature Biotechnology*, *26*(8), 909–915. doi:10.1038/nbt.1482 PMID:18660804

Bale, J. S., van Lenteren, J. C., & Bigler, F. (2008). Biological control and sustainable food production. *Philosophical Transactions of the Royal Society of London. Series B, Biological Sciences*, *363*(1492), 761–776. doi:10.1098/rstb.2007.2182 PMID:17827110

Batalla-Carrera, L., Morton, A., & Garcıa-del-Pino, F. (2010). Efficacy of nematodes against the tomato leafminer Tuta absoluta in laboratory and greenhouse conditions. *BioControl*, *55*, 523–530. doi:10.100710526-010-9284-z

Belair, G., Fournier, Y., & Dauphinais, N. (2003). Efficacy of steinernematid nematodes against three insect pests of crucifers in Quebec. *Journal of Nematology*, *35*, 259–265. PMID:19262759

Bender, S. F., & van der Heijden, M. G. A. (2015). Soil biota enhance agricultural sustainability by improving crop yield, nutrient uptake and reducing nitrogen leaching losses. *Journal of Applied Ecology*, *52*(1), 228–239. doi:10.1111/1365-2664.12351

Bongers, T., & Bongers, M. (1998). Functional diversity of nematodes. *Applied Soil Ecology*, *10*(3), 239–251. doi:10.1016/S0929-1393(98)00123-1 de Waal, J., Malan, A. P., & Addison, M. F. (2011). Evaluating mulches together with Heterorhabditis zealandica (Rhabditida: Heterorhabditidae) for the control of diapausing codling moth larvae, Cydia pomonella (L.) (Lepidoptera: Tortricidae). *Biocontrol Science and Technology*, *21*(3), 255–270. doi:10.1080/09583157.2010.540749

Deseo, K. V., & Docci, R. (1985). Microbiological control against Zeuzera pyrina L. (Lepidoptera: Cossidae) - Lotta microbiologica contro Zeuzera pyrina L. (Lepidoptera: Cossidae). *La Difesa delle Piante*, *8*, 285–291.

Doran, J. W., Sarrantonio, M., & Liebig, M. A. (1996). Soil health and sustainability. *Advances in Agronomy*, *56*, 1–54. doi:10.1016/S0065-2113(08)60178-9

Doran, J. W., & Zeiss, M. R. (2000). Soil health and sustainability: Managing the bioticcomponent of soil quality. *Applied Soil Ecology*, *15*(1), 3–11. doi:10.1016/S0929-1393(00)00067-6

Ebssa, L., & Koppenhöfer, A. M. (2012). Entomopathogenic nematodes for the management of Agrotis ipsilon: Effect of instar, nematode species and nematode production method. *Pest Management Science*, *68*(6), 947–957. doi:10.1002/ps.3259 PMID:22344709

Ehlers, R. U. (2011). Use of entomopathogenic nematodes in Mediterranean countries. Report on the 29th annual meeting of the working group "Beneficial Arthropods and Entomopathogenic Nematodes". *Journal of Plant Diseases and Protection*, *118*(2), 80–85. doi:10.1007/BF03356386

Ferris, H., & Bongers, T. (2006). Nematode indicators of organic enrichment. *Journal of Nematology*, *38*, 3–12. PMID:19259424

Grewal, P. S., De Nardo, E. A. B., & Aguillera, M. M. (2001). Entomopathogenic nematodes: Potential for exploration and use in South America. *Neotropical Entomology*, *30*, 191–205.

Grewal, P. S., Ehlers, R. U., & Shapiro-Ilan, D. I. (2005). *Nematodes as biological control agents*. CABI Publishing. doi:10.1079/9780851990170.0000

Griffin, C. T., Boemare, N. E., & Lewis, E. E. (2005). Biology and behaviour. In P. S. Grewal, R. U. Ehlers, & D. I. Shapiro-Ilan (Eds.), *Nematodes as biocontrol agents* (pp. 47–64). CABI Publishing. doi:10.1079/9780851990170.0047

Griffiths, B. S., Römbke, J., Schmelz, R. M., Scheffczyk, A., Faber, J. H., Bloem, J., Pérès, G., Cluzeau, D., Chabbi, A., Suhadolc, M., Sousa, J. P., Martins da Silva, P., Carvalho, F., Mendes, S., Morais, P., Francisco, R., Pereira, C., Bonkowski, M., Geisen, S., ... Stone, D. (2016). Selecting cost effective and policy-relevant biological indicators for European monitoring of soil biodiversity and ecosystem function. *Ecological Indicators*, *69*, 213–223. doi:10.1016/j.ecolind.2016.04.023

Head, J., & Walters, K. F. A. (2003). Augmentation biological control utilizing the entomopathogenic nematode, Steinernema feltiae, against the South American leafminer, Liriomyza huidobrensis. *Proceeding of the first International Symposium Biological control*, 136-140.

Jagdale, G. B., Casey, M. L., Grewal, P. S., & Lindquist, R. K. (2004). Effects of application rate and timing, potting medium and host plant on efficacy of Steinernema feltiae against fungus gnat, Bradysia coprophila, in floriculture. *Biological Control*, *29*(2), 296–305. doi:10.1016/S1049-9644(03)00164-6

Karlen, D. L., Mausbach, M. J., Doran, J. W., Cline, R. G., Harris, R. F., & Schuman, G. E. (1997). Soil quality: A concept, definition, and framework for evaluation. *Soil Science Society of America Journal*, *61*(1), 4–10. doi:10.2136ssaj1997.03615995 006100010001x

Lavelle, P., & Spain, A. (2001). *Soil ecology*. Kluwer Academic Publishers. doi:10.1007/978-94-017-5279-4

Liu, J. S., Chen, Y., Du, C., Liu, X. X., Ma, Q. H., Zhang, X., & Wang, D. L. (2019). Interactive effects of nitrogen addition and litter on soil nematodes in grassland. *European Journal of Soil Science*, *70*(3), 697–706. doi:10.1111/ejss.12779

Mahmoud, M. F. (2014a). Efficacy of entomopathogenic nematodes to certain Insect Pests infesting Oilseed Rape in the Laboratory and Greenhouse. *Egyptian Journal of Biological Pest Control*, *24*, 387–391.

Mahmoud, M. F. (2014b). Virulence of entomopathogenic nematodes against the Jasmine Moth, Palpita unionals Hb. (Lepidoptera: Pyralidae). Egypt. *J. Biol. Pest. Co.*, *24*, 393–397.

Mahmoud, M.F. (2016). Biology and use of entomopathogenic nematodes in insect pests biocontrol, a generic view. *Cercetări Agronomice în Moldova, 49*(4), 83-103. doi:10.1515/cerce-2016-0039

Mahmoud, M. F., & Osman, M. A. M. (2007). Use of the nematode Steinernema feltiae Cross N 33 as a biological control agent against the Peach Fruit Fly Bactrocera zonata. *Tunisian Journal of Plant Protection*, *2*, 109–115.

Mahmoud, M. F., & Osman, M. A. M. (2014). Activity of Psammotermes hypostoma (Desn.) and its biocontrol using entomopathogenic nematodes. Bull. Ent. Soc. Egypt. *Econ. Ser.*, *40*, 135–147.

Mahmoud, M. F., & Shebl, M. (2014). Insect fauna of canola and phenology of the diamondback moth, Plutella xylostella L. (Lepidoptera: Plutellidae) as a key pest. *Journal of Zoology (London, England)*, *97*, 125–132.

Mann, C., Lynch, D., Fillmore, S., & Mills, A. (2019). Relationships between field management, soil health, and microbial community composition. *Applied Soil Ecology*, *144*, 12–21. doi:10.1016/j.apsoil.2019.06.012

Martinez de Altube, M., Strauch, O., Fernandez De Castro, G., & Martinez Peña, A. (2008). Control of the flat-headed root borer Capnodis tenebrionis (Linné) (Coleoptera: Buprestidae) with the entomopathogenic nematode Steinernema carpocapsae (Weiser) (Nematoda: Steinernematidae) in a chitosan formulation in apricot orchards. *BioControl*, *53*(3), 531–539. doi:10.100710526-007-9094-0

Moebius-Clune, B. N. (2016). *Comprehensive assessment of soil health: the Cornell framework manual*. Cornell University.

Neher, D., & Olson, R. K. (1999). Nematode communities in soils of four farm cropping management systems. *Pedobiologia*, *43*, 430–438.

Neher, D. A. (2001). Role of nematodes in soil health and their use as indicators. *Journal of Nematology*, *33*(4), 161–168. https://www.ncbi.nlm.nih.gov/pubmed/19265875 PMID:19265875

Nguyen, C., Barker, B. M., Hoover, S., Nix, D. E., Ampel, N. M., Frelinger, J. A., Orbach, M. J., & Galgiani, J. N. (2013). Recent advances in our understanding of the environmental, epidemiological, immunological, and clinical dimensions of coccidioidomycosis. *Clinical Microbiology Reviews*, *26*(3), 505–525. doi:10.1128/CMR.00005-13 PMID:23824371

Nguyen, K. B., & Hunt, D. J. (2007). *Entomopathogenic nematodes: systematics, phylogeny and bacterial symbionts*. Brill. doi:10.1163/ej.9789004152939.i-816

Nielsen, U. N., Wall, D. H., & Six, J. (2015). Soil Biodiversity and the Environment. *Annual Review of Environment and Resources*, *40*(1), 63–90. doi:10.1146/annurev-environ-102014-021257

Nigon, V. M., & Félix, M. A. (2017). History of research on C. elegans and other free-living nematodes as model organisms. *WormBook*, *2017*, 1–84. https://www.ncbi.nlm.nih.gov/pubmed/28326696 PMID:28326696

Parkman, J. P., Hudson, G. W., Frank, J. H., Nguyen, K. B., & Smart, J. Jr. (1993). Establishment and persistence of Steinernema scapterisci (Rhabditida: Steinernematidae) in field populations of Scapteriscus spp. Mole crickets (Orthoptera: Gryllotalpidae). *Journal of Entomological Science*, *28*(2), 182–190. doi:10.18474/0749-8004-28.2.182

Parkman, J. P., & Smart, J. G. C. Jr. (1996). Entomopathogenic nematodes, a case study: Introduction of Steinernema scapterisci in Florida. *Biocontrol Science and Technology*, *6*(3), 413–419. doi:10.1080/09583159631389

Peterson, H., & Luxton, M. (1982). A comparative analysis of soil fauna populations and their role in decomposition processes. *Oikos*, *39*(3), 287–388. doi:10.2307/3544689

Poinar, G. O., Jr. (1990). Taxonomy and biology of Steinernematidae and Heterorhabditidae. In Entomopathogenic nematodes in biological control. CRC.

Reddy, P. (2021). Nematode Diseases of Crops and their Management. In Nematode Diseases of Crop Plants: An Overview. doi:10.1007/978-981-16-3242-6_1

Romig, D. E., Garlynd, M. J., Harris, R. F., & Mcsweeney, K. (1995). How farmers assess soil health and quality. *Journal of Soil and Water Conservation*, *50*(3), 229–236.

Sasser, J., & Freckman, D. (1987). A world perspective on nematology: the role of the society. In J. A. Veech & D. W. Dickson (Eds.), *Vistas on nematology* (pp. 7–14). Society of Nematologists.

Seesao, Y., Gay, M., Merlin, S., Viscogliosi, E., Aliouat-Denis, C. M., & Audebert, C. (2017). A review of methods for nematode identification. *Journal of Microbiological Methods, 138*, 37–49. . doi:10.1016/j.mimet.2016.05.030 PMID:27262374

Shamseldean, M. M., Hasanain, S. A., & Rezk, M. Z. A. 2009. Virulence of entomopathogenic nematodes against lepidopterous pests of horticultural crops in Egypt. *4th Conference on recent technologies in agriculture*, 74-84.

Shamseldean, M. S. M., Abd-Elbary, N. A., Shalaby, H., & Ibraheem, H. I. H. (2014). Entomopathogenic nematodes as biocontrol agents of the tomato leaf miner Tuta absoluta (Meyrick) (Lepidoptera: Gelechiidae) on tomato plants. Egypt. *J. Biol. Pest. Co.*, *24*, 503–513.

Shapiro-Ilan, D. I., Duncan, L. W., Lacey, L. A., & Han, R. (2005). Orchard crops. In P. Grewal, R.-U. Ehlers, & D. Shapiro-Ilan (Eds.), *Nematodes as biological control agents* (pp. 215–230). CABI. doi:10.1079/9780851990170.0215

Shapiro-Ilan, D. I., & Gaugler, R. (2010). Nematodes: Rhabditida: Steinernematidae & Heterorhabditidae. In A. Shelton (Ed.), *Biological Control: A Guide to Natural Enemies in North America*. Cornell University.

Thiele-Bruhn, S., Bloem, J., de Vries, F. T., Kalbitz, K., & Wagg, C. (2012). Linking soil biodiversity and agricultural soil management. *Current Opinion in Environmental Sustainability*, *4*(5), 523–528. doi:10.1016/j.cosust.2012.06.004

Tomalak, M., Piggott, S., & Jagdale, G. B. (2005). Glasshouse applications. In P. S. Grewal, R.-U. Ehlers, & D. I. Shapiro-Ilan (Eds.), *Nematodes as biological control agents* (pp. 147–166). CABI Publishing. doi:10.1079/9780851990170.0147

Torres-Barragan, A., Suazo, A., Buhler, W. G., & Cardoza, Y. J. (2011). Studies on the entomopathogenicity and bacterial associates of the nematode Oscheius carolinensis. *Biological Control*, *59*(2), 123–129. doi:10.1016/j.biocontrol.2011.05.020

Urzelai, A., Hernández, A. J., & Pastor, J. (2000). Biotic indices based on soil nematode communities for assessing soil quality in terrestrial ecosystems. *The Science of the Total Environment*, *247*(2–3), 253–261. . doi:10.1016/S0048-9697(99)00494-5 PMID:10803552

van Tol, R. W. H. M., & Raupp, M. J. (2006). Nursery and tree applications. In P. S. Grewal, R. U. Ehlers, & D. I. Shapiro-Ilan (Eds.), *Nematodes as biological control agents*. CABI Publishing.

Wall, D. H., Nielsen, U. N., & Six, J. (2015). Soil biodiversity and human health. *Nature*, *528*(7580), 69–76. doi:10.1038/nature15744 PMID:26595276

Wardle, D. (2002). *Communities and Ecosystems: Linking the Aboveground and Belowground Components*. Princeton University Press.

Williams, E. C., & Walters, K. F. A. (2000). Foliar application of the entomopathogenic nematode Steinernema feltiae against leafminers on vegetables. *Biocontrol Science and Technology*, *10*(1), 61–70. doi:10.1080/09583150029396

Ye, W. M., Torres-Barragan, A., & Cardoza, Y. J. (2010). Oscheius carolinensis n. sp (Nematoda: Rhabditidae), a potential entomopathogenic nematode from vermicompost. *Nematology*, *12*(1), 121–135. doi:10.1163/156854109X458464

Yeates, G., & Bongers, T. (1999). Nematode diversity in agroecosystems. *Agriculture, Ecosystems & Environment*, *74*(1-3), 113–135. doi:10.1016/S0167-8809(99)00033-X

Yeates, G., Bongers, T., De Goede, R., Freckman, D., & Georgieva, S. (1993). Feeding habits in soil nematode families and genera- An outline for soil ecologists. *Journal of Nematology*, *25*, 315–331. PMID:19279775

Compilation of References

Abad, P., Gouzy, J., Aury, J., Castagnone-Sereno, P., Danchin, E. G. J., Deleury, E., Perfus-Barbeoch, L., Anthouard, V., Artiguenave, F., Blok, V. C., Caillaud, M.-C., Coutinho, P. M., Dasilva, C., De Luca, F., Deau, F., Esquibet, M., Flutre, T., Goldstone, J. V., Hamamouch, N., ... Wincker, P. (2008). Genome sequence of the metazoan plant-parasitic nematode Meloidogyne incognita. *Nature Biotechnology*, *26*(8), 909–915. doi:10.1038/nbt.1482 PMID:18660804

Abdel Hamid, M. E., Ramadan, H. H., Salem, F. M., & Osman, G. Y. (1981). The susceptibility of different field crops to infestation by *Meloidogyne javanica* and *M. incognita acrita. Anzeiger für Schädlingskunde, Pflanzenschutz. Umweltschutz*, *54*, 81–82.

Abd-Elgawad, M. M. (2016). Use of Taylor's power law parameters in nematode sampling. *International Journal of Pharm Tech Research*, *9*(12), 999–1004.

Abd-Elgawad, M. M. (2020). Managing nematodes in Egyptian citrus orchards. *Bulletin of the National Research Center*, *44*(1), 1–15.

Abd-Elgawad, M. M. (2021). Optimizing safe approaches to manage plant-parasitic nematodes. *Plants*, *10*(9), 1911. doi:10.3390/plants10091911 PMID:34579442

Abd-Elgawad, M. M., Koura, F. F. H., Montasser, S. A., & Hammam, M. M. A. (2016). Distribution and losses of *Tylenchulus semipenetrans* in citrus orchards on reclaimed land in Egypt. *Nematology*, *18*(10), 1141–1150. doi:10.1163/15685411-00003020

Abobatta, W. F. (2019a). Nutritional Benefits of Citrus Fruits. *Am J Biomed Sci & Res. 3*(4), 303-306. doi:10.34297/AJBSR.2019.03.000681

Abobatta, W. F. (2019b). *Arbuscular Mycorrhizal* and Citrus Growth: Overview. *Acta Scientific Microbiology*, *2*(6), 14–17.

Abobatta, W. F. (2020). Citriculture and Climate Change. *Modern Concepts & Developments in Agronomy*, *6*(3), 649–650. doi:10.31031/MCDA.2020.06.000639

Abu Habib, A. H. A., Younes, H. A., Ibrahim, I. K. A., & Khalil, A. E. (2020). Plant parasitic nematodes associated with citrus trees and reaction of two citrus cultivars to *Tylenchulus semipenetrans* in northern Egypt. *Journal of the Advances in Agricultural Researches*, *25*(2), 166–175. doi:10.21608/jalexu.2020.161764

Adam, M., Westphal, A., Hallmann, J., & Heuer, H. (2014). Specific microbial attachment to root knot nematodes in suppressive soil. *Applied and Environmental Microbiology*, *80*(9), 2679–2686. doi:10.1128/AEM.03905-13 PMID:24532076

Agrios, G. N. (2005). *Plant pathology*. Elsevier.

Ahirwar, N. K., Singh, R., Chaurasia, S., Chandra, R., & Ramana, S. (2020). Effective role of beneficial microbes in achieving the sustainable agriculture and eco-friendly environment development goals: A review. *Frontiers in Microbiology*, *5*(6), 111–123. doi:10.11648/j.fem.20190506.12

Ahmad, N., & Jairajpuri, M. S. (1982). *Population fluctuations of predatory nematode* Parahadronchus shakili. *Proceedings of Symposium on Ecology and Animal Population in India*, *3*, 1–12.

Ahmad, G., Khan, A., Khan, A. A., Ali, A., & Mohhamad, H. I. (2021). Biological control: A novel strategy for the control of the plant parasitic nematodes. *Antonie van Leeuwenhoek*, *114*(7), 885–912. doi:10.100710482-021-01577-9 PMID:33893903

Ahmad, R., Khan, M. A., Mukhtar, T., & Javed, N. (2007). Infestation of citrus orchards by citrus nematode (*Tylenchulus semipenetrans*) and screening of some rootstocks for resistance or susceptibility. *Proceedings of the International Symposium on Prospects of Horticultural Industry in Pakistan*, 165-169.

Aït Hamza, M., Moukhli, A., Ferji, Z., Fossati-Gaschignard, O., Tavoillot, J., Ali, N., Boubaker, H., El Mousadik, A., & Mateille, T. (2018). Diversity of plant-parasitic nematode communities associated with olive nurseries in Morocco: Origin and environmental impacts. *Applied Soil Ecology*, *124*, 7–16. doi:10.1016/j.apsoil.2017.10.019

Akhtar, M., & Malik, A. (2000). Roles of organic soil amendments and soil organisms in the biological control of plant-parasitic nematodes: A review. *Bioresource Technology*, *74*(1), 35–47. doi:10.1016/S0960-8524(99)00154-6

Alabouvette, C., Olivain, C., & Steinberg, C. (2006). Biological Control of Plant Diseases: The European Situation. *European Journal of Plant Pathology*, *114*(3), 329–341. doi:10.100710658-005-0233-0

Al-Azzeh, T. K., & Gharbieh, W. A. (2005). Factors related to seasonal changes in numbers of *Tylenchulus semipenetrans* in the Jordan valley. *Pakistan Journal of Nematology*, *23*, 241–250.

Alfy, H., Ghareeb, R. Y., & Aziz, W. Z. (2020). Semi-field Application of a New Formulation Based on *Spodoptera litura* Nuclear Polyhedrosis Virus *and Bacillus thuringiensis* subsp. mexicanensis Against Cotton Leaf Worm, *Spodoptera littoralis* and Root Knot Nematode, *Meloidogyne incognita* in Egypt. *Current Applied Science and Technology*, 512-527.

Alheyalee, R. S., & Aljuboori, F. K. (2020). The first recording of root-knot nematodes Meloidogyne Javanica on the winter cherry Solanum pseudocapsicum and controlling it in vitro. *EurAsian Journal of BioSciences, 14*(2).

Ali, A., Matthews, W., Cavagnaro, P., Iorizzo, M., Roberts, P., & Simon, P. (2014). Inheritance and mapping of Mj-2, a new source of root-knot nematode (Meloidogyne javanica) resistance in carrot. *Journal NNN/of Heredity, 105*, 288-291.

Ali, M. A., Azeem, F., Li, H., & Bohlmann, H. (2017). Smart parasitic nematodes use multifaceted strategies to parasitize plants. *Frontiers in Plant Science, 8*, 1699. doi:10.3389/fpls.2017.01699 PMID:29046680

Ali, N., Chapuis, E., Tavoillot, J., & Mateille, T. (2014). Plant-parasitic nematodes associated with olive tree (Olea europaea L.) with a focus on the Mediterranean Basin: A review. *Comptes Rendus Biologies, 337*(7-8), 423–442. doi:10.1016/j.crvi.2014.05.006 PMID:25103828

Ali, Y., Abbas, G., Sajid, M., & Imran, M. (2018). Response of Citrus Slow Decline Causing Nematode (*T. semipenetrans*) against Different Cultivars (Kinnow, Musambi, Shakri, Feutrel Early, Red Blood) of Citrus. *Advances in Zoology and Botany., 6*(1), 1–18. doi:10.13189/azb.2018.060101

Almeida, N. O., Teixeira, R. A., Carneiro, F. A., de Oliveira, C. M., Ribeiro, V. A., Lobo Júnior, M., & da Rocha, M. R. (2018). Occurrence and correlations of nematodes, Fusarium oxysporum and edaphic factors on banana plantations. *Journal of Phytopathology, 166*(4), 265–272. doi:10.1111/jph.12683

Almohithef, A. H., Al-Yahya, F. A., Al-Hazmi, A. S., Dawabah, A. A. M., & Lafi, H. A. (2018). Prevalence of plant-parasitic nematodes associated with certain greenhouse vegetable crops in Riyadh region, Saudi Arabia. *Journal of the Saudi Society of Agricultural Sciences, 19*(1), 22–25. doi:10.1016/j.jssas.2018.05.001

Al-Qasem, M. S., & Abu-Gharbieh, W. I. (1995). Occurrence and distribution of citrus nematode, *Tylenchulus semipenetrans* in Jordan. *Nematologia Mediterranea, 23*, 335–339.

Al-Rehiayani, S., & Motawei, M. I. (2014) Detection and molecular characterization of Pasteuria penetrans isolates parasitizing root-knot nematodes in AL-Qassim area, Saudi Arabia. *Biotechnology*. http://www.omicsgroup.com/conferences/ ACS/conference/pdfs/2 113-Poster-Pdf.pdf

Alvandi, J., Karimi, J., Ghadamyari, M., Sharifi, M., & Asoodeh, A. (2017). Physiological defense of the white grub, *Polyphylla adspersa* Motschulsky (Col., Scarabaeidae) against entomopathogenic nematodes. *Journal of Asia-Pacific Entomology, 20*(3), 878–885. doi:10.1016/j.aspen.2017.06.003

Ami, S. N., Taher, I. E., Hussen, F. S., & Ahmed, A. I. (2019). First molecular identification of wheat seed gall nematode races parasitized on wheat in Iraq. *Acta Universitatis Sapientiae. Agriculture and Environment, 11*(1), 5–15.

Annapurna, M., Bhagawati, B., & Uday, K. (2018). In-vitro Efficacy of Native Fungal Bioagents against Meloidogyne incognita. *International Journal of Current Microbiology and Applied Sciences, 7*(11), 396–410. doi:10.20546/ijcmas.2018.711.048

Anonymous. (2019). *Floriculture Crops 2018 Summary* (May 2019). National Agricultural Statistics Service.

Ara, I., Khan, H., Syed, T., & Bhat, B. (2021). Prevalence and seasonal dynamics of gastrointestinal nematodes of domestic fowls (Gallus gallus domesticus) in Kashmir, India. *Journal of Advanced Veterinary and Animal Research*, *8*(3), 448. doi:10.5455/javar.2021.h533 PMID:34722743

Araya, M., & De Waele, D. (2011). Reaction of six Musa genotypes to root-parasitic nematodes. *International Journal of Pest Management*, *57*(3), 229–238. doi:10.1080/09670874.2011.581770

Archidona-Yuste, A., Cantalapiedra-Navarrete, C., Liebanas, G., Rapoport, H. F., Castillo, P., & Palomares-Rius, J. E. (2018). Diversity of root-knot nematodes of the genus *Meloidogyne* Göeldi, 1892 (Nematoda: Meloidogynidae) associated with olive plants and environmental cues regarding their distribution in southern Spain. *PLoS One*, *13*(6), e0198236. Advance online publication. doi:10.1371/journal.pone.0198236 PMID:29924813

Ardakani, S. A., Mafi, Z. T., Mokaram, A. H., & Goltappeh, E. M. (2014). Relationship between soil properties and abundance of *Tylenchulus semipenetrans* in citrus orchards, Kohgilouyeh va Boyerahmad Province. *Journal of Agricultural Science and Technology*, *16*(7), 1699–1710.

Asghar, M. N., Khan, S. A., Javed, N., & Arshad, M. (2021). Exploring Citrus Nematodes and its Antagonists in Citrus Growing Areas of Punjab, *Pakistan. J. Plant Environ*, *03*(02), 91–101. doi:10.33687/jpe.003.02.3714

Askary, T. H., Gani, M., & Wani, A. R. (2021). Nematodes Associated with Stone Fruits and Their Management Strategies. Production Technology of Stone Fruits, 423-436. doi:10.1007/978-981-15-8920-1_16

Askary, T. H. (2017). Diversity of plant parasitic nematodes in pulses. In A. A. Ansari, S. S. Gill, Z. K. Abbas, & M. Naeem (Eds.), *Plant diversity: present situation and future scenario* (pp. 232–274). CAB International. doi:10.1079/9781780646947.0239

Askary, T. H., & Abd-Elgawad, M. M. M. (2017). Beneficial nematodes in agroecosystems: a global perspective. In *Biocontrol agents: entomopathogenic and slug parasitic nematodes* (pp. 3–25). CABI. doi:10.1079/9781786390004.0003

Askary, T. H., Banday, S. A., Iqbal, U., Khan, A. A., Mir, M. M., & Waliullah, M. I. S. (2012). Plant Parasitic Nematode Diversity in Pome, Stone and Nut Fruits. In E. Lichtfouse (Ed.), *Agroecology and Strategies for Climate Change. Sustainable Agriculture Reviews* (Vol. 8). Springer. doi:10.1007/978-94-007-1905-7_10

Askary, T. H., Khalil, A., Nazir, N., Khan, A. A., & Banday, S. A. (2018). Nematode Parasites of Grapevines. In E. Lichtfouse (Ed.), *Sustainable Agriculture Reviews 31. Sustainable Agriculture Reviews* (Vol. 31). Springer. doi:10.1007/978-3-319-94232-2_7

Atkins, S., Hidalgo-Diaz, L., Kalisz, H., Mauchline, T., Hirsch, P., & Kerry, B. (2003). Development of a new management strategy for the control of root-knot nematodes (*Meloidogyne* spp.) in organic vegetable production. *Pest Management Science*, *59*(2), 183–189. doi:10.1002/ps.603 PMID:12587872

Avendaño, F., Pierce, F. J., & Melakeberhan, H. (2004). The relationship between soybean cyst nematode seasonal population dynamics and soil texture. *Nematology*, *6*(4), 511–525. doi:10.1163/1568541042665214

Aydınlı, G., Mennan, S., Devran, Z., Sirca, S., & Urek, G. (2013). First report of the root-knot nematode *Meloidogyne ethiopica* on tomato and cucumber in Turkey. *Plant Disease*, *97*(9), 1262. doi:10.1094/PDIS-01-13-0019-PDN PMID:30722413

Ayoub, S. M. (1980). *Plant Nematology: An agricultural training aid.* Nema Aid Publications.

Azmi, M. I. (1983). Predatory behaviour of nematodes. Biological control of *Helicotylenchus dihystera* through the predaceous nematodes, *Iotonchus monhystera. Indian Journal of Nematology*, *13*, 1–8.

Back, M. A., Haydock, P. P. J., & Jenkinson, P. (2002). Disease complexes involving plant parasitic nematodes and soilborne pathogens. *Plant Pathology*, *51*(6), 683–697. doi:10.1046/j.1365-3059.2002.00785.x

Bahadur, A. (2022). Nematodes Diseases of Fruits and Vegetables Crops in India. In C. Belle & T. Kaspary (Eds.), *Nematodes - Recent Advances, Management and New Perspectives* (pp. 1–18). IntechOpen. doi:10.5772/intechopen.98850

Baines, R. C., Cameron, J. W., & Soost, R. K. (1974). Four biotypes of *Tylenchulus semipenetrans* in California identified, and their importance in the development of resistant citrus rootstocks. *Journal of Nematology*, *6*, 63–66. PMID:19319369

Baines, R. c., Miyakawa, T., Cameron, J. W., & Small, R. H. (1969). Biotypes of the citrus nematode. *Proceedings of the lst International Citrus Symposium*, 955-956.

Baines, R. C., & Thorne, G. (1961). The olive tree as a host of the citrus-root nematode. *Phytopathology*, *42*, 77–78.

Baiocchi, T., Lee, G., Choe, D.-H., & Dillman, A. R. (2017). Host-seeking parasitic nematodes use specific odors to assess host resources. *Scientific Reports*, *7*(1), 1. Advance online publication. doi:10.103841598-017-06620-2 PMID:28740104

Bajaj, H.K., Kanwar, R.S., Baghel, P.P.S. & Bhatti, D.S. (1988). Motia (*Jasminum sambac*) - a new host record of citrus nematode, *Tylenchulus semipenetrans* Cobb, 1913. *Haryana Agricultural University Journal of Research*, *18*(2), 146-147.

Bale, J. S., van Lenteren, J. C., & Bigler, F. (2008). Biological control and sustainable food production. *Philosophical Transactions of the Royal Society of London. Series B, Biological Sciences*, *363*(1492), 761–776. doi:10.1098/rstb.2007.2182 PMID:17827110

Banage, W. B. (1963). The ecological importance of free-living soil nematodes with special reference to those of moorland soil. *Journal of Animal Ecology*, *32*(1), 133–140. doi:10.2307/2522

Banyer, R. J., & Fisher, J. M. (1971). Diagnostic and advisory programes. In Plant parasitic nematodes. Academic Press.

Baran, M. (2011). Nematodes reproductive system. *Parasitology*, *7*, 163.

Barker, K. R. (1998). Introduction. In K. R. Barker, G. A. Pederson, & G. L. Winhan (Eds.), *Plant and Nematode Interactions* (pp. 1–120). American Society of Agronomy. doi:10.2134/agronmonogr36

Barker, K. R., Nusbaum, C. J., & Nelson, L. A. (1969). Season population dynamics of selected plant-parasitic nematodes as measured by three extraction procedures. F. *Nematology*, *I*, 232.

Barnosky, A. D., Matzke, N., Tomiya, S., Wogan, G. O., Swartz, B., Quental, T. B., Marshall, C., McGuire, J. L., Lindsey, E. L., Maguire, K. C., Mersey, B., & Ferrer, E. A. (2011). Has the Earth's sixth mass extinction already arrived? *Nature*, *471*(7336), 51–57. doi:10.1038/nature09678 PMID:21368823

Basyoni, M. M. A., & Rizk, E. M. A. (2016). Nematodes ultrastructure: Complex systems and processes. *Journal of Parasitic Diseases : Official Organ of the Indian Society for Parasitology*, *40*(4), 1130–1140. doi:10.100712639-015-0707-8 PMID:27876901

Batalla-Carrera, L., Morton, A., & Garcıa-del-Pino, F. (2010). Efficacy of nematodes against the tomato leafminer Tuta absoluta in laboratory and greenhouse conditions. *BioControl*, *55*, 523–530. doi:10.100710526-010-9284-z

Bawa, J., Mohammed, I., & Liadi, S. (2014). Nematicidal effect of some plants extracts on root-knot nematodes (*Meloidogyne incognita*) of tomato (*Lycopersicon esculentum*). *World Journal of Life Science and Medical Research*, *3*, 81–87.

Belair, G., Fournier, Y., & Dauphinais, N. (2003). Efficacy of steinernematid nematodes against three insect pests of crucifers in Quebec. *Journal of Nematology*, *35*, 259–265. PMID:19262759

Bender, S. F., & van der Heijden, M. G. A. (2015). Soil biota enhance agricultural sustainability by improving crop yield, nutrient uptake and reducing nitrogen leaching losses. *Journal of Applied Ecology*, *52*(1), 228–239. doi:10.1111/1365-2664.12351

Benjlil, H., Elkassemi, K., Aït Hamza, M., Mateille, T., Furze, J. N., Cherifi, K., Mayad, E. H., & Ferji, Z. (2020). Plant-parasitic nematodes parasitizing saffron in Morocco: Structuring drivers and biological risk identification. *Applied Soil Ecology*, *147*, 103362. doi:10.1016/j.apsoil.2019.103362

Bernard, G. C., Egnin, M., & Bonsi, C. (2017). The Impact of Plant-Parasitic Nematodes on Agriculture and Methods of Control. In Nematology - Concepts, Diagnosis and Control. doi:10.5772/intechopen.68958

Bharali, A., Bhagawati, B., & Kurulkar, U. (2019). Nematicidal toxicity of native antagonists against *Meloidogyne incognita. International Journal of Current Microbiology and Applied Sciences*, *8*(2), 1385–1398. doi:10.20546/ijcmas.2019.802.163

Bhuiyan, S. A., Garlick, K., Anderson, J. M., Wickramasinghe, P., & Stirling, G. R. (2018). Biological control of root-knot nematode on sugarcane in soil naturally or artificially infested with Pasteuria penetrans. *Australasian Plant Pathology*, *47*(1), 45–52. doi:10.100713313-017-0530-z

Bilgrami, A. L. (2008). Biological control potentials of predatory nematodes. *Integrated management and biocontrol of vegetable and grain crops nematodes*, 3-28.

Bilgrami, L. A., & Brey, C. (2005). Potential of predatory nematodes to control plant-parasitic nematodes. In Nematodes as Biocontrol Agents. CAB International. doi:10.1079/9780851990170.0447

Bilgrami, A. L. (1992). Resistance and susceptibility of prey nematodes to predation and strike rate of the predators, *Mononchus aquaticus, Dorylaimus stagnalis*, and *Aquatides thornei. Fundamental and Applied Nematology*, *15*, 265–270.

Bilgrami, A. L., Ahmad, I., & Jairajpuri, M. S. (1985). Predatory behavior of *Aquatides thornei* (Nygolaimina: Nematoda). *Nematologica*, *30*, 457–462.

Bilgrami, A. L., Ahmad, I., & Jairajpuri, M. S. (1986). A study on the intestinal contents of some mononchs. *Review of Nematology*, *17*, 339–340.

Bilgrami, A. L., & Gaugler, R. (2005). Feeding behaviour of the predatory nematodes *Laimydorus baldus* and *Discolaimus major* (Nematoda: Dorylaimida). *Nematology*, *7*(1), 11–20. doi:10.1163/1568541054192207

Bilgrami, A. L., Gaugler, R., & Brey, C. (2005). Prey preference and feeding behaviour of diplogastrid predator *Mononchoides gaugleri* (Nematoda: Diplogasterida). *Nematology*, *7*(3), 333–342. doi:10.1163/156854105774355563

Bird, A. F., & Bird, J. (2012). *The structure of nematodes*. Academic Press.

Bird, G., Bishop, B., Grafius, E., Hausbeck, M., Lynnae, J., Williams, K., & Pett, W. (2004). Insect, disease and nematode control for commercial vegetables. *Michigan State University Extension Bulletin*, *312*, 81–82.

Bliss, C.I. (1958). Periodic regression in biology and climatology. *Connecticut Agri. Exp. Stn. Bull., 615*.

Böckenhoff, A., & Grundler, F. M. W. (1994). Studies on the Nutrient Uptake by the Beet Cyst Nematode Heterodera Schachtii by in Situ Microinjection of Fluorescent Probes into the Feeding Structures in Arabidopsis Thaliana. *Parasitology*, *109*(2), 249–255. doi:10.1017/S003118200007637X

Bodenheimer, F. S. (1958). *The history of biology: an introduction*. Dawson.

Bonfil, D., Dolgin, B., Mufradi, I., & Asido, S. (2004). Bioassay to forecast cereal cyst nematode damage to wheat in fields. *Precision Agriculture*, *5*(4), 329–344. doi:10.1023/B:PRAG.0000040804.97462.02

Bongers, T., & Bongers, M. (1998). Functional diversity of nematodes. *Applied Soil Ecology*, *10*(3), 239–251. doi:10.1016/S0929-1393(98)00123-1

Bonsi, C., & Phills, B. (1979). Reaction of twelve sweetpotato cultivars and breeding lines to root-knot species with three experimental methods. *Horticultural Science*, *14*, 539–541.

Boosalis, M. G., & Mankau, R. (1965). Parasitism and predation of soil microorganisms. In K. F. Baker & W. C. Snyder (Eds.), *Ecology of Soil-borne Plant Pathogens* (pp. 374–391). Berkeley University of California Press.

Bourne, J. M., & Kerry, B. R. (1999). Effect of the host plant on the efficacy of *Verticillium chlamydosporium* as a biological control agent of root-knot nematodes at different nematode densities and fungal application rates. *Soil Biology & Biochemistry*, *31*(1), 75–84. doi:10.1016/S0038-0717(98)00107-2

Bozbuga, R., Lilley, C. J., Knox, J. P., & Urwin, P. E. (2018). Host-specific signatures of the cell wall changes induced by the plant parasitic nematode. *Meloidogyne incognita. Scientific Reports*, *8*(1), 17302. doi:10.103841598-018-35529-7 PMID:30470775

BozbugaS.YildizS.YukselE.ÖzerG.DababatA. A.İmrenM. (2022). Nematode-citrus plant interactions: host preference, damage rate and molecular characterization of Citrus root nematode *Tylenchulus semipenetrans*. Research square. doi:10.21203/rs.3.rs-1502604/v1

Bravo, A., Gómez, I., Porta, H., García-Gómez, B. I., Rodriguez-Almazan, C., Pardo, L., & Soberón, M. (2012). Evolution of *Bacillus thuringiensis* Cry toxins insecticidal activity. *Microbial Biotechnology*, *6*(1), 17–26. doi:10.1111/j.1751-7915.2012.00342.x PMID:22463726

Bridge, J., Plowright, R., & Peng, D. (2005). Nematode parasites of rice. In Plant Parasitic Nematodes in Subtropical and Tropical Agriculture (2nd ed.). CABI Bioscience. doi:10.1079/9780851997278.0087

Brivio, M. F., & Mastore, M. (2018). Nematobacterial Complexes and Insect Hosts: Different Weapons for the Same War. *Insects, 9*(3), 117. doi:10.3390/insects9030117

Brlansky, R. H., Chung, K. R., & Rogers, M. E. (2012). *2006 Florida citrus pest management guide: Huanglongbing (citrus greening)*. UF/IFAS Extension.

Brodie, B. B. (1976). Vertical distribution of three nematode species in relation to certain soil properties. *Journal of Nematology*, *8*, 243–247. PMID:19308230

Brown, E. B. (1958). Pea root eelworm in the eastern countries of England. *Nematologica*, *3*(4), 257–268. doi:10.1163/187529258X00012

CABI. (2022). https://www.cabi.org/isc/datasheet/61059

Cai, W. (1999). Extension theory and its application. *Chinese Science Bulletin*, *44*(17), 1538–1548. doi:10.1007/BF02886090

Caldwell, M. M., Richards, J. H., & Beyschlag, W. (1991). Hydraulic lift: ecological implications of water efflux from roots. In D. Atkinson (Ed.), *Plant Root Growth: an Ecological Perspective* (pp. 423–436). Blackwell Scientific.

Calvet, C., Pinochet, J., Hernández-Dorrego, A., Estaún, V., & Camprubí, A. (2001). Field microplot performance of the peach-almond hybrid GF-677 after inoculation with arbuscular mycorrhizal fungi in a replant soil infested with root-knot nematodes. *Mycorrhiza*, *10*(6), 295–300. doi:10.1007/PL00009998

Carneiro, R. M. D. G., de Oliveira Lima, F. S., & Correia, V. R. (2017). *Methods and tools currently used for the identification of plant parasitic nematodes*. Nematology – Concepts, Diagnosis and Control. doi:10.5772/intechopen.69403

Cassidy, G. H. (1931). Some mononchs of Hawaii. *Hawaiian Planters' Rec*, *35*(3), 305–339.

Castillo, J. C., Reynolds, S. E., & Eleftherianos, I. (2011). Insect immune responses to nematode parasites. *Trends in Parasitology*, *27*(12), 537–547. doi:10.1016/j.pt.2011.09.001 PMID:21982477

Cervantes-Flores, J., Yencho, G., & Davis, E. (2008). Detection of quantitative trait loci and inheritance of root-knot nematode resistance in sweet potato. *Horticultural Science (Prague)*, *133*, 841–851.

Chaisson, K., & Hallem, E. (2012). Chemosensory behaviors of parasites. *Trends in Parasitology*, *28*(10), 427–436. doi:10.1016/j.pt.2012.07.004 PMID:22921895

Chapman, R. A. (1959). Development of *Pratylenchus penetrans* and *Tylenchorhynchus martini* on red clover Alfa Alfa. *Phytopathology*, *49*, 357.

Chawla, M. L., Samathanam, G. J., & Sharma, S. B. (1980). Occurrence of citrus nematode (*Tylenchulus semipenetrans* Cobb, 1913) on roots of mangos teen (*Garcinia mangostana*). *Indian Journal of Nematology*, *10*(2), 240–242.

Chen, Z. X., Chen, S. Y., & Dickson, D. W. (Eds.). (2004). Nematology: advances and perspectives. Academic Press.

Chen, Z., & Dickson, D. (1998). Review of *Pasteuria penetrans*: Biology, ecology, and biological control potential. *Journal of Nematology*, *30*, 313–340. PMID:19274225

Childs& Nathan. (2017). *U.S. 2016/17 Export Forecast Lowered 2.0 Million Cwt to 110.0 Million cwt. USDA-ERS Situation and Outlook*. RCS-17B.

Chirchir, A. K., Kimenju, J. W., Olubayo, F., & Mutua, G. K. (2008). Abundance and Distribution of Plant Parasitic Nematodes Associated With Sugarcane in Western Kenya. *Asian Journal of Plant Pathology*, *2*(1), 48–53. doi:10.3923/ajppaj.2008.48.53

Chitambar, J. J., Westerdahl, B. B., & Subbotin, S. A. (2018). Plant Parasitic Nematodes in California Agriculture. In Plant Parasitic Nematodes in Sustainable Agriculture of North America. Sustainability in Plant and Crop Protection. Springer. doi:10.1007/978-3-319-99585-4_6

Chitambar, J. J., & Noffsinger, E. M. (1989). Predaceous behaviour and life history of *Odontopharynx longicaudata* (Diplogasterida). *Journal of Nematology*, *21*, 284–291. PMID:19287610

Chitwood, D. J. (2003). Research on plant-parasitic nematode biology conducted by the United States Department of Agriculture–Agricultural Research Service. *Pest Management Science*, *59*(6-7), 748–753. doi:10.1002/ps.684 PMID:12846325

Christie, J. R. (1960). Biological control-predaceous nematodes. In J. M. Sasser & W. R. Jenkins (Eds.), *Nematology: Fundamentals and Recent Advances with Emphasis on Plant Parasitic and Soil Forms* (pp. 466–468). University of North Carolina Press.

Ciancio, A. (2018). Biocontrol potential of *Pasteuria* spp. for the management of plant parasitic nematodes. *Perspectives in Agriculture, Veterinary Science, Nutrition and Natural Resources*, *13*(013), 1–13. doi:10.1079/PAVSNNR201813013

Ciancio, A., & Mukerji, K. G. (Eds.). (2009). *Integrated management of fruit crops and forest nematodes* (Vol. 4). Springer Science & Business Media. doi:10.1007/978-1-4020-9858-1

CIP International Potato Center. (2014). *Sweetpotato Facts and Figures*. Available from: https://cipotato.org/sweetpotato/facts-2/

Clark, C., & Wright, V. (1983). Effect and reproduction of *Rotylenchulus reniformis* on sweet potato selections. *Journal of Nematology*, *15*, 198–203. PMID:19295791

Cobb, N. A. (1913). Notes on Mononchus and Tylenchulus. *Journal of the Washington Academy of Sciences*, *3*, 287–288.

Cobb, N. A. (1914). Citrus-root nematode. *Journal of Agricultural Research*, *2*, 217–230.

Cobb, N. A. (1917). The mononchs (Mononchus Bastian, 1865): A genus of free-living predatory nematodes. *Soil Science*, *3*(5), 431–486. doi:10.1097/00010694-191705000-00004

Cobb, N. A. (1920). Transfer of nematodes (mononchs) from place to place for economic purposes. *Science*, *51*(1330), 640–641. doi:10.1126cience.51.1330.640 PMID:17797737

Cohn, E. (1965). The development of the citrus nematode on some of its hosts. *Nematologica*, *11*(4), 593–600. doi:10.1163/187529265X00744

Cohn, E. (1969). The occurrence and distribution of species of *Xiphinema* and *Longidorus* in Israel. *Nematologica*, *15*(2), 179–192. doi:10.1163/187529269X00614

Cohn, E. (1972). Nematode diseases of citrus. In J. M. Webster (Ed.), *Economic Nematology* (pp. 215–244). Academic Press.

Cohn, E., & Mordechai, E. (1974). Experiments in suppressing citrus nematode populations by use of marigold and a predacious nematode. *Nematologia Mediterranea*, *2*, 43–53.

Coyne, D. L., Dalzell, J., Cortada, L., Claudius-Cole, A. O., Haukeland, S., Luambano, N., & Talwana, H. (2018). Plant-Parasitic Nematodesand Food Security in Sub-Saharan Africa. *Annual Review of Phytopathology*, *56*(1), 381–403. doi:10.1146/annurev-phyto-080417-045833 PMID:29958072

Coyne, D. L., Talwana, H. A. L., & Maslen, N. R. (2003). Plant-parasitic nematodes associated with root and tuber crops in Uganda. *African Plant Protection*, *9*, 87–98.

Creasy, G. L., & Creasy, L. L. (2018). *Grapes* (Vol. 27). Cabi. doi:10.1079/9781786391360.0000

Crickmore, N., Zeigler, D. R., Schnepf, E., Van Rie, J., Lereclus, D., Baum, J., Bravo, A., & Dean, D. H. (2011). *Bacillus thuringiensis toxin nomenclature*. Available at http://www. lifesci. sussex.ac.uk/Home/Neil_Crickmore/Bt/

Crickmore, N. (2005). Using worms to better understand how *Bacillus thuringiensis* kills insects. *Trends in Microbiology*, *13*(8), 347–350. doi:10.1016/j.tim.2005.06.002 PMID:15967665

Cunha, T. G. D., Visôtto, L. E., Lopes, E. A., Oliveira, C. M. G., & God, P. I. V. G. (2018). Diagnostic methods for identification of root-knot nematodes species from Brazil. *Ciência Rural*, *48*(2), 48. doi:10.1590/0103-8478cr20170449

Dababat, A. A., Sikora, R. A., & Hauschild, R. (2006). Use of Trichoderma harzianum and Trichoderma viride for the biological control of Meloidogyne incognita on tomato. *Communications in Agricultural and Applied Biological Sciences*, *71*(3 Pt B), 953–961. PMID:17390844

Dahlin, P., Eder, R., Consoli, E., Krauss, J., & Kiewnick, S. (2019). Integrated control of Meloidogyne incognita in tomatoes using fluopyram and Purpureocillium lilacinum strain 251. *Crop Protection (Guildford, Surrey)*, *124*, 104874. doi:10.1016/j.cropro.2019.104874

Das, S., Abdul Wadud, M., & Atiqur Rahman Khokon, M. (2021). Functional evaluation of culture filtrates of Bacillus subtilis and Pseudomonas fluorescens on the mortality and hatching of Meloidogyne javanica. *Saudi Journal of Biological Sciences*, *28*(2), 1318–1323. doi:10.1016/j. sjbs.2020.11.055 PMID:33613061

Davide, R. G. (1988). Nematode problems affecting agriculture in the Philippines. *Journal of Nematology*, *20*(2), 214–218. PMID:19290204

Dawabah, A. A., Al-Yahya, F. A., & Lafi, H. A. (2019). Integrated management of plant-parasitic nematodes on guava and fig trees under tropical field conditions. *Egyptian Journal of Biological Pest Control*, *29*(1), 1–9. doi:10.118641938-019-0133-9

De Ley, P., Blaxter, M., Vierstraete, A., & Lockington, D. (2006). Nematode Morphology: An Overview with Emphasis on Free-Living Species. *International Journal for Parasitology*, *36*(13), 1327–1358. doi:10.1016/j.ijpara.2006.06.003

de Waal, J., Malan, A. P., & Addison, M. F. (2011). Evaluating mulches together with Heterorhabditis zealandica (Rhabditida: Heterorhabditidae) for the control of diapausing codling moth larvae, Cydia pomonella (L.) (Lepidoptera: Tortricidae). *Biocontrol Science and Technology*, *21*(3), 255–270. doi:10.1080/09583157.2010.540749

Decraemer, W., & Hunt, D. (2006). Structure and classification. In R. Perry & M. Moens (Eds.), *Plant Nematology* (pp. 3–32). CAB International., doi:10.1079/9781845930561.0003

Decraemer, W., & Hunt, D. J. (2013). Structure and classification. In R. N. Perry & M. Moens (Eds.), *Plant nematology* (2nd ed., pp. 3–39). CABI Publishing. doi:10.1079/9781780641515.0003

Deng, X., Wang, X., & Li, G. (2022). Nematicidal effects of volatile organic compounds from microorganisms and plants on Plant-parasitic nematodes. *Microorganisms, 10*(6), 1201. doi:10.3390/microorganisms10061201 PMID:35744719

Deseo, K. V., & Docci, R. (1985). Microbiological control against Zeuzera pyrina L. (Lepidoptera: Cossidae) - Lotta microbiologica contro Zeuzera pyrina L. (Lepidoptera: Cossidae). *La Difesa delle Piante*, *8*, 285–291.

Devi, G., & George, J. (2018). Predatory nematodes as bio-control agent against plant-parasitic nematode-A review. *Agricultural Reviews (Karnal)*, *39*(1), 55–61. doi:10.18805/ag.R-1715

Devi, G., & Nath, D. (2017). Entomopathogenic nematodes: A tool in biocontrol of insect pests of vegetables-A review. *Agricultural Reviews (Karnal)*, *38*(2). Advance online publication. doi:10.18805/ag.v38i02.7945

Dias, W. P., Ribeiro, N. R., Lopes, I. O. N., Garcia, A., Carneiro, G. E. S., & Silva, J. F. V. (2007). Manejo de nematoides na cultura da soja. In *Congresso Brasileiro de Nematologia; 12 de Agosto; Goiânia, Brazil* (pp. 26–30). Sociedade Brasileira de Nematologia.

Dickson, D. W., & Hewlett, T. E. (1988). Horizontal and vertical migration of Meloidogyne arenaria in soil columns in the field. *Nematropica*, *18*, 4–5.

Doran, J. W., Sarrantonio, M., & Liebig, M. A. (1996). Soil health and sustainability. *Advances in Agronomy*, *56*, 1–54. doi:10.1016/S0065-2113(08)60178-9

Doran, J. W., & Zeiss, M. R. (2000). Soil health and sustainability: Managing the bioticcomponent of soil quality. *Applied Soil Ecology*, *15*(1), 3–11. doi:10.1016/S0929-1393(00)00067-6

Ducharme, E. P. (1967). Annual population periodicity of *Radopholus simils* in Florida citrus groves. *Dis. Reporter.*, *51*, 1031.

Duncan, L. W. (1999). Nematode diseases of citrus. In L. W. Timmer & L. W. Duncan (Eds.), *Citrus Health Management* (pp. 136–148). American Phytopathological Society Press.

Duncan, L. W. (2005). Nematode parasites of citrus. In M. Luc, R. A. Sikora, & J. Bridge (Eds.), *Plant Parasitic Nematodes in Subtropical and Tropical Agriculture* (pp. 437–466). CAB International. doi:10.1079/9780851997278.0437

Duncan, L. W., & Eissenstat, D. M. (1993). Response to *Tylenchulus semipenetrans* to citrus fruit removal: Implications for carbohydrates composition. *Journal of Nematology*, *25*, 7–14. PMID:19279735

Duncan, L. W., & El-Morshedy, M. M. (1996). Population changes of *Tylenchulus semipenetrans* under localized versus uniform drought in the citrus root zone. *Journal of Nematology*, *28*(3), 360. PMID:19277154

Duncan, L. W., El-Morshedy, M. M., & McSorley, R. (1994). Sampling citrus fibrous roots and *Tylenchulus semipenetrans*. *Journal of Nematology*, *26*, 442–451. PMID:19279914

Duncan, L. W., Graham, J. H., & Timmer, L. W. (1993). Seasonal pattern associated with *Tylenchulus semipenetrans* and *Phtophthora parasitic* in the citrus rhizosphere. *Phytopathology*, *83*(5), 573–581. doi:10.1094/Phyto-83-573

Duncan, L. W., Noling, J. W., & Inserra, R. N. (2022). *2022–2023 Florida Citrus Production Guide: Nematodes*. EDIS. doi:10.32473/edis-cg010-2022

Dutta, T. K., Khan, M. R., & Phani, V. (2019). Plant-parasitic nematode management via biofumigation using brassica and non-brassica plants: Current status and future prospects. *Current Plant Biology*, *17*, 17–32. doi:10.1016/j.cpb.2019.02.001

Ebrahimi, L., Niknam, G., Dunphy, G. B., & Toorchi, M. (2014). Side effects of immune response of Colorado potato beetle, *Leptinotarsa decemlineata* against the entomopathogenic nematode, *Steinernema carpocapsae* infection. *Invertebrate Survival Journal : ISJ*, *11*(1), 132–142.

Ebssa, L., & Koppenhöfer, A. M. (2012). Entomopathogenic nematodes for the management of Agrotis ipsilon: Effect of instar, nematode species and nematode production method. *Pest Management Science*, *68*(6), 947–957. doi:10.1002/ps.3259 PMID:22344709

Ehlers, R. U. (2011). Use of entomopathogenic nematodes in Mediterranean countries. Report on the 29th annual meeting of the working group "Beneficial Arthropods and Entomopathogenic Nematodes". *Journal of Plant Diseases and Protection*, *118*(2), 80–85. doi:10.1007/BF03356386

Eisenback, J. D., & Triantaphyllou, H. H. (2020). Root-knot nematodes: Meloidogyne species and races. In *Manual of agricultural nematology* (pp. 191–274). CRC Press. doi:10.1201/9781003066576-6

Eleftherianos, I., Joyce, S., Ffrench-Constant, R. H., Clarke, D. J., & Reynolds, S. E. (2010). Probing the tri-trophic interaction between insects, nematodes, and *Photorhabdus*. *Parasitology*, *137*(11), 1695–1706. doi:10.1017/S0031182010000508 PMID:20500922

Elhady, A., Giné, A., Topalovic, O., Jacquiod, S., Sørensen, S. J., Sorribas, F. J., & Heuer, H. (2017). Microbiomes associated with infective stages of root-knot and lesion nematodes in soil. *PLoS One*, *12*(5), e0177145. doi:10.1371/journal.pone.0177145 PMID:28472099

El-Saadony, M. T., Abuljadayel, D. A., Shafi, M. E., Albaqami, N. M., Desoky, E. S. M., El-Tahan, A. M., Mesiha, P. K., Elnahal, A. S. M., Almakas, A., Taha, A. E., Abd El-Mageed, T. A., Hassanin, A. A., Elrys, A. M., & Saad, A. M. (2021). Control of foliar phytoparasitic nematodes through sustainable natural materials: Current progress and challenges. *Saudi Journal of Biological Sciences*, *28*(12), 7314–7326. doi:10.1016/j.sjbs.2021.08.035 PMID:34867034

El-Saedy, M. A. M., Sandy, E. H., & Sherin, F. A. (2019). Nematicidal effect of abamectin, boron, chitosan, hydrogen peroxide and Bacillus thuringiensis against citrus nematode on Valencia orange trees. *Journal of Plant Science and Phytopathology*, *3*(3), 111–117. doi:10.29328/journal.jpsp.1001041

El-Sappah, A. H., MM, I., H. El-Awady, H., Yan, S., Qi, S., Liu, J., & Liang, Y. (2019). Tomato natural resistance genes in controlling the root-knot nematode. *Genes*, *10*(11), 925. doi:10.3390/genes10110925 PMID:31739481

Endo, B. Y. (1975). Pathogenesis of nematode-infected plants. *Annual Review of Phytopathology*, *13*(1), 213–238. doi:10.1146/annurev.py.13.090175.001241

Erez, A. (Ed.). (2013). *Temperate fruit crops in warm climates*. Springer Science & Business Media.

Escobar, C., Barcala, M., Cabrera, J. & Fenoll, C. (2015). Chapter One - Overview of Root-Knot Nematodes and Giant Cells. *Advances in Botanical Research*, (73), 1-32.

Esser, R. P. (1963). Nematode interactions in plates of no sterile water agar. *Proceedings - Soil and Crop Science Society of Florida*, *23*, 121–128.

Esser, R. P., & Sobers, E. K. (1964). Natural enemies of nematodes. *Proceedings - Soil and Crop Science Society of Florida*, *24*, 326–353.

Fan-xiang, M., Wen-ju, L., Wei, O., Yong, J., Qi, L., & Da-zhong, W. (2005). Vertical distribution of plant nematodes in an aquic brown soil under different land uses. *Journal of Forestry Research*, *16*(1), 39–42. doi:10.1007/BF02856852

FAO. Cereal Supply & Demand Brief. (2017). Available from: https://www.fao.org/worldfoodsituation/csdb/en/

FAOSTAT. (2015). *World production of potato*. Available from https://www.nationalpotatocouncil.org/files/1114/4223/8719/Pg._78_World_Potato_Production.pdf

FAOSTAT. (2020). *Citrus fruit fresh and processed – Statistical Bulletin*. Author.

FAOSTAT. (2022). *Markets and Trade*. Citrus.

Faulkner, L. R., & Bolander, W. J. (1966). Occurrence of large nematode populations in irrigation canals of South Central Washington. *Nematologica*, *12*(4), 591–600. doi:10.1163/187529266X00446

Fauzia, Jairajpuri, M. S. & Khan, Z. (1998). Biocontrol potential of *Mononchoides longicaudatus* on *Meloidogyne incognita* on tomato plants. *International Journal of Nematology*, *8*, 89–91.

Ferlian, O., Biere, A., Bonfante, P., Buscot, F., Eisenhauer, N., Fernandez, I., Hause, B., Herrmann, S., Krajinski-Barth, F., Meier, I. C., Pozo, M. J., Rasmann, S., Rillig, M. C., Tarkka, M. T., van Dam, N. M., Wagg, C., & Martinez-Medina, A. (2018). Growing Research Networks on Mycorrhizae for Mutual Benefits. *Trends in Plant Science*, *23*(11), 975–984. doi:10.1016/j.tplants.2018.08.008 PMID:30241736

Ferree, D. C., & Warrington, I. J. (Eds.). (2003). *Apples: botany, production, and uses*. CABI. doi:10.1079/9780851995922.0000

Ferreira, B. S., Santana, M., Macedo, R. S., Silva, J. O., Carneiro, M. A. C., & Rocha, M. R. (2018). Co-occurrence patterns between plant-parasitic nematodes and arbuscular mycorrhizal fungi are driven by environmental factors. *Agriculture, Ecosystems & Environment, 265*, 54–61. doi:10.1016/j.agee.2018.05.020

Ferris, H., & Bongers, T. (2006). Nematode indicators of organic enrichment. *Journal of Nematology, 38*, 3–12. PMID:19259424

Ferris, H., Zheng, L., & Walker, M. A. (2012). Resistance of Grape Rootstocks to Plant-parasitic Nematodes. *Journal of Nematology, 44*(4), 377–386. PMID:23482972

Ferris, J. M., & Ferris, V. R. (1998). Biology of Plant-Parasitic Nematodes. In K. R. Barker, G. A. Pederson, & Gl. Windham (Eds.), *Plant and Nematode Interactions* (pp. 21–35). Soil Science Society of America.

Ferris, R. V. (1968). Biometric analysis in the genus *Labronema* (Nematoda: Dorylaimida) with a description of *L. thornei* n. sp. *Nematologica, 14*(2), 276–284. doi:10.1163/187529268X00480

Ferris, V. R., & Bernard, R. L. (1971). Crop rotation effects on population densities of end parasitic nematodes. F. *Nematology, 3*, 119. PMID:19322357

Filgueiras, C. C., & Willett, D. S. (2021). Non-lethal effects of entomopathogenic nematode infection. *Scientific Reports, 11*(1), 1–8. doi:10.103841598-021-96270-2 PMID:34429457

Flegg, J. J. M. (1968). Life cycle studies of some *Xiphinema* and *Lohgidorus* species in South eastern England. *Nematologica, 14*(2), 197–210. doi:10.1163/187529268X00426

Fontana, J. D. (2011). *Biodiesel: para leitores de 9 a 90 anos.* UFPR.

Food and Agriculture Organization. (2022, December 23). *Citrus.* Https://Www.Fao.Org/Markets-and-Trade/Commodities/Citrus/En/

Forge, T. A., DeYoung, R., & Vrain, T. C. (1998). Temporal changes in the vertical distribution of Pratylenchus penetrans under raspberry. *Journal of Nematology, 30*, 179–183. PMID:19274208

Fosu-Nyarko, J., & Jones, M. G. (2016). Advances in understanding the molecular mechanisms of root lesion nematode host interactions. *Annual Review of Phytopathology, 54*(1), 253–278. doi:10.1146/annurev-phyto-080615-100257 PMID:27296144

Francl, L. J. (1993). Multivariate analysis of selected edaphic factors and their relationship to *Heterodera glycines* population density. *Journal of Nematology, 25*, 270–276. PMID:19279768

Franzener, G., Unfried, J. R., Stangarlin, J. R., & Furlanetto, C. (2005). Nematoides formadores de galhas e de cisto patogênicos a cultura da soja em municípios do Oeste do Paraná. *Nematologia Brasileira, 29*, 161–165.

Freckman, D. W., & Chapman, R. A. (1972). Infection of Red clover seedlings by *Heterodera trifolii* Goffart and *Pratylenchus penetrans* (Cobb). *Journal of Nematology, 4*(1), 23–28. PMID:19319241

Furlanetto, C., Seifert, K. E., Fenterseifer, C. E., Page, E. C., Davi, J. J. S., & Grabouski, M. M. S. (2010). Desenvolvimento das culturas de soja, milho e trigo cultivadas em áreas infestadas com o nematoide Tubixaba tuxaua no Oeste do Paraná. *Tropical Plant Pathology*, *35*, 295–302.

Galeano, M. (2002). Dinamica di popolazione di *Tylenchulus semipenetrans* e della nematofauna di vita libera nella rizosfera di agrumi in Spagna. *Supplemento Nematologia Mediterranea*, *30*, 49–53.

Gang, S. S., & Hallem, E. A. (2016). Mechanisms of host seeking by parasitic nematodes. *Molecular and Biochemical Parasitology*, *208*(1), 23–32. doi:10.1016/j.molbiopara.2016.05.007 PMID:27211240

Gan, Z., Yang, J., Tao, N., Liang, L., Mi, Q., Li, J., & Zhang, K. (2007). Cloning of the gene *Lecanicillium psalliotae* chitinase Lpchi1 and identification of its potential role in the biocontrol of root-knot nematode *Meloidogyne incognita*. *Applied Microbiology and Biotechnology*, *76*(6), 1309–1317. doi:10.100700253-007-1111-9 PMID:17665191

Gao, H., Qi, G., Yin, R., Zhang, H., Li, C., & Zhao, X. (2016). *Bacillus cereus* strain S2 shows high nematicidal activity against *Meloidogyne incognita* by producing sphingosine. *Scientific Reports*, *6*(1), 28756. doi:10.1038rep28756 PMID:27338781

Garriga, A., Morton, A., & Garcia-del-Pino, F. (2018). Is *Drosophila suzukii* as susceptible to entomopathogenic nematodes as *Drosophila melanogaster? Journal of Pest Science*, *91*(2), 789–798. doi:10.100710340-017-0920-6

Gaur, H. S. (1988). Dissemination and mode of survival of nematodes in dust storms. *Indian Journal of Nematology*, *18*(1), 94–98.

Gaur, H. S. (2004). Nematology: advances and perspectives, volume-i: nematode morphology, physiology and ecology. *Indian Journal of Nematology*, *34*(2), 240–241.

Gerber, K., & Giblin-Davis, R. M. (1990). Association of the red ring nematode and other nematode species with the palm weevil, Rhynchophorus palmarum. *Journal of Nematology*, *22*(2), 143. PMID:19287703

Ghini, R., & Morandi, M. A. B. (2006). Biotic and abiotic factors associated with soil suppressiveness to Rhizoctonia solani. *Scientia Agrícola*, *63*(2), 153–160. doi:10.1590/S0103-90162006000200007

Giannakou, I. O., & Panopoulou, S. (2019). The use of fluensulfone for the control of rootknot nematodes in greenhouse cultivated crops: Efficacy and phytotoxicity effects. *Cogent Food & Agriculture*, *5*(1), 1643819. doi:10.1080/23311932.2019.1643819

Goffart, H. (1951). *Nematoden der Kulturpflanzen Europas*. Paul Parey.

Gozel, U., & Gozel, C. (2016). Entomopathogenic Nematodes in Pest Management. In Integrated Pest Management (IPM): Environmentally Sound Pest Management. IntechOpen. https://doi.org/ doi:10.5772/63894

Grewal, P. S., De Nardo, E. A. B., & Aguillera, M. M. (2001). Entomopathogenic nematodes: Potential for exploration and use in South America. *Neotropical Entomology*, *30*, 191–205.

Grewal, P. S., Ehlers, R. U., & Shapiro-Ilan, D. I. (2005). *Nematodes as biological control agents*. CABI Publishing. doi:10.1079/9780851990170.0000

Griffin, C. T., Boemare, N. E., & Lewis, E. E. (2005). Biology and behaviour. In P. S. Grewal, R. U. Ehlers, & D. I. Shapiro-Ilan (Eds.), *Nematodes as biocontrol agents* (pp. 47–64). CABI Publishing. doi:10.1079/9780851990170.0047

Griffin, G. D., & Darling, H. M. (1964). An ecological study of *Xiphinema americanum* Cobb in an ornamental spruce nursery. *Nematologica*, *10*(3), 471–479. doi:10.1163/187529264X00493

Griffiths, B. S., Römbke, J., Schmelz, R. M., Scheffczyk, A., Faber, J. H., Bloem, J., Pérès, G., Cluzeau, D., Chabbi, A., Suhadolc, M., Sousa, J. P., Martins da Silva, P., Carvalho, F., Mendes, S., Morais, P., Francisco, R., Pereira, C., Bonkowski, M., Geisen, S., ... Stone, D. (2016). Selecting cost effective and policy-relevant biological indicators for European monitoring of soil biodiversity and ecosystem function. *Ecological Indicators*, *69*, 213–223. doi:10.1016/j.ecolind.2016.04.023

Grootaert, P., & Maertens, D. (1976). Cultivation and life cycle of *Mononchus aquaticus*. *Nematologica*, *22*(2), 173–181. doi:10.1163/187529276X00265

Gullino, M. L., Albajes, R., & Nicot, P. C. (2019). *Integrated pest and disease management in grreenhouse crops*. Springer Int Publishing., doi:10.1007/978-3-030-22304-5

Hague, N. G. M., & Gowen, S. (1987). Chemical control of nematodes. In R. H. Brown & B. R. Kerry (Eds.), *Principles and practice of nematode control in crops* (pp. 131–178). Academic Press.

Ha, J., Won, J. C., Jung, Y. H., Yang, J. W., Lee, H. U., Nam, K. J., Park, S.-C., Jeong, J. C., Lee, S.-W., Lee, D. W., Chung, J.-S., Lee, J. J., & Kim, Y. H. (2017). Comparative proteomic analysis of the response of fibrous roots of nematode-resistant and-sensitive sweet potato cultivars to root-knot nematode Meloidogyne incognita. *Acta Physiologiae Plantarum*, *39*(12), 1–11. doi:10.100711738-017-2560-0

Hajihassani, A. (2018). *Chemical nematicides for control of plant-parasitic nematodes in Georgia vegetable crops*. University of Georgia.

Hammam, M. M. A., Abd-Elgawad, M. M. M., Ruan, W., & El-bahrawy, A. (2021). Management of pests and pathogens affecting citrus yield in Egypt with special emphasis on nematodes. *Egyptian Journal of Agronematology*, *20*(2), 64–84. doi:10.21608/ejaj.2021.183231

Hannon, C. I. (1962). The occurrence and Distribution of the citrus-root nematode, *Tylenchulus semipentrans* Cobb., in Florida. *Pl. Dis. Reptr.*, *46*, 451–455.

Haougui, A., Doumma, A., Toufique, B. M., & Kollo, I. A. (2011). Survey of plant parasitic nematodes associated with sweet potato in Niger. *Asian Journal of Agricultural Sciences.*, *3*, 32–36.

Head, J., & Walters, K. F. A. (2003). Augmentation biological control utilizing the entomopathogenic nematode, Steinernema feltiae, against the South American leafminer, Liriomyza huidobrensis. *Proceeding of the first International Symposium Biological control*, 136-140.

Heald, C. M. (1994). *Citrus Declines Caused by Nematodes in Florida V. Slow Decline. University of Florida Cooperative Extension Service, Institute of Food and Agriculture Sciences*. EDIS.

Hechler, H. C. (1963). Description, developmental biology, and feeding habits of *Seinura tenuicaudata* (De Man), J.B. Goodey, 1960 (Nematoda: Aphelenchoididae), a nematode predator. *Proceedings of the Helminthological Society of Washington, 30*, 183–195.

Hechler, H. C., & Taylor, D. P. (1966). The life histories of *Seinura celeris, S. olivierae, S. oxura* and *S. steineri* (Nematoda: Aphelenchoididae). *Proceedings of the Helminthological Society of Washington, 33*, 71–83.

Helms, A. M., Ray, S., Matulis, N. L., Kuzemchak, M. C., Grisales, W., Tooker, J. F., & Ali, J. G. (2019). Chemical cues linked to risk: Cues from below-ground natural enemies enhance plant defences and influence herbivore behaviour and performance. *Functional Ecology, 33*(5), 798–808. doi:10.1111/1365-2435.13297

Hemlata, C., Bagyaraj, G., Selvakumar, G., & Sundaramc, S. P. (2015). Novel plant growth promoting rhizobacterial. App. *Soil Ecology, 95*, 38–53. doi:10.1016/j.apsoil.2015.05.011

Hermosa, R., Viterbo, A., Chet, I., & Monte, E. (2012). Plant-beneficial effects of Trichoderma and of its genes. *Microbiology, 158*(1), 17–25. doi:10.1099/mic.0.052274-0 PMID:21998166

Hill, L., & Perry, L. (2012). *The Fruit Gardener's Bible: A Complete Guide to Growing Fruits and Nuts in the Home Garden*. Storey Publishing, LLC.

Hu, C., & Qi, Y. (2013). Effective microorganisms and compost favor nematodes in wheat crops. *Agronomy for Sustainable Development, 33*(3), 573–579. doi:10.100713593-012-0130-9

Hunt, D. J., Palomares-Rius, J. E., & Manzanilla-López, R. H. (2018). Identification, morphology and biology of plant parasitic nematodes. In *Plant parasitic nematodes in subtropical and tropical agriculture* (pp. 20–61). CAB International. doi:10.1079/9781786391247.0020

Hunter, J. J., Archer, E., Van Schalkwyk, D., Strever, A. E., & Volschenk, C. G. (2016). Grapevine roots: Interaction with natural factors and agronomic practices. *Acta Horticulturae*, (1136), 63–80. doi:10.17660/ActaHortic.2016.1136.10

Hussey, R. S., Barker, K. R., & Starr, J. L. (2003). Plant-parasitic nematodes. In J. J. Burdon & J. G. Ellis (Eds.), *Plant pathology* (pp. 551–569). Blackwell Science Ltd.

Huston, M. A. (1997). Hidden treatments in ecological experiments: Re-evaluating the ecosystem function of biodiversity. *Oecologia, 110*(4), 449–460. doi:10.1007004420050180 PMID:28307235

Ingham, R., Hamm, P., Williams, R., & Swanson, W. (2000). Control of *Meloidogyne chitwoodi* in potato with fumigant and nonfumigant nematicides. *Journal of Nematology, 32*, 556–565. PMID:19271010

Inserra, R. N., & Vovlas, N. (1978). *Tylenchulus semipenetrans* on olive trees in Northern Italy. *3rd International Congress of Plant Pathology, München, GFR, 16-23 Aug., 1978. Abstracts of Papers: l'5I.*

Inserra, R. N., Duncan, L. W., O'Bannon, I. H., & Fuller, S. A. (1994). Citrus nematode biotypes and resistance of citrus rootstocks in Florida. Florida. Dept. Agric. & Consumer Serv. Div. Pl. Ind.

Inserra, R. N., Vovlas, N., & O'Bannon, J. H. (1980). A classification of *Tylenchulus semipenetrans* biotypes. *Journal of Nematology*, *12*, 283–287. PMID:19300704

Inserra, R. N., Vovlas, N., O'Bannon, J. H., & Esser, R. P. (1988). Tylenchulus graminis n. sp. and T. palustris n. sp. (Tylenchulidae) from native flora of Florida, with notes on *T. semipenetrans* and *T. furcus*. *Journal of Nematology*, *20*, 266–287. PMID:19290211

Iqbal, M. A., Ahmad, R., Mukhtar, T., & Khan, H. U. (2004). Interrelationship of citrus nematode (*Tylenchulus semipenetrans* Cobb) population and seasonal fluctuations in Sargodha district. *Pakistan Journal of Phytopathology*, *16*, 60–62.

Iqbal, M. A., Mukhtar, T., & Ahmad, R. (2007). Effect of soil types on the movement and infection of citrus nematode, *Tylenchulus semipenetrans*. *Pakistan Journal of Nematology*, *25*, 69–72.

Istiranca, R. (1940). *Narenciye Nematodu*. Seyhan Ziraat Mücadele İstasyonu.

Jaffee, B. A. (1993). Density-dependent parasitism in biological control of soil borne insects, nematodes, fungi and bacteria. *Biocontrol Science and Technology*, *3*(3), 235–246. doi:10.1080/09583159309355279

Jagdale, G. B., Casey, M. L., Grewal, P. S., & Lindquist, R. K. (2004). Effects of application rate and timing, potting medium and host plant on efficacy of Steinernema feltiae against fungus gnat, Bradysia coprophila, in floriculture. *Biological Control*, *29*(2), 296–305. doi:10.1016/S1049-9644(03)00164-6

Jain, R., Mathur, K., & Singh, R. (2007). Estimation of losses due to plant parasitic nematodes on different crops in India. *Indian Journal of Nematology*, *37*, 217–219.

Jairajpuri, M. S. (2019). *Nematode pests of rice*. CRC Press. doi:10.1201/9780429034749

Jaiswal, A. K. (Ed.). (2020). *Nutritional composition and antioxidant properties of fruits and vegetables*. Elsevier Academic Press. doi:10.1016/C2016-0-04117-7

Jatala, P., Jansson, R. K., & Raman, K. V. (2019). Sweet potato pest management. New York: CRC Press.

Jeyanthi, V., & Kanimozhi, S. (2018). Plant Growth Promoting Rhizobacteria (PGPR) - Prospective and Mechanisms: A Review. *Journal of Pure & Applied Microbiology*, *12*(2), 733–749. doi:10.22207/JPAM.12.2.34

Jonathan, E. I. (2010). Nematology: Fundamentals and applications. Pitam Pura: New India Publishing Agency.

Jonathan, E. I. (2022). *Nematology Fundamentals & Applications (2nd Revised & Enlarged Edition)*. New India Publishing Agency. doi:10.59317/9789395319461

Jones, F. G. W. (1974). Control of nematode pest, background and outlook for biological control. In *Proceedings of the 13th Symposium British on Biology in Pest and Disease Control.* Ecology Society Oxford.

Jones, D. R. (Ed.). (2018). *Handbook of diseases of banana, abaca and enset.* CABI.

Jones, J. T., Haegeman, A., Danchin, E. G. J., Gaur, H. S., Helder, J., Jones, M. G. K., Kikuchi, T., Manzanilla-López, R., Palomares-Rius, J. E., Wesemael, W. M. L., & Perry, R. N. (2013). Top 10 plant-parasitic nematodes in molecular plant pathology. *Molecular Plant Pathology, 14*(9), 946–961. doi:10.1111/mpp.12057 PMID:23809086

Kampfe, L. (1960). Die raumliche vertelung des primarbefaals von *Heterodera shachtii* Schmidt in den Wirtswurzeln. *Nematologica, 5*(1), 18–26. doi:10.1163/187529260X00208

Kanwar, R. S., Patil, J. A., & Yadav, S. (2021). Prospects of using predatory nematodes in biological control for plant parasiticnematodes—A review. *Biological Control, 160*, 104668. doi:10.1016/j.biocontrol.2021.104668

Kaplan, D. T., & O'Bannon, J. H. (1981). Evaluation and nature of citrus nematode resistance in Swingle citrumelo. *Proceedings of the Annual Meeting of the Florida State Horticultural Society, 94*, 33–36.

Karki, P. B. (1997). Status of nematode problems and research in Nepal. Diagnosis of key nematode pests of chickpea and pigeonpea and their management. *Proceedings of a Regional Training Course*, 112.

Karlen, D. L., Mausbach, M. J., Doran, J. W., Cline, R. G., Harris, R. F., & Schuman, G. E. (1997). Soil quality: A concept, definition, and framework for evaluation. *Soil Science Society of America Journal, 61*(1), 4–10. doi:10.2136ssaj1997.03615995006100010001x

Karuri, H. W., Olago, D., Neilson, R., Njeri, E., Opere, A., & Ndegwa, P. (2017). Plant Parasitic Nematode Assemblages Associated with Sweet Potato in Kenya and Their Relationship with Environmental Variables. *Tropical Plant Pathology, 42*(1), 1–12. doi:10.100740858-016-0114-4

Keçici, A. İ., Bozbuğa, R., Öcal, A., Yüksel, E., Özer, G., Yildiz, Ş., Lahlali, R., Slaats, B., Dababat, A. A., & İmren, M. (2022). Diversity and identification of plant-parasitic nematodes in wheat-growing ecosystems. *Microorganisms, 10*(8), 1534. doi:10.3390/microorganisms10081534 PMID:36013952

Kenney, E., & Eleftherianos, I. (2016). Entomopathogenic and plant pathogenic nematodes as opposing forces in agriculture. *International Journal for Parasitology, 46*(1), 13–19. doi:10.1016/j.ijpara.2015.09.005 PMID:26527129

Keshari, N., & Mallikarjun, G. (2022). Plant Parasitic Nematodes: A Major Constraint in Fruit Production. In C. Belle & T. Kaspary (Eds.), *Nematodes - Recent Advances, Management and New Perspectives* (pp. 1–33). IntechOpen. doi:10.5772/intechopen.101696

Khambay, B. P., Bourne, J. M., Cameron, S., Kerry, B. R., & Zaki, M. J. (2000). A nematicidal metabolite fromVerticillium chlamydosporium. *Pest Management Science*, *56*(12), 1098–1099. doi:10.1002/1526-4998(200012)56:12<1098::AID-PS244>3.0.CO;2-H

Khan, A., Sayed, M., Shaukat, S., & Handoo, Z. (2008). Efficacy of four plant extracts on nematodes associated with papaya in Sindh, Pakistan. *Nematologia Mediterranea*, *36*, 93–98.

Khan, M. R., Ahmad, I., & Ahamad, F. (2018). Effect of pure culture and culture filtrates of Trichoderma species on root-knot nematode, Meloidogyne incognita infesting tomato. *Indian Phytopathology*, *71*(2), 265–274. doi:10.100742360-018-0031-1

Khan, M. R., Khan, M. M., Anwer, M. A., & Haque, Z. (2012). Laboratory and field performance of some soil bacteria used as seed treatments on Meloidogyne incognita in chickpea. *Nematologia Mediterranea*, *40*, 143–151.

Khanna, K., Jamwal, V. L., Sharma, A., Gandhi, S. G., Ohri, P., Bhardwaj, R., Al-Huqail, A. A., Siddiqui, M. H., Marraiki, N., & Ahmad, P. (2019). Evaluation of the role of rhizobacteria in controlling root-knot nematode infection in *Lycopersicon esculentum* plants by modulation in the secondary metabolite profiles. *AoB Plants*, *11*(6), plz069. doi:10.1093/aobpla/plz069

Khan, Z. R., & .... (2017). Predatory Nematodes As Biological Control Agents Of Plant Parasitic Nematodes. *Frontiers in Microbiology*, *8*, 1–14.

Khan, Z., Bilgrami, A. L., & Jairajpuri, M. S. (1991). Some observations on the predation ability of *Aporcelaimellus nivalis* (Altherr, 1952) Heyns 1966 (Nematoda: Dorylaimida). *Nematologica*, *37*(1-4), 333–342. doi:10.1163/187529291X00330

Khan, Z., Bilgrami, A. L., & Jairajpuri, M. S. (1994). Attraction and food preference behaviour of predatory nematodes, *Allodorylaimus americanus* and *Discolaimus silvicolus* (Nematoda: Dorylaimida). *Indian Journal of Nematology*, *24*, 168–175.

Khan, Z., Bilgrami, A. L., & Jairajpuri, M. S. (1995a). A comparative study on predation by *Allodorylaimus americanus* and *Discolaimus silvicolus* (Nematoda: Dorylaimida) on different species of plant parasitic nematodes. *Indian Journal of Nematology*, *25*, 94–103.

Khan, Z., Bilgrami, A. L., & Jairajpuri, M. S. (1995b). Observations on the predation abilities of *Neoactinolaimus agilis* (Dorylaimida: Actinolaimoidea). *Indian Journal of Nematology*, *25*, 129–135.

Khan, Z., & Jairajpuri, M. S. (1997). Predatory behavior of *Paractinolaimus elongatus* on plant parasitic nematodes. *International Journal of Nematology*, *8*, 178–181.

Khan, Z., & Kim, Y. H. (2005). The predatory nematode, *Mononchoides fortidens* (Nematoda: Diplogasterida), suppresses the root-knot nematode, *Meloidogyne arenaria*, in potted field soil. *Biological Control*, *35*(1), 78–82. doi:10.1016/j.biocontrol.2005.05.015

Kim, E., Seo, Y., Kim, Y. S., Park, Y., & Kim, Y. H. (2017). Effects of soil textures on infectivity of root-knot nematodes on carrot. *The Plant Pathology Journal*, *33*(1), 66–74. doi:10.5423/PPJ.OA.07.2016.0155 PMID:28167889

Kim, Y.-H., & Yang, J.-W. (2019). Recent research on enhanced resistance to parasitic nematodes in sweet potato. *Plant Biotechnology Reports*, *13*(6), 559–566. doi:10.100711816-019-00557-w

Kinloch, R. A., & Allen, M. W. (1972). Interaction of *Meloidogyne hapla* and *M. javanica* infecting tomato.F. *Nematology*, *4*, 7. PMID:19319239

Kiptoo, J., Mutisya, D., Ndegwa, P., Amata, R., Irungu, L., & Godfrey, R. (2022). Influence of soil types on citrus nematode species diversity and abundance on varied ecological conditions in Kenya. *Pakistan Journal of Nematology*, *40*(1), 12–21. doi:10.17582/journal.pjn/2022/40.1.12.21

Koenning, S. R., Overstreet, C., Noling, J. W., Donald, P. A., Becker, J. O., & Fortnum, B. A. (1999). Survey of crop losses in response to phytoparasitic nematodes in the United States for 1994. *Journal of Nematology*, *31*, 587–618. PMID:19270925

Kokalis-Burelle, N. (2015). Pasteuria penetrans for control of *Meloidogyne incognita* on tomato and cucumber and *M. arenaria* on snapdragon. *Journal of Nematology*, *47*, 207–213. PMID:26527842

Koppenhöfer, A. M., Shapiro-Ilan, D. I., & Hiltpold, I. (2020). Advances in the use of entomopathogenic nematodes (EPNs) as biopesticides in suppressing crop insect pests. In *Biopesticides for sustainable agriculture*. Burleigh Dodds Science Publishing. doi:10.19103/AS.2020.0073.10

Korayem, A. M., & Hassabo, S. A. A. (2005). Citrus yield in relation to Tylenchulus semipanetrans in silty loam soil. *International Journal of Nematology*, *15*(2), 179–182.

Korthals, G. W., Alexiev, A. D., Lexmond, T. M., Kammenga, J. E., & Bongers, T. (1996). Long-term effects of copper and ph on the nematode community in an agroecosystem. *Environmental Toxicology and Chemistry*, *15*(6), 979–985. doi:10.1002/etc.5620150621

Kreiter, S. (2018). Pest management in organic grape production. In *Handbook of pest management in organic farming* (pp. 173–217). CAB International. doi:10.1079/9781780644998.0173

Krusberg, L. R. (1963). Host response to nematode infection. *Annual Review of Phytopathology*, *1*(1), 219–240. doi:10.1146/annurev.py.01.090163.001251

Kumar, K. K., & Arthurs, S. (2021). Recent advances in the biological control of citrus nematodes: A review. *Biological Control*, *157*, 104593. doi:10.1016/j.biocontrol.2021.104593

Kumar, S., & Singh, A. (2015). Biopesticides: Present Status and the Future Prospects. *Journal of Fertilizers & Pesticides*, *06*(02). Advance online publication. doi:10.4172/2471-2728.1000e129

Kumar, V., Khan, M. R., & Walia, R. K. (2020). Crop loss estimations due to plant-parasitic nematodes in major crops in India. *National Academy Science Letters*, *43*(5), 409–412. doi:10.100740009-020-00895-2

Kyndt, T., Fernandez, D., & Gheysen, G. (2014). Plant-parasitic nematode infections in rice: Molecular and cellular insights. *Annual Review of Phytopathology*, *52*(1), 135–153. doi:10.1146/annurev-phyto-102313-050111 PMID:24906129

Labaude, S., & Griffin, C. T. (2018). Transmission Success of Entomopathogenic Nematodes Used in Pest Control. *Insects*, *9*(2), 72. doi:10.3390/insects9020072 PMID:29925806

Labiadh, M., Mhamdi, B., Loulou, A., & Sadreddine, K. (2023). Impact of rhizobacteria community of citrus root on Tylenchulus semipenetrans and on Citrus plant growth. *Biocontrol Science and Technology*, *33*(3), 1–17. doi:10.1080/09583157.2023.2175785

Lahlali, R., Ezrari, S., Radouane, N., Kenfaoui, J., Esmaeel, Q., el Hamss, H., Belabess, Z., & Barka, E. A. (2022). Biological Control of Plant Pathogens: A Global Perspective. *Microorganisms*, *10*(3), 596. doi:10.3390/microorganisms10030596 PMID:35336171

Lamovšek, J., Urek, G., & Trdan, S. (2013). Biological control of root-knot nematodes (Meloidogyne spp.): Microbes against the pests. *Acta Agriculturae Slovenica*, *101*(2), 263–275. doi:10.2478/acas-2013-0022

Lasserre, F., Rivoal, R., & Cook, R. (1994). Interactions between *Heterodera avenae* and *Pratylenchus neglectus* on wheat. *Journal of Nematology*, *26*, 336–344. PMID:19279901

Lavelle, P., & Spain, A. (2001). *Soil ecology*. Kluwer Academic Publishers. doi:10.1007/978-94-017-5279-4

Lawrence, G. W., Clark, C. A., & Wright, V. L. (1986). Influence of Meloidogyne incognita on resistant and susceptible sweet potato cultivars. *Journal of Nematology*, *18*, 59–65. PMID:19294141

Layne, D., & Bassi, D. (Eds.). (2008). *The peach: botany, production and uses*. Cabi. doi:10.1079/9781845933869.0000

Le Roux, H. F., Pretorius, M. C., & Huisman, L. (2000). Citrus nematode IPM in Southern Africa. *Proc. Int. Soc. Citricult.*, *2*, 823–827.

Lewis, E. E., & Clarke, D. J. (2012). *Nematode parasites and entomopathogens*. Academic Press.

Liao, D., Wang, S., Cui, M., Liu, J., Chen, A., & Xu, G. (2018). Phytohormones Regulate the Development of Arbuscular Mycorrhizal Symbiosis. *International Journal of Molecular Sciences*, *19*(10), 3146. doi:10.3390/ijms19103146 PMID:30322086

Lilley, C. J., Maqbool, A., Wu, D., Yusup, H. B., Jones, L. M., Birch, P. R., Banfield, M. J., Urwin, P. E., & Eves-van den Akker, S. (2018). Effector gene birth in plant parasitic nematodes: Neofunctionalization of a housekeeping glutathione synthetase gene. *PLOS Genetics*, *14*(4), e1007310. doi:10.1371/journal.pgen.1007310 PMID:29641602

Linford, M. B., & Oliviera, J. M. (1937). The feeding of hollow-spear nematodes on other nematodes. *Science*, *85*(2203), 295–297. doi:10.1126cience.85.2203.295 PMID:17787453

Liu, J. S., Chen, Y., Du, C., Liu, X. X., Ma, Q. H., Zhang, X., & Wang, D. L. (2019). Interactive effects of nitrogen addition and litter on soil nematodes in grassland. *European Journal of Soil Science*, *70*(3), 697–706. doi:10.1111/ejss.12779

Liu, T., Hu, F., & Li, H. (2019). Spatial ecology of soil nematodes: Perspectives from global to micro scales. *Soil Biology & Biochemistry*, *137*, 107565. doi:10.1016/j.soilbio.2019.107565

Li, X., Zhu, H., Geisen, S., Bellard, C., Hu, F., Li, H., Chen, X., & Liu, M. (2020). Agriculture erases climate constraints on soil nematode communities across large spatial scales. *Global Change Biology*, *26*(2), 919–930. doi:10.1111/gcb.14821 PMID:31479174

Li, Z., Tao, J., & Zong, S. (2022). Cold tolerance in pinewood nematode Bursaphelenchus xylophilus promoted multiple invasion events in mid-temperate zone of China. *Forests*, *13*(7), 1100. doi:10.3390/f13071100

Lopez, E. A., Gapasin, R. M., & Palomar, M. K. (1981). Effect of different levels of *Helicotylenchus* nematode infestation on the growth and yield of sweet potato. *Annals of Tropical Research*, *3*, 275–280.

Lu, D., Macchietto, M., Chang, D., Barros, M. M., Baldwin, J., Mortazavi, A., & Dillman, A. R. (2017). Activated entomopathogenic nematode infective juveniles release lethal venom proteins. *PLoS Pathogens*, *13*(4), e1006302. doi:10.1371/journal.ppat.1006302 PMID:28426766

Luo, J., Zhang, X., Kou, X., Xie, H., Bao, X., Mahamood, M., & Liang, W. (2021). Effects of residue mulching amounts on metabolic footprints based on production and respiration of soil nematodes in a long-term no-tillage system. *Land Degradation & Development*, *32*(7), 2383–2392. doi:10.1002/ldr.3918

Macaron, J. (1972). *Contribution to the study of the plant-parasitic nematode* Tylenchulus semipenetrans *Cobb 1913 (Nematoda-Tylenchida)* [PhD Thesis]. University of Science and Technology of Languedoc.

Maegher, J. W. (1970). Seasonal fluctuations in numbers of larvae of the cereal cyst nematode (*Heterodera avenae*) and of *Pratylenchus minyus* and *Tylechorhynchus brevidens* in soil. *Nematologica*, *16*(3), 333–347. doi:10.1163/187529270X00018

Maertens, D. (1975). Observations on life cycle of *Prionchulus punctatus* (Cobb, 1917) and culture conditions. *Biologische Jaarboek Dodonaea*, *43*, 197–218.

Mahmoud, M.F. (2016). Biology and use of entomopathogenic nematodes in insect pests biocontrol, a generic view. *Cercetări Agronomice în Moldova, 49*(4), 83-103. doi:10.1515/cerce-2016-0039

Mahmoud, M. F. (2014a). Efficacy of entomopathogenic nematodes to certain Insect Pests infesting Oilseed Rapc in the Laboratory and Greenhouse. *Egyptian Journal of Biological Pest Control*, *24*, 387–391.

Mahmoud, M. F. (2014b). Virulence of entomopathogenic nematodes against the Jasmine Moth, Palpita unionals Hb. (Lepidoptera: Pyralidae). Egypt. *J. Biol. Pest. Co.*, *24*, 393–397.

Mahmoud, M. F., & Osman, M. A. M. (2007). Use of the nematode Steinernema feltiae Cross N 33 as a biological control agent against the Peach Fruit Fly Bactrocera zonata. *Tunisian Journal of Plant Protection*, *2*, 109–115.

Mahmoud, M. F., & Osman, M. A. M. (2014). Activity of Psammotermes hypostoma (Desn.) and its biocontrol using entomopathogenic nematodes. Bull. Ent. Soc. Egypt. *Econ. Ser.*, *40*, 135–147.

Mahmoud, M. F., & Shebl, M. (2014). Insect fauna of canola and phenology of the diamondback moth, Plutella xylostella L. (Lepidoptera: Plutellidae) as a key pest. *Journal of Zoology (London, England)*, *97*, 125–132.

Mangat, B. P. S., & Sharma, N. K. (1981). Influence of Host Nutrition on Multiplication and Development of Citrus Nematode. *Indian Phytopathology*, *34*, 90–91.

Mankau, R. (1980). Biological control of *Meloidogyne* populations by *Bacillus penetrans* in West Africa. *Journal of Nematology*, *12*, 23.

Mann, C., Lynch, D., Fillmore, S., & Mills, A. (2019). Relationships between field management, soil health, and microbial community composition. *Applied Soil Ecology*, *144*, 12–21. doi:10.1016/j.apsoil.2019.06.012

Marín-Buzón, C., Pérez-Romero, A., Tucci-Álvarez, F., & Manzano-Agugliaro, F. (2020). Assessing the Orange Tree Crown Volumes Using Google Maps as a Low-Cost Photogrammetric Alternative. *Agronomy (Basel)*, *10*(6), 893. doi:10.3390/agronomy10060893

Martinelli, P. R. P., Santos, J. M., & Barbosa, J. C. (2012). Efficacy of formulations containing five nematophagous fungi for the management of Pratylenchus jaehni in Citrus. *Nematologia Brasileira*, *36*, 1–8.

Martinez de Altube, M., Strauch, O., Fernandez De Castro, G., & Martinez Peña, A. (2008). Control of the flat-headed root borer Capnodis tenebrionis (Linné) (Coleoptera: Buprestidae) with the entomopathogenic nematode Steinernema carpocapsae (Weiser) (Nematoda: Steinernematidae) in a chitosan formulation in apricot orchards. *BioControl*, *53*(3), 531–539. doi:10.100710526-007-9094-0

Martínez-Medina, A., Fernandez, I., Lok, G. B., Pozo, M. J., Pieterse, C. M. J., & van Wees, S. C. M. (2017). Shifting from priming of salicylic acid- to jasmonic acid-regulated defences by *Trichoderma* protects tomato against the root knot nematode *Meloidogyne incognita*. *The New Phytologist*, *213*(3), 1363–1377. doi:10.1111/nph.14251 PMID:27801946

Martinez, O. V., Jorquera, M. A., Crowley, D. E., Gajardo, G., & Mora, M. L. (2010). Mechanisms and practical considerations involved in plant growth promotion by rhizobacterial. *Journal of Soil Science and Plant Nutrition*, *3*, 293–319.

Mckeen, C. D., & Mountain, W. B. (1960). Synergism between *Pratylenchus pentrans* (Cobb) and *Verticillium albo-atrum* R and B in eggplant wilt. Canad.f. *Bot.*, *38*, 789.

McSorley, R., & Dickson, D. W. (1990). Vertical distribution of plant-parasitic nematodes in sandy soil under soybean. *Journal of Nematology*, *22*, 90–96. PMID:19287693

Mehl, H., & Kleczewski, N. (2017). Disease and nematode management in field crops. *Pest Management Guide: Field Crops*, 456-016.

Mendoza-Mendoza, A., Zaid, R., Lawry, R., Hermosa, R., Monte, E., Horwitz, B. A., & Mukherjee, P. K. (2018). Molecular dialogues between Trichoderma and roots: Role of the fungal secretome. *Fungal Biology Reviews*, *32*(2), 62–85. doi:10.1016/j.fbr.2017.12.001

Mesa-Valle, C. M., Garrido-Cardenas, J. A., Cebrian-Carmona, J., & Miguel, T. F. M.-A. (2020). Global Research on Plant Nematodes. *Agronomy (Basel)*, *10*(8), 1148. doi:10.3390/agronomy10081148

Migunova, V. D., & Sasanelli, N. (2021). Bacteria as Biocontrol Tool against Phytoparasitic Nematodes. *Plants*, *10*(2), 389. doi:10.3390/plants10020389 PMID:33670522

Miliute, I., Buzaite, O., Baniulis, D., & Stanys, V. (2015). Bacterial endophytes in agricultural crops and their role in stress tolerance: A review. *Zemdirbyste*, *102*(4), 465–478. doi:10.13080/z-a.2015.102.060

Miller, H. N., & Di Edwardo, A. A. (1962). Leaf galls on *Siderasis fuscata* caused by the root knot nematode, *Meloiodgyne incognita incognita. Phytopathology*, *52*, 1070.

Minton, N. A., & Sayre, R. M. (1989). Suppressive influence of *Pasteuria penetrans* in Georgia soils on reproduction of *Meloidogyne arenaria. Journal of Nematology*, *21*, 574–575.

Mitiku, M. (2018). Plant-parasitic nematodes and their management: A review. *Agricultural Research & Technology: Open Access Journal*, *8*(2), 30–38. doi:10.19080/ARTOAJ.2018.16.55580

Mkandawire, T. T., Grencis, R. K., Berriman, M., & Duque-Correa, M. A. (2022). Hatching of parasitic nematode eggs: A crucial step determining infection. *Trends in Parasitology*, *38*(2), 174–187. doi:10.1016/j.pt.2021.08.008 PMID:34538735

Moebius-Clune, B. N. (2016). *Comprehensive assessment of soil health: the Cornell framework manual*. Cornell University.

Mohammed, S.H., El Saedy, M.A., Enan, M.R., Ibrahim, N.E., Ghareeb, A. and Moustafa, S.A., (2008). Biocontrol efficiency of *Bacillus thuringiensis* toxins against root-knot nematode, *Meloidogyne incognita. Journal of Cell and Molecular Biology, 7*(1), pp.57-66.

Mokrini, F., Laasli, S. E., Iraqui, D., Wifaya, A., Mimouni, A., Erginbas-Orakci, G., Imren, M., & Dababat, A. A. (2019). Distribution and occurrence of plant-parasitic nematodes associated with raspberry (*Rubus Idaeus*) in Souss-Massa region of Morocco: Relationship with soil physico-chemical factors. *Russian Journal of Nematology*, *27*, 107–121. doi:10.24411/0869-6918-2019-10011

Mountain, W. B., & Mckeen, C. D. (1962). Effect of *Verticillium dahliae* on the population of *Pratylenchus penetrans. Nematologica*, *7*(4), 261–266. doi:10.1163/187529262X00549

Mulvey, R. H. (1961). The Mononchidae: a family of predaceous nematodes Genus *Mylonchulus* (Enoplida: Mononchidae). *Canadian Journal of Zoology*, *39*(5), 665–696. doi:10.1139/z61-070

Nagachandrabose, S., & Baidoo, R. (2018). *A guide to introductory nematology*. Nova Science Publishers, Incorporated.

Naik, B. H., & Thippesh, D. (2014). *Fundamentals of Horticulture & Production Technology of Fruit Crops*. UAHS.

Nand, S., Meher, H. C., & Singh, G. (1992). Clerodendron inerme (L.) Gaertn: A new host of *Meloidogyne incognita* and Tylenchulus semipenetrans. *Indian Journal of Nematology*, *22*(2), 149.

National root and tuber crops development strategy 2019-2022. (2019). *Ministry of Agriculture, Livestock and Fisheries*. https://kilimo.go.ke/wp-content/uploads/2021/04/Roots-and-Tuber-Crops-Strategy-2019-2022.pdf

Navas, A., Nombela, G., & Bello, A. (1992). Caracterizaciǿn de la modalidad de dis-tribuciǿn de *Tylenchulus semipenetrans* en el levante español. *Nematropica*, *22*, 205–216.

Neher, D. A. (2001). Role of nematodes in soil health and their use as indicators. *Journal of Nematology*, *33*(4), 161–168. https://www.ncbi.nlm.nih.gov/pubmed/19265875 PMID:19265875

Neher, D., & Olson, R. K. (1999). Nematode communities in soils of four farm cropping management systems. *Pedobiologia*, *43*, 430–438.

Ngalimat, M. S., Mohd Hata, E., Zulperi, D., Ismail, S. I., Ismail, M. R., Mohd Zainudin, N. A. I., Saidi, N. B., & Yusof, M. T. (2021). Plant Growth-Promoting Bacteria as an Emerging Tool to Manage Bacterial Rice Pathogens. *Microorganisms*, *9*(4), 682. doi:10.3390/microorganisms9040682 PMID:33810209

Nguyen, C., Barker, B. M., Hoover, S., Nix, D. E., Ampel, N. M., Frelinger, J. A., Orbach, M. J., & Galgiani, J. N. (2013). Recent advances in our understanding of the environmental, epidemiological, immunological, and clinical dimensions of coccidioidomycosis. *Clinical Microbiology Reviews*, *26*(3), 505–525. doi:10.1128/CMR.00005-13 PMID:23824371

Nguyen, K. B., & Hunt, D. J. (2007). *Entomopathogenic nematodes: systematics, phylogeny and bacterial symbionts*. Brill. doi:10.1163/ej.9789004152939.i-816

Nguyen, S. V., Nguyen, P. K. T., Araki, M., Perry, R. N., Tran, L. B., Chau, K. M., Min, Y. Y., & Toyota, K. (2020). Seasonal variation and vertical distribution of nematode communities and their relation to soil chemical property and rice productivity under triple rice cultivation in the Mekong Delta, Vietnam. *Nematology*, *23*, 33–46. doi:10.1163/15685411-bja10028

Nico, A. I., Jimenez-Diaz, R. M., & Castillo, P. (2003). Solarization of soil in piles for the control of Meloidogyne incognita in olive nurseries in southern Spain. *Plant Pathology*, *52*(6), 770–778. doi:10.1111/j.1365-3059.2003.00927.x

Nicol, J., Turner, D., Coyne, L., den Nijs, L., Hockland, S., & Maafi, Z. (2011). Current nematode threats to world agriculture. In J. Jones, G. Gheysen, & C. Fenoll (Eds.), *Genomics and Molecular Genetics of Plant-Nematode Interactions* (pp. 21–43). Springer Science Business Media. doi:10.1007/978-94-007-0434-3_2

Nielsen, U. N., Wall, D. H., & Six, J. (2015). Soil Biodiversity and the Environment. *Annual Review of Environment and Resources*, *40*(1), 63–90. doi:10.1146/annurev-environ-102014-021257

Nigon, V. M., & Félix, M. A. (2017). History of research on C. elegans and other free-living nematodes as model organisms. *WormBook, 2017*, 1–84. https://www.ncbi.nlm.nih.gov/pubmed/28326696 PMID:28326696

Nisa, R. U., Tantray, A. Y., Kouser, N., Allie, K. A., Wani, S. M., Alamri, S. A., Alyemeni, M. N., Wijaya, L., & Shah, A. A. (2021). Influence of ecological and edaphic factors on biodiversity of soil nematodes. *Saudi Journal of Biological Sciences*, *28*(5), 3049–3059. doi:10.1016/j.sjbs.2021.02.046 PMID:34025181

Njuguna, L., & Bridge, J. (1998). Plant parasitic nematodes of Irish potatoes (Solanum tuberosum) in Central Province and sweet potatoes (Ipomoea batatas) in Central, Nyanza and Coast Provinces of Kenya. *International Journal of Nematology*, *8*, 21–26.

Norton, D. (1983). Maize nematode problems. *Plant Disease*, *67*(3), 253–256. doi:10.1094/PD-67-253

Norton, D. C., & Niblack, T. L. (2020). Biology and ecology of nematodes. In *Manual of agricultural nematology* (pp. 47–72). CRC Press. doi:10.1201/9781003066576-2

Norton, D., & Hoffmann, J. (1975). *Longidorus breviannulatus* n. sp. (Nematoda: Longidoridae) associated with stunted corn in Iowa. *Journal of Nematology*, *7*, 168–171. PMID:19308150

Noweer, E.M.A. (2018). Effect of the Nematode-Trapping Fungus Dactylaria Brochopaga and the Nematode Egg Parasitic Fungus Verticilium Chlamydosporium in Controlling Citrus Nematode Infesting Mandarin, And Interrelationship with the Co Inhabitant Fungi. *International Journal of Engineering & Technology, 7*(3.32), 19. doi:10.14419/ijet.v7i3.32.18383

Nxitywa, A., & Malan, A. P. (2023). Formulation of *Steinernema yirgalemense* in gel for long-term storage at room temperature. *Journal of Plant Diseases and Protection*, *130*(4), 809–816. doi:10.100741348-023-00764-2

O'Bannon, J. H., & Esser, R. P. (1985). *Citrus declines caused by nematodes in Florida. II. Physiological races. Nematology Circular No. 116*. Florida Department of Agriculture and Consumer Services.

O'Bannon, J. H., Radewald, J. D., & Tomerlin, A. T. (1972). Population fluctuation of three parasitic nematodes in Florida citrus. *Journal of Nematology*, *4*, 194–199. PMID:19319266

O'Bannon, J. H., Reynolds, H. W., & Leathers, C. R. (1966). Effects of temperature on penetration, development, and reproduction of *Tylenchulus semipenetrans*. *Nematologica*, *12*(4), 483–487. doi:10.1163/187529266X00301

Oka, Y., Koltai, H., Bar-Eyal, M., Mor, M., Sharon, E., Chet, I., & Spiegel, Y. (2000). New strategies for the control of plant-parasitic nematodes. *Pest Management Science: Formerly. Pest Management Science*, *56*(11), 983–988. doi:10.1002/1526-4998(200011)56:11<983::AID-PS233>3.0.CO;2-X

Oka, Y., Tkachi, N., Shuker, S., Rosenberg, R., Suriano, S., & Fine, P. (2006). Laboratory Studies on the Enhancement of Nematicidal Activity of Ammonia-releasing Fertilizers by Alkaline Amendments. *Nematology*, *8*(3), 335–346. doi:10.1163/156854106778493466

Okechalu, O. B., & Wonang, D. L. (2015). The response of eleven sweet potato (*Ipomoea batatas* (L.) Lam) cultivars to infection by *Meloidogyne* spp. in Jos, Nigeria. *IOSR Journal of Pharmacy and Biological Sciences*, *10*, 42–49.

Okeniyi, M., Afolami, S., Fademi, O., & Oduwaye, O. (2013). Effect of botanical extracts on root-knot nematode (*Meloidogyne incognita*) infection and growth of cashew (*Anacardium occidentale*) seedlings. *Academia Journal of Biotechnology*, *1*, 81–86.

Ollat, N., Peccoux, A., Papura, D., Esmenjaud, D., Marguerit, E., Tandonnet, J.-P., … Delrot, S. (2016). Rootstocks as a component of adaptation to environment. *Grapevine in a Changing Environment*, 68–108. doi:10.1002/9781118735985.ch4

Olthof, T. H. A. (1971). Seasonal fluctuations in population densities of *Pratylechus pentrans* under a rye-tobacco rotation in Ontario. *Nematologica*, *17*(3), 453–459. doi:10.1163/187529271X00693

Ono, M., & Yoshiga, T. (2019). Cellular immunity in the insect *Galleria mellonella* against insect non-parasitic nematodes. *Parasitology*, *146*(6), 708–715. doi:10.1017/S003118201800210X PMID:30567609

Orf, O. M. H., Soad, Y. S. E.-S., & Sahar, H. A.-B. (2019). Impact Of Plant Growth Promoting Rhizobacteria As Bio-Control Agents On Citrus Nematode, *Tylenchulus Semipenetrans* Infecting Balady Orange (*Citrus Sinensis* L.) And Improving Its Productivity. *The Future of Agriculture*, (4), 73–87.

Osei, K., Gowen, S., Pembroke, B., Brandenburg, R., & Jordan, D. (2010). Potential of leguminous cover crops in management of a mixed population of root-knot nematodes (*Meloidogyne* spp.). *Journal of Nematology*, *42*, 173–178. PMID:22736854

Overstreet, C., Loebenstein, G., & Thottappilly, G. (2009). The sweet potato. Dordrecht: Springer.

Ozakman, Y., & Eleftherianos, I. (2021). Nematode infection and antinematode immunity in Drosophila. *Trends in Parasitology*, *37*(11), 1002–1013. doi:10.1016/j.pt.2021.06.001 PMID:34154933

Ozarslandan, A., & Tanriver, E. (2018). Evaluation of some stone fruit rootstocks against resistance to root knot nematode (Meloidogyne incognita). *International Journal of Agricultural and Natural Science*, *1*, 137–141.

Palomares-Rius, J. E., Escobar, C., Cabrera, J., Vovlas, A., & Castillo, P. (2017). Anatomical alterations in plant tissues induced by plant-parasitic nematodes. *Frontiers in Plant Science*, *8*, 1987. doi:10.3389/fpls.2017.01987 PMID:29201038

Pandey, P. (2022). A Review on Major Diseases of Citrus in Nepal and their Management. *Journal of Tikapur Multiple Campus*, *5*(June), 152–167. doi:10.3126/jotmc.v5i1.46522

Pandey, R. K., Goswami, B. K., & Singh, S. (2004). Influence of soil pH on population dynamics of *Tylenchulus semipenetrans* infesting citrus and its bio-management using VAM fungi. *International Journal of Nematology*, *14*, 174–176.

Parkman, J. P., Hudson, G. W., Frank, J. H., Nguyen, K. B., & Smart, J. Jr. (1993). Establishment and persistence of Steinernema scapterisci (Rhabditida: Steinernematidae) in field populations of Scapteriscus spp. Mole crickets (Orthoptera: Gryllotalpidae). *Journal of Entomological Science*, *28*(2), 182–190. doi:10.18474/0749-8004-28.2.182

Parkman, J. P., & Smart, J. G. C. Jr. (1996). Entomopathogenic nematodes, a case study: Introduction of Steinernema scapterisci in Florida. *Biocontrol Science and Technology*, *6*(3), 413–419. doi:10.1080/09583159631389

Parvatha Reddy, P., Khan, R. M., & Rao, M. S. (1991). Integrated management of the citrus nematode *Tylenchulus semipenetrans* using oil-cakes and *Paecilomyces lilacinus*. *Afro-Asian Journal of Nematology*, *1*, 221–222.

Paschapur, A. U., Lakshmi, K. V., & Pawar, B. S. V. (2018). Survey for the presence of entomopathogenic nematodes (EPNs) in sugarcane growing regions of Telangana state. *Journal of Entomology and Zoology Studies*, *6*(6), 320–326.

Peng, D., Chai, L., Wang, F., Zhang, F., Ruan, L., & Sun, M. (2011). Synergistic activity between Bacillus thuringiensis Cry6Aa and Cry55Aa toxins against *Meloidogyne incognita*. *Microbial Biotechnology*, *4*(6), 794–798. doi:10.1111/j.1751-7915.2011.00295.x PMID:21923640

Pérez,, E, Weingartner, D, & McSorley,, R. (2000). Distribution of *Paratrichodorus minor* and *Belonolaimus longicaudatus* following fumigation on potato and cabbage. *Journal of Nematology*, *32*, 343–348. PMID:19270987

Pérez, E., Weingartne, D., & McSorley, R. (2000). Niche distribution of *Paratrichodorus minor* and *Belonolaimus longicaudatus* following fumigation on potato and cabbage. *Journal of Nematology*, *32*, 343–348. PMID:19270987

Peterson, H., & Luxton, M. (1982). A comparative analysis of soil fauna populations and their role in decomposition processes. *Oikos*, *39*(3), 287–388. doi:10.2307/3544689

Pettigrew, W. T., Meredith, W. R. Jr, & Young, L. D. (2005). Potassium Fertilization Effects on Cotton Lint Yield, Yield Components, and Reniform Nematode Populations. *Agronomy Journal*, *7*(4), 1245–1251. doi:10.2134/agronj2004.0321

Phani, V., Khan, M. R., & Dutta, T. K. (2021). Plant-parasitic nematodes as a potential threat to protected agriculture: Current status and management options. *Crop Protection (Guildford, Surrey)*, *144*, 105573. doi:10.1016/j.cropro.2021.105573

Philis, J. (1989). Yield loss assessment caused by the citrus nematode *Tylenchulus semipenetrans* on Valencia oranges in Cyprus. *Nematologia Mediterranea*, *17*, 5–6.

Poinar, G. O., Jr. (1990). Taxonomy and biology of Steinernematidae and Heterorhabditidae. In Entomopathogenic nematodes in biological control. CRC.

Pollock, M. (Ed.). (2012). *Fruit & Vegetable Gardening*. DK Publishing.

Poveda, J. (2020). Trichoderma parareesei Favors the Tolerance of Rapeseed (Brassica napus L.) to Salinity and Drought Due to a Chorismate Mutase. *Agronomy (Basel)*, *10*(1), 118. doi:10.3390/agronomy10010118

Poveda, J., Abril-Urias, P., & Escobar, C. (2020). Biological Control of Plant-Parasitic Nematodes by Filamentous Fungi Inducers of Resistance: Trichoderma, Mycorrhizal and Endophytic Fungi. *Frontiers in Microbiology*, *11*, 992. Advance online publication. doi:10.3389/fmicb.2020.00992 PMID:32523567

Prasad, S. K., & Chawla, M. L. (1965). Observation on the population fluctuation of citrus nematode, *Tylenchulus semipenetrans* Cobb, 1913. *Indian Journal of Entomology*, *27*, 450–454.

Ptatscheck, C., Gansfort, B., & Traunspurger, W. (2018). The extent of wind-mediated dispersal of small metazoans, focusing nematodes. *Scientific Reports*, *8*(1), 6814. doi:10.103841598-018-24747-8 PMID:29717144

Pudasaini, M. P., Schomaker, C. H., Been, T. H., & Moens, M. (2006). Vertical distribution of the plant-parasitic nematode, *Pratylenchus penetrans*, under four field crops. *Phytopathology*, *96*(3), 226–233. doi:10.1094/PHYTO-96-0226 PMID:18944436

Qing, X., Wang, M., Karssen, G., Bucki, P., Bert, W., & Braun-Miyara, S. (2020). PPNID: A reference database and molecular identification pipeline for plant-parasitic nematodes. *Bioinformatics (Oxford, England)*, *36*(4), 1052–1056. doi:10.1093/bioinformatics/btz707 PMID:31529041

Rajinikanth, R., Sreenivasa, R. M., & Pavani, K. V. (2016). Evaluation of potential bio-control agents on root-knot nematode Meloidogyne incognita and wilt causing fungus Fusarium oxysporum f.sp. conglutinans in vitro. *African Journal of Biotechnology*, *15*(19), 798–805. doi:10.5897/AJB2015.15235

Rama, K., & Dasgupta, M. K. (1998). Biocontrol of nematodes associated with mandarin orange decline by the promotion of predatory nematode *Iotonchus tenuicaudatus* (Kreis, 1924). *Indian Journal of Nematology*, *28*, 118–124.

Rashidifard, M., Shokoohi, E., Hoseinipour, A., & Jamali, S. (2015). *Tylenchulus semipenetrans* (Nematoda: Tylenchulidae) on pomegranate in Iran. *Australasian Plant Disease Notes, Australasian Plant Pathology Society*, *10*(1), 2. doi:10.100713314-014-0157-7

Ravari, S. B., & Moghaddam, E. M. (2015). Efficacy of Bacillus thuringiensis Cry14 toxin against root knot nematode, *Meloidogyne javanica. Plant Protection Science*, *51*(1), 46–51. doi:10.17221/93/2013-PPS

Rebois, V. R., & Cairns, E. J. (1968). Nematodes associated with soybeans in Alabama, Florida and Georgia. *Pl.Dis.Rptr.*, *52*, 40.

Reddy, P. (2021). Nematode Diseases of Crops and their Management. In Nematode Diseases of Crop Plants: An Overview. doi:10.1007/978-981-16-3242-6_1

Reddy, P. P. (2021). *Nematode diseases of crops and their management*. Springer Singapore. doi:10.1007/978-981-16-3242-6

Rehab, Elnouby, Hasan, Ticu, Popa, Bungau, & Hafez. (2019). New Trend for Using the Reduced Graphene Oxide as Effective and Eco-friendly Nematicide. *Materiale Plastice*, *56*(1).

Reitz, M., Oger, P., Meyer, A., Niehaus, K., Farrand, S. K., Hallman, J., Sikora, R., & Richard, S. A. (2002). Importance of the Oantigen, core-region and lipid A of rhizobial induction of systemic resistance in potato to *Globodera pallida*. *Nematology*, *4*(1), 73–79. doi:10.1163/156854102760082221

Rijal, J., & Bergh, J. (2016). Food-Finding Capability of Grape Root Borer (Lepidoptera: Sesiidae) Neonates in Soil Column Bioassays. *Journal of Entomological Science, 51*. . doi:10.18474/JES15-21.1

Riley, T., & Reardon, T. (1995). Isolation and characterization of *Clavibacter tritici* associated with *Anguina tritici* in wheat from Western Australia. *Plant Pathology*, *44*(5), 805–810. doi:10.1111/j.1365-3059.1995.tb02739.x

Risède, J. M., & Simoneau, P. (2004). Pathogenic and genetic diversity of soilborne isolates of Cylindrocladium from banana cropping systems. *European Journal of Plant Pathology*, *110*(2), 139–154. doi:10.1023/B:EJPP.0000015337.54178.c0

Rishbeth, J. (1963). Stump protection against *Fomes annosus*.3.Inoculation with *Peniophora gigantean*. *Annals of Applied Biology*, *52*(1), 63–77. doi:10.1111/j.1744-7348.1963.tb03728.x

Robinson, A. F., Akridge, R., Bradford, J. M., Cook, C. G., Gazaway, W. S., Kirkpatrick, T. L., Lawrence, G. W., Lee, G., McGawley, E. C., Overstreet, C., & Padgett, B. (2005). Vertical distribution of *Rotylenchulus reniformis* in cotton fields. *Journal of Nematology*, *37*, 265–271. PMID:19262871

Robinson, J. C., & Saúco, V. G. (2010). *Bananas and plantains* (Vol. 19). Cabi. doi:10.1079/9781845936587.0000

Roccuzzo, G., Ciancio, A., & Lo Giudice, V. (1992). Some observations on the ecology of the citrus nematode *Tylenchulus semipenetrans* Cobb in Southern Italy. *Proceedings of the VII International Citrus Congress*, *3*, 950–952.

Rodriguez-Kábana, R. (1986). Organic and Inorganic Nitrogen Amendments to Soil as Nematode Suppressants. *Journal of Nematology*, *18*, 129–135. PMID:19294153

Rodriguez-Kabana, R., & Robertson, D. (1987). Vertical distribution of *Meloidogyne arenaria* juvenile populations in a peanut field. *Nematropica*, *17*, 199–208.

Romanenko, N. D., Zaets, V. G., Kozereva, N. I., Popov, I. O., & Tabolin, S. B. (2008). Biological agents for plant protection against plant parasitic nematodes and other pathogens, and perspectives of their use in the 21st century. *RUDN J Agronomy Animal Industries.*, *2*, 39–51.

Romig, D. E., Garlynd, M. J., Harris, R. F., & Mcsweeney, K. (1995). How farmers assess soil health and quality. *Journal of Soil and Water Conservation*, *50*(3), 229–236.

Ross, J. P. (1960). *Heterodera trifolii* a foliage pathogen of white clover. *Phytopathogy*, *50*, 866.

Roussos, P. A., Denaxa, N.-K., Tsafouros, A., Efstathios, N., & Intidhar, B. (2016). Apricot (Prunus armeniaca L.). *Nutritional Composition of Fruit Cultivars*, 19–48. doi:10.1016/B978-0-12-408117-8.00002-7

Rueda-Ramírez, D., Palevsky, E., & Ruess, L. (2023). Soil Nematodes as a Means of Conservation of Soil Predatory Mites for Biocontrol. *Agronomy (Basel)*, *13*(1), 32. doi:10.3390/agronomy13010032

Ruiu, L. (2020). Plant-Growth-Promoting Bacteria (PGPB) against Insects and Other Agricultural Pests. *Agronomy (Basel)*, *10*(6), 861. doi:10.3390/agronomy10060861

Saadi, I., De Luca, F., Fanelli, E., Troccoli, A., Greco, N., & Benbouza, H. (2019). Morpho-molecular characterization of Ditylenchus gigas and D. oncogenus parasitizing broad bean, Vicia faba, in Algeria. *European Journal of Plant Pathology*, *155*(2), 505–513. doi:10.100710658-019-01784-1

Saeed, M., Mukhtar, T., & Rehman, M. A. (2019). Temporal fluctuations in the population of citrus nematode (*Tylenchulus semipenetrans*) in the Pothowar region of Pakistan. *Pakistan Journal of Zoology*, *51*(6), 2257–2263. doi:10.17582/journal.pjz/2019.51.6.2257.2263

Sahebani, N., & Hadavi, N. (2008). Biological control of the root-knot nematode Meloidogyne javanica by Trichoderma harzianum. *Soil Biology & Biochemistry*, *40*(8), 2016–2020. doi:10.1016/j.soilbio.2008.03.011

Salahi Ardakani, A., Tanha Mafi, Z., Mokaram Hesar, A., & Mohammadi Goltappeh, E. (2014). Relationship between soil properties and abundance of *Tylenchulus semipenetrans* in citrus orchards, Kohgilouyeh va Boyerahmad Province. *J. Agric. Sci. Technol.*, *16*, 1699–1710.

Samak, A. A., & Abo-Korah, M. S. (2021). Impact of Irrigation Systems and Applied Water on Crop Yield, WUE, Certain Nematicides Efficiency, and Plant-Parasitic Nematodes Activity Infected Tomato Plants. *Egyptian Academic Journal of Biological Sciences. F, Toxicology and Pest Control*, *13*(2), 301–316. doi:10.21608/eajbsf.2021.244427

Santana, H. P. E., Comerlato, A. P., Nasu, E. G. C., & Furlanetto, C. (2009). Variabilidade genética em populações de campo do nematoide de cisto da soja provenientes dos estados do Paraná e Rio Grande do Sul. *Tropical Plant Pathology*, *34*(4), 261–264. doi:10.1590/S1982-56762009000400010

Santo, G., O'Bannon, J., Finley, A., & Golden, A. (1980). Occurrence and host range of a new root-knot nematode (*Meloidogyne chitwoodi*) in the Pacific Northwest. *Plant Disease*, *64*(10), 951–952. doi:10.1094/PD-64-951

Sasser, J., & Freckman, D. (1987). A world perspective on nematology: the role of the society. In J. A. Veech & D. W. Dickson (Eds.), *Vistas on nematology* (pp. 7–14). Society of Nematologists.

Sasser, J., & Freckman, D. (1987). A world perspective on nematology: The role of the society. In J. Veech & D. Dickson (Eds.), *Vistas on Nematology* (pp. 7–14). Society of Nematologists, Inc.

Schmitt, D. P., & Barker, K. R. (1981). Damage and reproduction potentials of *Pratylenchus brachuyrus* and *P. penetrans* on soybean. *Journal of Nematology*, *13*(3), 327–332. PMID:19300771

Schoonbeek, H., van Nistelrooy, J. G., & de Waard, M. A. (2003). Functional Analysis of ABC Transporter Genes from *Botrytis cinerea* Identifies BcatrB as a Transporter of Eugenol. *European Journal of Plant Pathology*, *109*(9), 1003–1011. doi:10.1023/B:EJPP.0000003936.61182.14

Schouteden, N., de Waele, D., Panis, B., & Vos, C. M. (2015). Arbuscular Mycorrhizal Fungi for the Biocontrol of Plant-Parasitic Nematodes: A Review of the Mechanisms Involved. *Frontiers in Microbiology*, *6*. Advance online publication. doi:10.3389/fmicb.2015.01280 PMID:26635750

Schurig, J., Ipach, U., Hahn, M., & Winterhagen, P. (2021). Evaluating nematode resistance of grapevine rootstocks based on Xiphinema index reproduction rates in a fast screening assay. *European Journal of Plant Pathology*, *160*(1), 233–238. doi:10.100710658-021-02227-6

Seesao, Y., Gay, M., Merlin, S., Viscogliosi, E., Aliouat-Denis, C. M., & Audebert, C. (2017). A review of methods for nematode identification. *Journal of Microbiological Methods*, *138*, 37–49. . doi:10.1016/j.mimet.2016.05.030 PMID:27262374

Sehgal, M., & Gaur, H. S. (1999). *Important nematode problem of crops in India and their management*. National Centre for Integrated Pest Management.

Seinhorst, J. W. (1965). The relation between nematode density and damage to plants. *Nematologica*, *11*(1), 137–154. doi:10.1163/187529265X00582

Seinhorst, J. W. (1986). Effects of Nematode Attack on the Growth and Yield of Crop Plants. In *Cyst Nematodes* (pp. 191–209). Springer US. doi:10.1007/978-1-4613-2251-1_11

Shamseldean, M. M., Hasanain, S. A., & Rezk, M. Z. A. 2009. Virulence of entomopathogenic nematodes against lepidopterous pests of horticultural crops in Egypt. *4th Conference on recent technologies in agriculture*, 74-84.

Shamseldean, M. S. M., Abd-Elbary, N. A., Shalaby, H., & Ibraheem, H. I. H. (2014). Entomopathogenic nematodes as biocontrol agents of the tomato leaf miner Tuta absoluta (Meyrick) (Lepidoptera: Gelechiidae) on tomato plants. Egypt. *J. Biol. Pest. Co.*, *24*, 503–513.

Shan, S., Wang, W., Song, C., Wang, M., Sun, B., Li, Y., Fu, Y., Gu, X., Ruan, W., & Rasmann, S. (2019). The symbiotic bacteria Alcaligenes faecalis of the entomopathogenic nematodes *Oscheius* spp. Exhibit potential biocontrol of plant- and entomopathogenic fungi. *Microbial Biotechnology*, *12*(3), 459–471. doi:10.1111/1751-7915.13365 PMID:30618110

Shapiro-Ilan, D. I., Hiltpold, I., & Lewis, E. E. (2017). Nematodes. In Ecology of Invertebrate Diseases (pp. 415–440). John Wiley & Sons, Ltd. doi:10.1002/9781119256106.ch11

Shapiro-Ilan, D. I., Bruck, D. J., & Lacey, L. A. (2012). Principles of epizootiology and microbial control. *Insect Pathology*, *2*, 29–72. doi:10.1016/B978-0-12-384984-7.00003-8

Shapiro-Ilan, D. I., Cottrell, T. E., Mizell, R. F. III, Horton, D. L., Behle, R. W., & Dunlap, C. A. (2010). Efficacy of *Steinernema carpocapsae* for control of the lesser peachtree borer, *Synanthedon pictipes*: Improved aboveground suppression with a novel gel application. *Biological Control, 54*(1), 23–28. doi:10.1016/j.biocontrol.2009.11.009

Shapiro-Ilan, D. I., Duncan, L. W., Lacey, L. A., & Han, R. (2005). Orchard crops. In P. Grewal, R.-U. Ehlers, & D. Shapiro-Ilan (Eds.), *Nematodes as biological control agents* (pp. 215–230). CABI. doi:10.1079/9780851990170.0215

Shapiro-Ilan, D. I., & Gaugler, R. (2002). Production technology for entomopathogenic nematodes and their bacterial symbionts. *Journal of Industrial Microbiology & Biotechnology, 28*(3), 137–146. doi:10.1038j.jim.7000230 PMID:12074087

Shapiro-Ilan, D. I., & Gaugler, R. (2010). Nematodes: Rhabditida: Steinernematidae & Heterorhabditidae. In A. Shelton (Ed.), *Biological Control: A Guide to Natural Enemies in North America*. Cornell University.

Shapiro-Ilan, D. I., Gouge, D. H., Piggott, S. J., & Fife, J. P. (2006). Application technology and environmental considerations for use of entomopathogenic nematodes in biological control. *Biological Control, 38*(1), 124–133. doi:10.1016/j.biocontrol.2005.09.005

Shapiro-Ilan, D. I., Han, R., & Qiu, X. (2014). Production of Entomopathogenic Nematodes. In J. A. Morales-Ramos, M. G. Rojas, & D. I. Shapiro-Ilan (Eds.), *Mass Production of Beneficial Organisms* (pp. 321–355). Academic Press. doi:10.1016/B978-0-12-391453-8.00010-8

Sharma, G. C., & Kashyap, A. S. (2009). Effect of different intercrops on the nematode populations and yield of apricot var. New Castale. *Indian Journal of Horticulture, 66*, 420–421.

Sharma, G. C., Thakur, B. S., & Kashyap, A. S. (2005). Impact of NPK on the Nematode Populations and Yield of Plum (*Prunus salicina*). *Acta Horticulturae*, (696), 433–436. doi:10.17660/ActaHortic.2005.696.77

Sharma, S., Waliyar, F., Subrahmanyam, P., & Ndunguru, B. (1992). Role of *Scutellonema clathricaudatum* in etiology of groundnut growth variability in Niger. *Plant and Soil, 143*(1), 133–139. doi:10.1007/BF00009138

Sharmishtha, T., & Sharma M. K. (2020). Population Dynamics of Horizontal and Vertical Population of Root-Knot Nematode, Meloidogyne incognita under Protected Cultivation. *International Journal of Current Microbiology and Applied Sciences*, (11), 329-336.

Shokoohi, E., & Duncan, L. W. (2018). Nematode parasites of citrus. In R. Sikora, P. Timper, & D. Coyne (Eds.), *Plant-parasitic nematodes in tropical & subtropical agriculture* (3rd ed., pp. 446–476). CAB International. doi:10.1079/9781786391247.0446

Siddiqi, M. R. (1996). Occurrence of the citrus nematode, *Tylenchulus semipenetrans* Cobb, 1913, and the reniform nematode, *Rotylenchulus reniformis* in India (Abstr.). *Proceedings of 48th Indian Science Congress,* 504.

Siddiqi, M. (2000). *Topley & Wilson's Microbiology & Microbial Infections: Parasitology* (10th ed.). Arnold Publishers Ltd.

Siddiqui, M. A. (2007). Seasonal fluctuation in nematode population associated with mango, *Mangifera indica* L. *Archiv für Phytopathologie und Pflanzenschutz, 40*(6), 389–394. doi:10.1080/03235400600627601

Sikandar, A., Zhang, M. Y., Wang, Y. Y., Zhu, X. F., Liu, X. Y., Fan, H. Y., Xuan, Y. H., Chen, L. J., & Duan, Y. X. (2020). Review article: *Meloidogyne incognita* (root-knot nematode) a risk to agriculture. *Applied Ecology and Environmental Research, 18*(1), 1679–1690. doi:10.15666/aeer/1801_16791690

Sikandar, A., Zhang, M., Wang, Y., Zhu, X., Liu, X., Fan, H., Xuan, Y., Chen, L., & Duan, Y. (2020). In vitro evaluation of *Penicillium chrysogenum* Snef1216 against *Meloidogyne incognita* (root-knot nematode). *Scientific Reports, 10*(1), 8342. doi:10.103841598-020-65262-z PMID:32433576

Sikora, R. A., & Roberts, P. A. (2018). Management practices: an overview of integrated nematode management technologie. *Plant parasitic nematodes in subtropical and tropical agriculture*, 795-838.

Sikora, R., & Fernandez, E. (2005). Nematode parasites of vegetables. In Plant parasitic nematodes in subtropical and tropical. Wallingford: CABI. doi:10.1079/9780851997278.0319

Sikora, R., Coyne, D., Hallman, J., & Timper, P. (2018). Plant Parasitic Nematodes—The World's Most Important Crop pathogen? *CABIBlog*. Available online: https://blog.cabi.org/2018/07/24/plant-parasitic-nematodes/

Silva, M. S. O., Cardoso, J. F. M., Ferreira, M. E. P., Baldo, F. B., Silva, R. S. A., Chacon-Orozco, J. G., Shapiro-Ilan, D. I., Hazir, S., Bueno, C. J., & Leite, L. G. (2021). An Assessment of *Steinernema rarum* as a Biocontrol Agent in Sugarcane with Focus on *Sphenophorus levis*, Host-Finding Ability, Compatibility with Vinasse and Field Efficacy. *Agriculture, 11*(6), 6. Advance online publication. doi:10.3390/agriculture11060500

Singh, B. (1999). Occurrence and distribution of *Tylenchulus semipenetrans* in Nagpur mandarin orchards of Vidarbha, Maharashtra. *Indian J. Nematol, 29*, 149–153.

Singh, S., Singh, B., & Singh, A. P. (2015). Nematodes: A threat to sustainability of agriculture. *Procedia Environmental Sciences, 29*, 215–216. doi:10.1016/j.proenv.2015.07.270

Sinha, N. K., Sidhu, J., Barta, J., Wu, J., & Cano, M. P. (Eds.). (2012). *Handbook of fruits and fruit processing* (2nd ed.). John Wiley & Sons. doi:10.1002/9781118352533

Small, R. W. (1979). The effects of predatory nematodes on populations of plant parasitic nematodes in pots. *Nematologica, 25*(1), 94–103. doi:10.1163/187529279X00424

Smith, B. P., Morales, N. B., Thomas, M. R., Smith, H. M., & Clingeleffer, P. R. (2016). Grapevine rootstocks resistant to the root-knot nematodeMeloidogyne javanica. *Australian Journal of Grape and Wine Research, 23*(1), 125–131. doi:10.1111/ajgw.12242

Somvanshi, V. S., Tathode, M., Shukla, R. N., & Rao, U. (2018). Nematode Genome Announcement: A draft genome for Rice root-knot nematode. *Journal of Nematology*, *50*(2), 111–116. doi:10.21307/jofnem-2018-018 PMID:30451432

Song Z., Chen, X. Z. & Huang T. Z. (1994). Studies on soil micro-organisms and biochemical activities of Chinese fir woodland infected by *Tylenchulus semipenetrans* II. Soil enzyme activity and soil chemical properties. *Journal of Fujian College of Forestry*, *14*(1), 81-85.

Song, D., Tariq, A., Pan, K., Chen, W., Zhang, A., Sun, X., Ran, Y., & Zeng, F. (2020). Effects of straw mulching practices on soil nematode communities under walnut plantation. *Scientific Reports*, *10*(1), 1–10. doi:10.103841598-020-72530-5 PMID:32948816

Soriano, I., Prot, J., & Matias, D. (2000). Expression of tolerance for *Meloidogyne graminicola* in rice cultivars as affected by soil type and flooding. *Journal of Nematology*, *32*, 309–317. PMID:19270982

Sorribas, F. J., Verdejo-Lucas, S., Forner, J. B., Alcaide, A., Pons, J., & Ornat, C. (2000). Seasonality of *Tylenchulus semipenetrans* Cobb and *Pasteuria* sp., in citrus orchards of spain. *Journal of Nematology*, *32*, 622–632. PMID:19271018

Sorribas, F. J., Verdejo-Lucas, S., Galeano, M., Pastor, J., & Ornat, C. (2003). Effect of 1,3-dichloropropene and rootstocks alone and in combination on *Tylenchulus semipenetrans* and citrus tree growth in a replant management program. *Nematropica*, *34*, 149–158.

Sorribas, F. J., Verdejo-Lucas, S., Pastor, J., Ornat, C., Pons, J., & Valero, J. (2008). Population densities of *Tylenchulus semipenetrans* related to physicochemical properties of soil and yield of clementine mandarin in Spain. *Plant Disease*, *92*(3), 445–450. doi:10.1094/PDIS-92-3-0445 PMID:30769676

Southwood, T. R. E. (1968). Ecological methods. Methuen.

Steiner, G., & Heinly, H. (1922). The possibility of control of *Heterodera radicicola* and other plant injurious nematodes by means of predatory nemas, especially *Mononchus aquaticus*. *Journal of the Washington Academy of Sciences*, *12*, 367–385.

Stirling, G. (1991). *Biological Control of Plant-parasitic Nematodes*. CAB International.

Stirling, G. R. (2014). *Biological control of plant-parasitic nematodes: soil ecosystem management in sustainable agriculture*. CABI. doi:10.1079/9781780644158.0000

Stirling, G. R. (2018). Biological control of plant-parasitic nematodes. In *Diseases of nematodes* (pp. 103–150). CRC Press.

Stirling, G. R. (2021). Surrounding the swollen roots of sweet potato with a decomposing band of an organic amendment enhances nematode-suppressive services and reduces damage caused by root-knot nematode. *Australasian Plant Pathology*, *50*(2), 151–168. doi:10.100713313-020-00751-3

Stirling, G. R., Stirling, A. M., & Prichard, M. (2020). Sustainable sweet potato farming systems to improve soil health and reduce losses caused by root-knot nematode. *Australasian Plant Pathology*, *49*(6), 591–604. doi:10.100713313-020-00731-7

Storey, S. G., Malan, A. P., & Hugo, H. J. (2017). Nematode Pests of Grapevine. In H. Fourie, V. Spaull, R. Jones, M. Daneel, & D. De Waele (Eds.), *Nematology in South Africa: A View from the 21st Century*. Springer. doi:10.1007/978-3-319-44210-5_14

Subbotin, S.A. (2019). *Plant Pathogenic Nematode Species: Identification Keys to Genera from Russia and Adjacent Countries.* Springer International Publishing AG.

Subbotin, S.A. (2020). *Plant Pathogenic Nematode Species: Identification Keys to Genera from Russia and Adjacent Countries.* Springer International Publishing AG.

Subbotin, S. A., Madani, M., Krall, E., Sturhan, D., & Moens, M. (2005). Molecular diagnostics, taxonomy, and phylogeny of the stem nematode Ditylenchus dipsaci species complex based on the sequences of the internal transcribed spacer-rDNA. *Phytopathology*, *95*(11), 1308–1315. doi:10.1094/PHYTO-95-1308 PMID:18943362

Sweelam, M. E., Abokorah, M. S., & Elsebaay, W. A. (2022). Plant parasitic nematodes associated with fruit trees in the newly reclaimed soils at El-Sadat province, Menoufia governorate. *Egyptian Journal of Agronematology*, *21*(1), 1–13. doi:10.21608/ejaj.2022.218676

Sweelam, M. E., Abokorah, M. S., Taka, S. M. A., & Ellatief, S. M. A. (2019). Biological control of the citrus nematode, Tylenchulus semipenetrans attacking Citrus sinensis Navel orange trees. *Menoufia Journal of Plant Protection*, *4*(5), 179–188. doi:10.21608/mjapam.2019.123021

Tafesh-Edwards, G., & Eleftherianos, I. (2022). Entomopathogenic Nematodes as a Model for Probing the Insect Immune System. *Nematodes as Model Organisms*, 42–57. https://doi.org/doi:10.1079/9781789248814.0003

Tankam-Chedjou, I., Touzeau, S., Mailleret, L., Tewa, J. J., & Grognard, F. (2020). Modelling and control of a banana soilborne pest in a multi-seasonal framework. *Mathematical Biosciences*, *322*, 108324. doi:10.1016/j.mbs.2020.108324 PMID:32092468

Tarasco, E., Fanelli, E., Salvemini, C., El-Khoury, Y., Troccoli, A., Vovlas, A., & De Luca, F. (2023). Entomopathogenic nematodes and their symbiotic bacteria: From genes to field uses. *Front. Insect Sci.*, *3*, 1195254. doi:10.3389/finsc.2023.1195254

Thakur, N., Tomar, P., Kaur, S., & Kumari, P. (2022a). Virulence of native entomopathogenic nematodes against major lepidopteran insect species of tomato (*Solanum lycopersicum* L.). *Journal of Applied Biology and Biotechnology*, *10*(1), 6–14. doi:10.7324/JABB.2022.10s102

Thakur, N., Tomar, P., Sharma, S., Kaur, S., Sharma, S., Yadav, A. N., & Hesham, A. E.-L. (2022b). Synergistic effect of entomopathogens against *Spodoptera litura* (Fabricius) under laboratory and greenhouse conditions. *Egyptian Journal of Biological Pest Control*, *32*(1), 1–10. doi:10.118641938-022-00537-3

Therios, I. (2009). *Crop Production Science in Horticulture 18, OLIVES* (1st ed.). CABI.

Thiele-Bruhn, S., Bloem, J., de Vries, F. T., Kalbitz, K., & Wagg, C. (2012). Linking soil biodiversity and agricultural soil management. *Current Opinion in Environmental Sustainability*, *4*(5), 523–528. doi:10.1016/j.cosust.2012.06.004

Thomas, R. J., & Clark, C. A. (1983). Effects of concomitant development on reproduction of *Meloidogyne incognita* and *Rotylenchulus reniformis* on sweet potato. *Journal of Nematology*, *15*, 215–221. PMID:19295794

Thorne, G. (1927). The life history, habits and economic importance of some mononchs. *Journal of Agricultural Research*, *34*, 265–286.

Thorne, G. (1961). *Principles of nematology*. McGraw-Hill Book Company.

Tomalak, M., Piggott, S., & Jagdale, G. B. (2005). Glasshouse applications. In P. S. Grewal, R.-U. Ehlers, & D. I. Shapiro-Ilan (Eds.), *Nematodes as biological control agents* (pp. 147–166). CABI Publishing. doi:10.1079/9780851990170.0147

Tomar, P., Thakur, N., & Yadav, A. N. (2022). Endosymbiotic microbes from entomopathogenic nematode (EPNs) and their applications as biocontrol agents for agro-environmental sustainability. *Egyptian Journal of Biological Pest Control*, *32*(1), 80. doi:10.118641938-022-00579-7

Topalovic, O., Hussain, M., & Heuer, H. (2020). Plants and associated soil microbiota cooperatively suppress plant-parasitic nematodes. *Frontiers in Microbiology*, *11*, 313. doi:10.3389/fmicb.2020.00313 PMID:32184773

Torres-Barragan, A., Suazo, A., Buhler, W. G., & Cardoza, Y. J. (2011). Studies on the entomopathogenicity and bacterial associates of the nematode Oscheius carolinensis. *Biological Control*, *59*(2), 123–129. doi:10.1016/j.biocontrol.2011.05.020

Treonis, A. M., Unangst, S. K., Kepler, R. M., Buyer, J. S., Cavigelli, M. A., Mirsky, S. B., & Maul, J. E. (2018). Characterization of soil nematode communities in three cropping systems through morphological and DNA metabarcoding approaches. *Scientific Reports*, *8*(1), 2004. doi:10.103841598-018-20366-5 PMID:29386563

Tsai, B. Y., & Van Gundy, S. D. (1988). Comparison of anhydrobiotic ability of the citrus nematode with other plant parasitic nematodes. *Proceedings of the International Society of Citriculture*, *2*, 983–992.

Turlings, T. C. J., Hiltpold, I., & Rasmann, S. (2012). The importance of root-produced volatiles as foraging cues for entomopathogenic nematodes. *Plant and Soil*, *358*(1), 51–60. doi:10.100711104-012-1295-3

Turner, S., & Evans, K. (1998). The origins, global distribution and biology of potato cyst nematodes (*Globodera rostochiensis* (Woll.) and *G. pallida* Stone. In R. Marks & B. Brodie (Eds.), *Potato Cyst Nematodes: Biology, Distribution, and Control* (pp. 7–26). University Press, pp.

Tyśkiewicz, R., Nowak, A., Ozimek, E., & Jaroszuk-Ściseł, J. (2022). Trichoderma: The Current Status of Its Application in Agriculture for the Biocontrol of Fungal Phytopathogens and Stimulation of Plant Growth. *International Journal of Molecular Sciences*, *23*(4), 2329. doi:10.3390/ijms23042329 PMID:35216444

Udin, N. (2014). Organic Farming Impact on Sustainable Livelihoods of Marginal Farmers in Shimoga District of Karnataka. *American Journal of Rural Development.*, *2*(4), 81–88. doi:10.12691/ajrd-2-4-4

Urzelai, A., Hernández, A. J., & Pastor, J. (2000). Biotic indices based on soil nematode communities for assessing soil quality in terrestrial ecosystems. *The Science of the Total Environment*, *247*(2–3), 253–261. . doi:10.1016/S0048-9697(99)00494-5 PMID:10803552

Van Gundy, S. D. (1958). The life history of the citrus nematode Tylenchulus semipenetrans Cobb. *Nematologica*, *3*(4), 283–294. doi:10.1163/187529258X00049

Van Gundy, S. D., & Kirkpatrick, J. D. (1964). Nature of resistance in certain citrus rootstocks to citrus nematode. *Phytopathology*, *54*, 419–427.

Van Gundy, S. D., & Martin, G. P. (1962). Soil texture, PH and moisture effect on the development of citrus nematode (*Tylenchulus Semipenetrans*). *Phytopathology*, *52*, 31.

van Tol, R. W. H. M., & Raupp, M. J. (2006). Nursery and tree applications. In P. S. Grewal, R. U. Ehlers, & D. I. Shapiro-Ilan (Eds.), *Nematodes as biological control agents*. CABI Publishing.

Van, Z. C., & Malan, A. P. (2014). The role of entomopathogenic nematodes as biological control agents of insect pests, with emphasis on the history of their mass culturing and in vivo production: Review article. *African Entomology*, *22*(2), 235–249. doi:10.4001/003.022.0222

Vashisth, S., Chandel, Y. S., & Sharma, P. K. (2013). Entomopathogenic nematodes—A review. *Agricultural Reviews (Karnal)*, *34*(3), 163. Advance online publication. doi:10.5958/j.0976-0741.34.3.001

Verdejo-Lucas, S. (2022). *Pre-planting solutions for the slow decline of citrus caused by Tylenchulus semipenetrans*. https://cabidigitallibrary.org

Verdejo-Lucas, S., & Talavera, M. (2009). Integrated Managemenr of Nematodes Parasitic on *Prunus SPP*. *Integrated Management of Fruit Crops Nematodes*, 177-194.

Verdejo-Lucas, S. (1992). On the occurrence of the Mediterranean biotype of *Tylenehulus semipenetrans* in Spain. Fundam. appl. *NemaLOI.*, *15*, 475–477.

Verdejo-Lucas, S., Galeano, M., Sorribas, F. J., Forner, F. B., & Alcaide, A. (2003). Effect on resistance to *Tylenchulus semipenetrans* of hybrid citrus rootstocks subjected to continuous exposure to high population densities of the nematode. *European Journal of Plant Pathology*, *109*(5), 427–433. doi:10.1023/A:1024220417737

Verdejo-Lucas, S., & Kaplan, D. T. (2002). The citrus nematode: *Tylenchulus semipenetrans*. In J. L. Starr, R. Cook, & J. Bridge (Eds.), *Plant resistance to parasitic nematodes* (pp. 207–219). CABI. doi:10.1079/9780851994666.0207

Verdejo-Lucas, S., & Mckenry, M. V. (2004). Management of the Citrus Nematode, *Tylenchulus semipenetrans. Journal of Nematology*, *36*, 424–432. PMID:19262822

Verdejo-Lucas, S., Sorribas, F. J., Pons, J., Forner, J. B., & Alcaide, A. (1997). The mediterranean biotypes of *Tylenchulus semipenetrans* in Spanish citrus orchards. *Fundamental and Applied Nematology*, *20*(4), 399–404.

Visagie, M., Mienie, C. M., Marais, M., Daneel, M., Karssen, G., & Fourie, H. (2018). Identification of Meloidogyne spp. associated with agri-and horticultural crops in South Africa. *Nematology*, *20*(4), 397–401. doi:10.1163/15685411-00003160

Vurukonda, S. S. K. P., Giovanardi, D., & Stefani, E. (2018). Plant growth promoting and biocontrol activity of Streptomyces spp. as endophytes. *International Journal of Molecular Sciences*, *19*(4), 952. doi:10.3390/ijms19040952 PMID:29565834

Waisen, P., Cheng, Z., Sipes, B. S., DeFrank, J., Marahatta, S. P., & Wang, K. (2020). Effects of biofumigant crop termination methods on suppression of plant-parasitic nematodes. *Applied Soil Ecology*, *154*, 103595. doi:10.1016/j.apsoil.2020.103595

Walia, R. K., & Bajaj, H. K. (2003). *Text book on introductory plant nematology. Directorate of Information and Publications of Agriculture*. ICAR.

Walker, G. E., & Morey, B. G. (1999). Effects of chemicals and microbial antagonists on nematodes and fungal pathogens of citrus roots. *Australian Journal of Experimental Agriculture*, *39*(5), 629–637. doi:10.1071/EA99003

Wallace, H. R. (1973). *Nematode ecology and plant disease*. Academic Press.

Wallace, H. R., & Dongcaster, C. C. (1964). A comparative study of the movement of some microphagus, plant parasitic and animal parasitic nematodes. *Parasitology*, *54*(2), 313–326. doi:10.1017/S0031182000067949 PMID:14170216

Wall, D. H., Nielsen, U. N., & Six, J. (2015). Soil biodiversity and human health. *Nature*, *528*(7580), 69–76. doi:10.1038/nature15744 PMID:26595276

Wallis, C. M. (2020). Grapevine (*Vitis* spp.) rootstock stilbenoid associations with host resistance to and induction by root knot nematodes, Meloidogyne incognita. *BMC Research Notes*, *13*(1), 360. doi:10.118613104-020-05201-3 PMID:32727572

Wang, C., Bruening, G., & Williamson, V. M. (2009). Determination of preferred ph for root-knot nematode aggregation using pluronic F-127 gel. *Journal of Chemical Ecology*, *35*(10), 1242–1251. doi:10.100710886-009-9703-8 PMID:19838866

Wang, X., Meyers, D., Yan, Y., Baum, T., Smant, G., Hussey, R., & Davis, E. (1999). In planta localization of a β-1,4-endoglucanase se-creted by Heterodera glycines. *Molecular Plant-Microbe Interactions*, *12*(1), 64–67. doi:10.1094/MPMI.1999.12.1.64 PMID:9885195

Wardle, D. (2002). *Communities and Ecosystems: Linking the Aboveground and Belowground Components*. Princeton University Press.

Warrior, P., Linda, A., Beach, M., Grau, P. A., Kirfman, G. W., & Conley, J. M. (1999). Commercial development and introduction of DiTera™, a new nematicide. *Pesticide Science*, *55*(3), 343–389. doi:10.1002/(SICI)1096-9063(199903)55:3<376::AID-PS918>3.0.CO;2-S

Webster, J. M. (1972). Nematodes and biocontrol. In J. M. Webster (Ed.), *Economic Nematology* (pp. 469–496). Academic Press.

Webster, J. M. (2008). *An anecdotal history of nematology*. Pensoft Publishers.

Weibelzahl-Fulton, E., Dickson, D., & Whitty, E. (1996). Suppression of *Meloidogyne incognita* and M. *javanica* by *Pasteuria penetrans* in field soil. *Journal of Nematology*, *28*, 43–49. PMID:19277344

Wei, J., Hale, K., Carta, L., Platzer, E., Wong, C., Fang, S., & Aroian, V. (2003). *Bacillus thurigiensis* crystal proteins that target nematode. *Microbiology*, *100*, 2760–2765. PMID:12598644

Wesemael, W. M. L., & Moens, M. (2008). Vertical distribution of the plant-parasitic nematode, *Meloidogyne chitwoodi*, under field crops. *European Journal of Plant Pathology*, *120*(3), 249–257. doi:10.100710658-007-9213-x

Westwood, M. N. (1993). *Temperate-zone pomology: Physiology and Culture*. Timber press.

Wheeler, L., & Crow, W. T. (2020). Foliar Nematode Aphelenchoides spp. (Nematoda: Aphelenchida: Aphelenchoididae): EENY749/IN1279, 2/2020. *EDIS, 2020*(3).

Whistler, C. A., Stockwell, V. O., & Loper, J. E. (2000). Lon protease influences antibiotic production and UV tolerance of Pseudomonas fluorescens Pf-5. *Applied and Environmental Microbiology*, *66*(7), 2718–2725. doi:10.1128/AEM.66.7.2718-2725.2000 PMID:10877760

Whitehead, A. G. (1997). *Plant nematode control*. CAB International.

Williams, C. D., Dillon, A. B., Ennis, D., Hennessy, R., & Griffin, C. T. (2015). Differential susceptibility of pine weevil, *Hylobius abietis* (Coleoptera: Curculionidae), larvae and pupae to entomopathogenic nematodes and death of adults infected as pupae. *BioControl*, *60*(4), 537–546. doi:10.100710526-015-9658-3

Williams, E. C., & Walters, K. F. A. (2000). Foliar application of the entomopathogenic nematode Steinernema feltiae against leafminers on vegetables. *Biocontrol Science and Technology*, *10*(1), 61–70. doi:10.1080/09583150029396

Winfield, A. L., Coppock, L. J., & Whitlock, A. J. (1970). A survey of brassica cyst eelworm in Bedfordshire. *Plant Pathology*, *19*(2), 91–92. doi:10.1111/j.1365-3059.1970.tb00988.x

Wong, C. L., & Willetts, H. J. (1969). Gall formation in aerial parts of plants inoculated with *Meloidogyne javanica. Nematologica, 15*(3), 425–428. doi:10.1163/187529269X00506

Wood, F. H. (1973). Nematode feeding relationships. Relationships of soil dwelling nematodes. *Soil Biology & Biochemistry, 5*(5), 593–601. doi:10.1016/0038-0717(73)90049-7

Woolfe, J. A. (1992). *Sweet potato: An untapped food resource*. Cambridge University Press.

Wu, S. Y., Tang, H., Zhang, C., Tang, F., Lin, J., Wang, Y., Chen, L., & Hou, Y. (2023). Potential of entomopathogenic nematode-infected insect cadavers for the biocontrol of the red imported fire ant *Solenopsis invicta. Pest Management Science*, ps.7637. Advance online publication. doi:10.1002/ps.7637 PMID:37384581

Wyss, U., & Grootaert, P. (1977). Feeding mechanisms in *Labronema* sp. *Med. Fac. Landbouww. Rijskuniv. Gent., 42*, 1521–1527.

Xi, W., & Lei, Y. (2020). Apricot. *Nutritional Composition and Antioxidant Properties of Fruits and Vegetables*, 613–629. doi:10.1016/B978-0-12-812780-3.00038-6

Xiong, J., Zhou, Q., Luo, H., Xia, L., Li, L., Sun, M., & Yu, Z. (2015). Systemic nematicidal activity and biocontrol efficacy of *Bacillus firmus* against the root-knot nematode *Meloidogyne incognita. World Journal of Microbiology & Biotechnology, 31*(4), 661–667. doi:10.100711274-015-1820-7 PMID:25672545

Xu, J., Zhang, Y., Zhang, P., Trivedi, P., Riera, N., Wang, Y., Liu, X., Fan, G., Tang, J., Coletta-Filho, H. D., Cubero, J., Deng, X., Ancona, V., Lu, Z., Zhong, B., Roper, M. C., Capote, N., Catara, V., Pietersen, G., ... Wang, N. (2018). The structure and function of the global citrus rhizosphere microbiome. *Nature Communications, 9*(1), 4894. doi:10.103841467-018-07343-2 PMID:30459421

Xu, Z., Zhao, Y., Yang, D., Sun, H., Zhang, C., & Xie, Y. (2015). Attractant and repellent effects of sweet potato root exudates on the potato rot nematode, *Ditylenchus destructor. Nematology, 17*(1), 117–124. doi:10.1163/15685411-00002856

Yahia, E. M. (2019). Classification of Horticultural Commodities. *Postharvest Technology of Perishable Horticultural Commodities*, 71–97. doi:10.1016/B978-0-12-813276-0.00003-1

Yeates, G. W. (1969). Predation by *Mononchoides potohikus* (Nematoda: Diplogasteridae) in laboratory culture. *Nematologica, 15*(1), 1–9. doi:10.1163/187529269X00029

Yeates, G., & Bongers, T. (1999). Nematode diversity in agroecosystems. *Agriculture, Ecosystems & Environment, 74*(1-3), 113–135. doi:10.1016/S0167-8809(99)00033-X

Yeates, G., Bongers, T., De Goede, R., Freckman, D., & Georgieva, S. (1993). Feeding habits in soil nematode families and genera- An outline for soil ecologists. *Journal of Nematology, 25*, 315–331. PMID:19279775

Ye, W. M., Torres-Barragan, A., & Cardoza, Y. J. (2010). Oscheius carolinensis n. sp (Nematoda: Rhabditidae), a potential entomopathogenic nematode from vermicompost. *Nematology*, *12*(1), 121–135. doi:10.1163/156854109X458464

Yokoo, T. (1964). Studies on the citrus nematode *Tylenchulus semipenetrans* Cobb, 1913) in Japan. *Agric. Bull. Saga Univ.*, *20*, 71–109.

Zasada, I. A., Howland, A. D., Peetz, A. B., East, K., & Moyer, M. (2018). *Vitis* spp. Rootstocks Are Poor Hosts for Meloidogyne hapla, a Nematode Commonly Found in Washington Winegrape Vineyards. *American Journal of Enology and Viticulture*. doi:10.5344/ajev.2018.18027

Zhang, F., Peng, D., Ye, P., Yu, Z., Hu, Z., Ruan, L., & Sun, M. (2012). *In vitro* uptake of 140 kDa *Bacillus thuringiensis* nematicidal crystal proteins by the second stage juvenile of *Meloidogyne hapla*. *PLoS One*, *7*(6), e38534. doi:10.1371/journal.pone.0038534 PMID:22737212

Zhang, S., Gan, Y., Ji, W., Xu, B., Hou, B., & Liu, J. (2017). Mechanisms and Characterization of Trichoderma longibrachiatum T6 in Suppressing Nematodes (Heterodera avenae) in Wheat. *Frontiers in Plant Science*, *8*, 1491. Advance online publication. doi:10.3389/fpls.2017.01491 PMID:28966623

Zhang, Y., Li, S., Li, H., Wang, R., Zhang, K. Q., & Xu, J. (2020). Fungi–nematode interactions: Diversity, ecology, and biocontrol prospects in agriculture. *Journal of Fungi (Basel, Switzerland)*, *6*(4), 206. doi:10.3390/jof6040206 PMID:33020457

Zhao, J. H., Tu, G. J., Wu, X. B., & Li, C. P. (2018). *Characterization of the complete mitochondrial genome of Ortleppascaris sinensis (Nematoda: Heterocheilidae) and comparative mitogenomic analysis of eighteen Ascaridida nematodes*. Academic Press.

Zhao, L. L., Wei, W., Kang, L., & Sun, J. H. (2007). Chemotaxis of the pinewood nematode, *Bursaphelenchus xylophilus*, to volatiles associated with host pine, Pinus massoniana, and its vector *Monochamus alternatus*. *Journal of Chemical Ecology*, *33*(6), 1207–1216. doi:10.100710886-007-9289-y PMID:17447123

Zheng, G. D., Shi, L. B., Wu, H. Y., & Peng, D. L. (2012). Nematode communities in continuous tomato-cropping field soil infested by root-knot nematodes. *Acta Agriculturæ Scandinavica. Section B, Soil and Plant Science*, *62*(3), 216–223. doi:10.1080/09064710.2011.598545

Zhong, S., Zeng, H., & Jin, Z. (2015). Responses of soil nematode abundance and diversity to long-term crop rotations in tropical China. *Pedosphere*, *25*(6), 844–852. doi:10.1016/S1002-0160(15)30065-5

Zhou, L., Yuen, G., Wang, Y., Wei, L., & Ji, G. (2016). Evaluation of bacterial biological control agents for control of root-knot nematode disease on tomato. *Crop Protection (Guildford, Surrey)*, *84*, 8–13. doi:10.1016/j.cropro.2015.12.009

Zoubi, B., Mokrini, F., Dababat, A. A., Amer, M., Ghoulam, C., Lahlali, R., Laasli, S.-E., Khfif, K., Imren, M., Akachoud, O., Benkebboura, A., Housseini, A. I., & Qaddoury, A. (2022). Occurrence and Geographic Distribution of Plant-Parasitic Nematodes Associated with Citrus in Morocco and Their Interaction with Soil Patterns. *Life (Chicago, Ill.)*, *12*(5), 637. doi:10.3390/life12050637 PMID:35629306

Related References

To continue our tradition of advancing academic research, we have compiled a list of recommended IGI Global readings. These references will provide additional information and guidance to further enrich your knowledge and assist you with your own research and future publications.

Abbasnejad, B., Moeinzadeh, S., Ahankoob, A., & Wong, P. S. (2021). The Role of Collaboration in the Implementation of BIM-Enabled Projects. In J. Underwood & M. Shelbourn (Eds.), *Handbook of Research on Driving Transformational Change in the Digital Built Environment* (pp. 27–62). IGI Global. https://doi.org/10.4018/978-1-7998-6600-8.ch002

Abdulrahman, K. O., Mahamood, R. M., & Akinlabi, E. T. (2022). Additive Manufacturing (AM): Processing Technique for Lightweight Alloys and Composite Material. In K. Kumar, B. Babu, & J. Davim (Ed.), *Handbook of Research on Advancements in the Processing, Characterization, and Application of Lightweight Materials* (pp. 27-48). IGI Global. https://doi.org/10.4018/978-1-7998-7864-3.ch002

Agrawal, R., Sharma, P., & Saxena, A. (2021). A Diamond Cut Leather Substrate Antenna for BAN (Body Area Network) Application. In V. Singh, V. Dubey, A. Saxena, R. Tiwari, & H. Sharma (Eds.), *Emerging Materials and Advanced Designs for Wearable Antennas* (pp. 54–59). IGI Global. https://doi.org/10.4018/978-1-7998-7611-3.ch004

Ahmad, F., Al-Ammar, E. A., & Alsaidan, I. (2022). Battery Swapping Station: A Potential Solution to Address the Limitations of EV Charging Infrastructure. In M. Alam, R. Pillai, & N. Murugesan (Eds.), *Developing Charging Infrastructure and Technologies for Electric Vehicles* (pp. 195–207). IGI Global. doi:10.4018/978-1-7998-6858-3.ch010

Aikhuele, D. (2018). A Study of Product Development Engineering and Design Reliability Concerns. *International Journal of Applied Industrial Engineering*, *5*(1), 79–89. doi:10.4018/IJAIE.2018010105

Al-Khatri, H., & Al-Atrash, F. (2021). Occupants' Habits and Natural Ventilation in a Hot Arid Climate. In R. González-Lezcano (Ed.), *Advancements in Sustainable Architecture and Energy Efficiency* (pp. 146–168). IGI Global. https://doi.org/10.4018/978-1-7998-7023-4.ch007

Al-Shebeeb, O. A., Rangaswamy, S., Gopalakrishan, B., & Devaru, D. G. (2017). Evaluation and Indexing of Process Plans Based on Electrical Demand and Energy Consumption. *International Journal of Manufacturing, Materials, and Mechanical Engineering*, *7*(3), 1–19. doi:10.4018/IJMMME.2017070101

Amuda, M. O., Lawal, T. F., & Akinlabi, E. T. (2017). Research Progress on Rheological Behavior of AA7075 Aluminum Alloy During Hot Deformation. *International Journal of Materials Forming and Machining Processes*, *4*(1), 53–96. doi:10.4018/IJMFMP.2017010104

Amuda, M. O., Lawal, T. F., & Mridha, S. (2021). Microstructure and Mechanical Properties of Silicon Carbide-Treated Ferritic Stainless Steel Welds. In L. Burstein (Ed.), *Handbook of Research on Advancements in Manufacturing, Materials, and Mechanical Engineering* (pp. 395–411). IGI Global. https://doi.org/10.4018/978-1-7998-4939-1.ch019

Anikeev, V., Gasem, K. A., & Fan, M. (2021). Application of Supercritical Technologies in Clean Energy Production: A Review. In L. Chen (Ed.), *Handbook of Research on Advancements in Supercritical Fluids Applications for Sustainable Energy Systems* (pp. 792–821). IGI Global. https://doi.org/10.4018/978-1-7998-5796-9.ch022

Arafat, M. Y., Saleem, I., & Devi, T. P. (2022). Drivers of EV Charging Infrastructure Entrepreneurship in India. In M. Alam, R. Pillai, & N. Murugesan (Eds.), *Developing Charging Infrastructure and Technologies for Electric Vehicles* (pp. 208–219). IGI Global. https://doi.org/10.4018/978-1-7998-6858-3.ch011

Araujo, A., & Manninen, H. (2022). Contribution of Project-Based Learning on Social Skills Development: An Industrial Engineer Perspective. In A. Alves & N. van Hattum-Janssen (Eds.), *Training Engineering Students for Modern Technological Advancement* (pp. 119–145). IGI Global. https://doi.org/10.4018/978-1-7998-8816-1.ch006

Armutlu, H. (2018). Intelligent Biomedical Engineering Operations by Cloud Computing Technologies. In U. Kose, G. Guraksin, & O. Deperlioglu (Eds.), *Nature-Inspired Intelligent Techniques for Solving Biomedical Engineering Problems* (pp. 297–317). Hershey, PA: IGI Global. doi:10.4018/978-1-5225-4769-3.ch015

Atik, M., Sadek, M., & Shahrour, I. (2017). Single-Run Adaptive Pushover Procedure for Shear Wall Structures. In V. Plevris, G. Kremmyda, & Y. Fahjan (Eds.), *Performance-Based Seismic Design of Concrete Structures and Infrastructures* (pp. 59–83). Hershey, PA: IGI Global. doi:10.4018/978-1-5225-2089-4.ch003

Attia, H. (2021). Smart Power Microgrid Impact on Sustainable Building. In R. González-Lezcano (Ed.), *Advancements in Sustainable Architecture and Energy Efficiency* (pp. 169–194). IGI Global. https://doi.org/10.4018/978-1-7998-7023-4.ch008

Aydin, A., Akyol, E., Gungor, M., Kaya, A., & Tasdelen, S. (2018). Geophysical Surveys in Engineering Geology Investigations With Field Examples. In N. Ceryan (Ed.), *Handbook of Research on Trends and Digital Advances in Engineering Geology* (pp. 257–280). Hershey, PA: IGI Global. doi:10.4018/978-1-5225-2709-1.ch007

Ayoobkhan, M. U. D., Y., A., J., Easwaran, B., & R., T. (2021). Smart Connected Digital Products and IoT Platform With the Digital Twin. In P. Vasant, G. Weber, & W. Punurai (Ed.), Research Advancements in Smart Technology, Optimization, and Renewable Energy (pp. 330-350). IGI Global. https://doi.org/ doi:10.4018/978-1-7998-3970-5.ch016

Baeza Moyano, D., & González Lezcano, R. A. (2021). The Importance of Light in Our Lives: Towards New Lighting in Schools. In R. González-Lezcano (Ed.), *Advancements in Sustainable Architecture and Energy Efficiency* (pp. 239–256). IGI Global. https://doi.org/10.4018/978-1-7998-7023-4.ch011

Bagdadee, A. H. (2021). A Brief Assessment of the Energy Sector of Bangladesh. *International Journal of Energy Optimization and Engineering*, *10*(1), 36–55. doi:10.4018/IJEOE.2021010103

Baklezos, A. T., & Hadjigeorgiou, N. G. (2021). Magnetic Sensors for Space Applications and Magnetic Cleanliness Considerations. In C. Nikolopoulos (Ed.), *Recent Trends on Electromagnetic Environmental Effects for Aeronautics and Space Applications* (pp. 147–185). IGI Global. https://doi.org/10.4018/978-1-7998-4879-0.ch006

Bas, T. G. (2017). Nutraceutical Industry with the Collaboration of Biotechnology and Nutrigenomics Engineering: The Significance of Intellectual Property in the Entrepreneurship and Scientific Research Ecosystems. In T. Bas & J. Zhao (Eds.), *Comparative Approaches to Biotechnology Development and Use in Developed and Emerging Nations* (pp. 1–17). Hershey, PA: IGI Global. doi:10.4018/978-1-5225-1040-6.ch001

Bazeer Ahamed, B., & Periakaruppan, S. (2021). Taxonomy of Influence Maximization Techniques in Unknown Social Networks. In P. Vasant, G. Weber, & W. Punurai (Eds.), *Research Advancements in Smart Technology, Optimization, and Renewable Energy* (pp. 351-363). IGI Global. https://doi.org/10.4018/978-1-7998-3970-5.ch017

Beale, R., & André, J. (2017). *Design Solutions and Innovations in Temporary Structures*. Hershey, PA: IGI Global. doi:10.4018/978-1-5225-2199-0

Behnam, B. (2017). Simulating Post-Earthquake Fire Loading in Conventional RC Structures. In P. Samui, S. Chakraborty, & D. Kim (Eds.), *Modeling and Simulation Techniques in Structural Engineering* (pp. 425–444). Hershey, PA: IGI Global. doi:10.4018/978-1-5225-0588-4.ch015

Ben Hamida, I., Salah, S. B., Msahli, F., & Mimouni, M. F. (2018). Distribution Network Reconfiguration Using SPEA2 for Power Loss Minimization and Reliability Improvement. *International Journal of Energy Optimization and Engineering*, *7*(1), 50–65. doi:10.4018/IJEOE.2018010103

Bentarzi, H. (2021). Fault Tree-Based Root Cause Analysis Used to Study Mal-Operation of a Protective Relay in a Smart Grid. In A. Recioui & H. Bentarzi (Eds.), *Optimizing and Measuring Smart Grid Operation and Control* (pp. 289–308). IGI Global. https://doi.org/10.4018/978-1-7998-4027-5.ch012

Beysens, D. A., Garrabos, Y., & Zappoli, B. (2021). Thermal Effects in Near-Critical Fluids: Piston Effect and Related Phenomena. In L. Chen (Ed.), *Handbook of Research on Advancements in Supercritical Fluids Applications for Sustainable Energy Systems* (pp. 1–31). IGI Global. https://doi.org/10.4018/978-1-7998-5796-9.ch001

Bhaskar, S. V., & Kudal, H. N. (2017). Effect of TiCN and AlCrN Coating on Tribological Behaviour of Plasma-nitrided AISI 4140 Steel. *International Journal of Surface Engineering and Interdisciplinary Materials Science*, *5*(2), 1–17. doi:10.4018/IJSEIMS.2017070101

Bhuyan, D. (2018). Designing of a Twin Tube Shock Absorber: A Study in Reverse Engineering. In K. Kumar & J. Davim (Eds.), *Design and Optimization of Mechanical Engineering Products* (pp. 83–104). Hershey, PA: IGI Global. doi:10.4018/978-1-5225-3401-3.ch005

Blumberg, G. (2021). Blockchains for Use in Construction and Engineering Projects. In J. Underwood & M. Shelbourn (Eds.), *Handbook of Research on Driving Transformational Change in the Digital Built Environment* (pp. 179–208). IGI Global. https://doi.org/10.4018/978-1-7998-6600-8.ch008

Burstein, L. (2021). Simulation Tool for Cable Design. In L. Burstein (Ed.), *Handbook of Research on Advancements in Manufacturing, Materials, and Mechanical Engineering* (pp. 54–74). IGI Global. https://doi.org/10.4018/978-1-7998-4939-1.ch003

Calderon, F. A., Giolo, E. G., Frau, C. D., Rengel, M. G., Rodriguez, H., Tornello, M., ... Gallucci, R. (2018). Seismic Microzonation and Site Effects Detection Through Microtremors Measures: A Review. In N. Ceryan (Ed.), *Handbook of Research on Trends and Digital Advances in Engineering Geology* (pp. 326–349). Hershey, PA: IGI Global. doi:10.4018/978-1-5225-2709-1.ch009

Ceryan, N., & Can, N. K. (2018). Prediction of The Uniaxial Compressive Strength of Rocks Materials. In N. Ceryan (Ed.), *Handbook of Research on Trends and Digital Advances in Engineering Geology* (pp. 31–96). Hershey, PA: IGI Global. doi:10.4018/978-1-5225-2709-1.ch002

Ceryan, S. (2018). Weathering Indices Used in Evaluation of the Weathering State of Rock Material. In N. Ceryan (Ed.), *Handbook of Research on Trends and Digital Advances in Engineering Geology* (pp. 132–186). Hershey, PA: IGI Global. doi:10.4018/978-1-5225-2709-1.ch004

Chen, H., Padilla, R. V., & Besarati, S. (2017). Supercritical Fluids and Their Applications in Power Generation. In L. Chen & Y. Iwamoto (Eds.), *Advanced Applications of Supercritical Fluids in Energy Systems* (pp. 369–402). Hershey, PA: IGI Global. doi:10.4018/978-1-5225-2047-4.ch012

Chen, H., Padilla, R. V., & Besarati, S. (2021). Supercritical Fluids and Their Applications in Power Generation. In L. Chen (Ed.), *Handbook of Research on Advancements in Supercritical Fluids Applications for Sustainable Energy Systems* (pp. 566–599). IGI Global. https://doi.org/10.4018/978-1-7998-5796-9.ch016

Chen, L. (2017). Principles, Experiments, and Numerical Studies of Supercritical Fluid Natural Circulation System. In L. Chen & Y. Iwamoto (Eds.), *Advanced Applications of Supercritical Fluids in Energy Systems* (pp. 136–187). Hershey, PA: IGI Global. doi:10.4018/978-1-5225-2047-4.ch005

Chen, L. (2021). Principles, Experiments, and Numerical Studies of Supercritical Fluid Natural Circulation System. In L. Chen (Ed.), *Handbook of Research on Advancements in Supercritical Fluids Applications for Sustainable Energy Systems* (pp. 219–269). IGI Global. https://doi.org/10.4018/978-1-7998-5796-9.ch007

Chiba, Y., Marif, Y., Henini, N., & Tlemcani, A. (2021). Modeling of Magnetic Refrigeration Device by Using Artificial Neural Networks Approach. *International Journal of Energy Optimization and Engineering, 10*(4), 68–76. https://doi.org/10.4018/IJEOE.2021100105

Clementi, F., Di Sciascio, G., Di Sciascio, S., & Lenci, S. (2017). Influence of the Shear-Bending Interaction on the Global Capacity of Reinforced Concrete Frames: A Brief Overview of the New Perspectives. In V. Plevris, G. Kremmyda, & Y. Fahjan (Eds.), *Performance-Based Seismic Design of Concrete Structures and Infrastructures* (pp. 84–111). Hershey, PA: IGI Global. doi:10.4018/978-1-5225-2089-4.ch004

Codinhoto, R., Fialho, B. C., Pinti, L., & Fabricio, M. M. (2021). BIM and IoT for Facilities Management: Understanding Key Maintenance Issues. In J. Underwood & M. Shelbourn (Eds.), *Handbook of Research on Driving Transformational Change in the Digital Built Environment* (pp. 209–231). IGI Global. doi:10.4018/978-1-7998-6600-8.ch009

Cortés-Polo, D., Calle-Cancho, J., Carmona-Murillo, J., & González-Sánchez, J. (2017). Future Trends in Mobile-Fixed Integration for Next Generation Networks: Classification and Analysis. *International Journal of Vehicular Telematics and Infotainment Systems, 1*(1), 33–53. doi:10.4018/IJVTIS.2017010103

Costa, H. G., Sheremetieff, F. H., & Araújo, E. A. (2022). Influence of Game-Based Methods in Developing Engineering Competences. In A. Alves & N. van Hattum-Janssen (Eds.), *Training Engineering Students for Modern Technological Advancement* (pp. 69–88). IGI Global. https://doi.org/10.4018/978-1-7998-8816-1.ch004

Cui, X., Zeng, S., Li, Z., Zheng, Q., Yu, X., & Han, B. (2018). Advanced Composites for Civil Engineering Infrastructures. In K. Kumar & J. Davim (Eds.), *Composites and Advanced Materials for Industrial Applications* (pp. 212–248). Hershey, PA: IGI Global. doi:10.4018/978-1-5225-5216-1.ch010

Dalgıç, S., & Kuşku, İ. (2018). Geological and Geotechnical Investigations in Tunneling. In N. Ceryan (Ed.), *Handbook of Research on Trends and Digital Advances in Engineering Geology* (pp. 482–529). Hershey, PA: IGI Global. doi:10.4018/978-1-5225-2709-1.ch014

Dang, C., & Hihara, E. (2021). Study on Cooling Heat Transfer of Supercritical Carbon Dioxide Applied to Transcritical Carbon Dioxide Heat Pump. In L. Chen (Ed.), *Handbook of Research on Advancements in Supercritical Fluids Applications for Sustainable Energy Systems* (pp. 451–493). IGI Global. https://doi.org/10.4018/978-1-7998-5796-9.ch013

Daus, Y., Kharchenko, V., & Yudaev, I. (2021). Research of Solar Energy Potential of Photovoltaic Installations on Enclosing Structures of Buildings. *International Journal of Energy Optimization and Engineering*, *10*(4), 18–34. https://doi.org/10.4018/IJEOE.2021100102

Daus, Y., Kharchenko, V., & Yudaev, I. (2021). Optimizing Layout of Distributed Generation Sources of Power Supply System of Agricultural Object. *International Journal of Energy Optimization and Engineering*, *10*(3), 70–84. https://doi.org/10.4018/IJEOE.2021070104

de la Varga, D., Soto, M., Arias, C. A., van Oirschot, D., Kilian, R., Pascual, A., & Álvarez, J. A. (2017). Constructed Wetlands for Industrial Wastewater Treatment and Removal of Nutrients. In Á. Val del Río, J. Campos Gómez, & A. Mosquera Corral (Eds.), *Technologies for the Treatment and Recovery of Nutrients from Industrial Wastewater* (pp. 202–230). Hershey, PA: IGI Global. doi:10.4018/978-1-5225-1037-6.ch008

Deb, S., Ammar, E. A., AlRajhi, H., Alsaidan, I., & Shariff, S. M. (2022). V2G Pilot Projects: Review and Lessons Learnt. In M. Alam, R. Pillai, & N. Murugesan (Eds.), *Developing Charging Infrastructure and Technologies for Electric Vehicles* (pp. 252–267). IGI Global. https://doi.org/10.4018/978-1-7998-6858-3.ch014

Dekhandji, F. Z., & Rais, M. C. (2021). A Comparative Study of Power Quality Monitoring Using Various Techniques. In A. Recioui & H. Bentarzi (Eds.), *Optimizing and Measuring Smart Grid Operation and Control* (pp. 259–288). IGI Global. https://doi.org/10.4018/978-1-7998-4027-5.ch011

Deperlioglu, O. (2018). Intelligent Techniques Inspired by Nature and Used in Biomedical Engineering. In U. Kose, G. Guraksin, & O. Deperlioglu (Eds.), *Nature-Inspired Intelligent Techniques for Solving Biomedical Engineering Problems* (pp. 51–77). Hershey, PA: IGI Global. doi:10.4018/978-1-5225-4769-3.ch003

Dhurpate, P. R., & Tang, H. (2021). Quantitative Analysis of the Impact of Inter-Line Conveyor Capacity for Throughput of Manufacturing Systems. *International Journal of Manufacturing, Materials, and Mechanical Engineering, 11*(1), 1–17. https://doi.org/10.4018/IJMMME.2021010101

Dinkar, S., & Deep, K. (2021). A Survey of Recent Variants and Applications of Antlion Optimizer. *International Journal of Energy Optimization and Engineering, 10*(2), 48–73. doi:10.4018/IJEOE.2021040103

Dixit, A. (2018). Application of Silica-Gel-Reinforced Aluminium Composite on the Piston of Internal Combustion Engine: Comparative Study of Silica-Gel-Reinforced Aluminium Composite Piston With Aluminium Alloy Piston. In K. Kumar & J. Davim (Eds.), *Composites and Advanced Materials for Industrial Applications* (pp. 63–98). Hershey, PA: IGI Global. doi:10.4018/978-1-5225-5216-1.ch004

Drabecki, M. P., & Kułak, K. B. (2021). Global Pandemics on European Electrical Energy Markets: Lessons Learned From the COVID-19 Outbreak. *International Journal of Energy Optimization and Engineering, 10*(3), 24–46. https://doi.org/10.4018/IJEOE.2021070102

Dutta, M. M. (2021). Nanomaterials for Food and Agriculture. In M. Bhat, I. Wani, & S. Ashraf (Eds.), *Applications of Nanomaterials in Agriculture, Food Science, and Medicine* (pp. 75–97). IGI Global. doi:10.4018/978-1-7998-5563-7.ch004

Dutta, M. M., & Goswami, M. (2021). Coating Materials: Nano-Materials. In S. Roy & G. Bose (Eds.), *Advanced Surface Coating Techniques for Modern Industrial Applications* (pp. 1–30). IGI Global. doi:10.4018/978-1-7998-4870-7.ch001

Elsayed, A. M., Dakkama, H. J., Mahmoud, S., Al-Dadah, R., & Kaialy, W. (2017). Sustainable Cooling Research Using Activated Carbon Adsorbents and Their Environmental Impact. In T. Kobayashi (Ed.), *Applied Environmental Materials Science for Sustainability* (pp. 186–221). Hershey, PA: IGI Global. doi:10.4018/978-1-5225-1971-3.ch009

Ercanoglu, M., & Sonmez, H. (2018). General Trends and New Perspectives on Landslide Mapping and Assessment Methods. In N. Ceryan (Ed.), *Handbook of Research on Trends and Digital Advances in Engineering Geology* (pp. 350–379). Hershey, PA: IGI Global. doi:10.4018/978-1-5225-2709-1.ch010

Faroz, S. A., Pujari, N. N., Rastogi, R., & Ghosh, S. (2017). Risk Analysis of Structural Engineering Systems Using Bayesian Inference. In P. Samui, S. Chakraborty, & D. Kim (Eds.), *Modeling and Simulation Techniques in Structural Engineering* (pp. 390–424). Hershey, PA: IGI Global. doi:10.4018/978-1-5225-0588-4.ch014

Fekik, A., Hamida, M. L., Denoun, H., Azar, A. T., Kamal, N. A., Vaidyanathan, S., Bousbaine, A., & Benamrouche, N. (2022). Multilevel Inverter for Hybrid Fuel Cell/PV Energy Conversion System. In A. Fekik & N. Benamrouche (Eds.), *Modeling and Control of Static Converters for Hybrid Storage Systems* (pp. 233–270). IGI Global. https://doi.org/10.4018/978-1-7998-7447-8.ch009

Fekik, A., Hamida, M. L., Houassine, H., Azar, A. T., Kamal, N. A., Denoun, H., Vaidyanathan, S., & Sambas, A. (2022). Power Quality Improvement for Grid-Connected Photovoltaic Panels Using Direct Power Control. In A. Fekik & N. Benamrouche (Eds.), *Modeling and Control of Static Converters for Hybrid Storage Systems* (pp. 107–142). IGI Global. https://doi.org/10.4018/978-1-7998-7447-8.ch005

Fernando, P. R., Hamigah, T., Disne, S., Wickramasingha, G. G., & Sutharshan, A. (2018). The Evaluation of Engineering Properties of Low Cost Concrete Blocks by Partial Doping of Sand with Sawdust: Low Cost Sawdust Concrete Block. *International Journal of Strategic Engineering*, *1*(2), 26–42. doi:10.4018/IJoSE.2018070103

Ferro, G., Minciardi, R., Parodi, L., & Robba, M. (2022). Optimal Charging Management of Microgrid-Integrated Electric Vehicles. In M. Alam, R. Pillai, & N. Murugesan (Eds.), *Developing Charging Infrastructure and Technologies for Electric Vehicles* (pp. 133–155). IGI Global. https://doi.org/10.4018/978-1-7998-6858-3.ch007

Flumerfelt, S., & Green, C. (2022). Graduate Lean Leadership Education: A Case Study of a Program. In A. Alves & N. van Hattum-Janssen (Eds.), *Training Engineering Students for Modern Technological Advancement* (pp. 202–224). IGI Global. https://doi.org/10.4018/978-1-7998-8816-1.ch010

Galli, B. J. (2021). Implications of Economic Decision Making to the Project Manager. *International Journal of Strategic Engineering*, *4*(1), 19–32. https://doi.org/10.4018/IJoSE.2021010102

Gento, A. M., Pimentel, C., & Pascual, J. A. (2022). Teaching Circular Economy and Lean Management in a Learning Factory. In A. Alves & N. van Hattum-Janssen (Eds.), *Training Engineering Students for Modern Technological Advancement* (pp. 183–201). IGI Global. https://doi.org/10.4018/978-1-7998-8816-1.ch009

Ghosh, S., Mitra, S., Ghosh, S., & Chakraborty, S. (2017). Seismic Reliability Analysis in the Framework of Metamodelling Based Monte Carlo Simulation. In P. Samui, S. Chakraborty, & D. Kim (Eds.), *Modeling and Simulation Techniques in Structural Engineering* (pp. 192–208). Hershey, PA: IGI Global. doi:10.4018/978-1-5225-0588-4.ch006

Gil, M., & Otero, B. (2017). Learning Engineering Skills through Creativity and Collaboration: A Game-Based Proposal. In R. Alexandre Peixoto de Queirós & M. Pinto (Eds.), *Gamification-Based E-Learning Strategies for Computer Programming Education* (pp. 14–29). Hershey, PA: IGI Global. doi:10.4018/978-1-5225-1034-5.ch002

Gill, J., Ayre, M., & Mills, J. (2017). Revisioning the Engineering Profession: How to Make It Happen! In M. Gray & K. Thomas (Eds.), *Strategies for Increasing Diversity in Engineering Majors and Careers* (pp. 156–175). Hershey, PA: IGI Global. doi:10.4018/978-1-5225-2212-6.ch008

Godzhaev, Z., Senkevich, S., Kuzmin, V., & Melikov, I. (2021). Use of the Neural Network Controller of Sprung Mass to Reduce Vibrations From Road Irregularities. In P. Vasant, G. Weber, & W. Punurai (Ed.), *Research Advancements in Smart Technology, Optimization, and Renewable Energy* (pp. 69-87). IGI Global. https://doi.org/10.4018/978-1-7998-3970-5.ch005

Gomes de Gusmão, C. M. (2022). Digital Competencies and Transformation in Higher Education: Upskilling With Extension Actions. In A. Alves & N. van Hattum-Janssen (Eds.), *Training Engineering Students for Modern Technological Advancement* (pp. 313–328). IGI Global. https://doi.org/10.4018/978-1-7998-8816-1.ch015A

Goyal, N., Ram, M., & Kumar, P. (2017). Welding Process under Fault Coverage Approach for Reliability and MTTF. In M. Ram & J. Davim (Eds.), *Mathematical Concepts and Applications in Mechanical Engineering and Mechatronics* (pp. 222–245). Hershey, PA: IGI Global. doi:10.4018/978-1-5225-1639-2.ch011

Gray, M., & Lundy, C. (2017). Engineering Study Abroad: High Impact Strategy for Increasing Access. In M. Gray & K. Thomas (Eds.), *Strategies for Increasing Diversity in Engineering Majors and Careers* (pp. 42–59). Hershey, PA: IGI Global. doi:10.4018/978-1-5225-2212-6.ch003

Güler, O., & Varol, T. (2021). Fabrication of Functionally Graded Metal and Ceramic Powders Synthesized by Electroless Deposition. In S. Roy & G. Bose (Eds.), *Advanced Surface Coating Techniques for Modern Industrial Applications* (pp. 150–187). IGI Global. https://doi.org/10.4018/978-1-7998-4870-7.ch007

Guraksin, G. E. (2018). Internet of Things and Nature-Inspired Intelligent Techniques for the Future of Biomedical Engineering. In U. Kose, G. Guraksin, & O. Deperlioglu (Eds.), *Nature-Inspired Intelligent Techniques for Solving Biomedical Engineering Problems* (pp. 263–282). Hershey, PA: IGI Global. doi:10.4018/978-1-5225-4769-3.ch013

Hamida, M. L., Fekik, A., Denoun, H., Ardjal, A., & Bokhtache, A. A. (2022). Flying Capacitor Inverter Integration in a Renewable Energy System. In A. Fekik & N. Benamrouche (Eds.), *Modeling and Control of Static Converters for Hybrid Storage Systems* (pp. 287–306). IGI Global. https://doi.org/10.4018/978-1-7998-7447-8.ch011

Hasegawa, N., & Takahashi, Y. (2021). Control of Soap Bubble Ejection Robot Using Facial Expressions. *International Journal of Manufacturing, Materials, and Mechanical Engineering, 11*(2), 1–16. https://doi.org/10.4018/IJMMME.2021040101

Hejazi, T., & Akbari, L. (2017). A Multiresponse Optimization Model for Statistical Design of Processes with Discrete Variables. In M. Ram & J. Davim (Eds.), *Mathematical Concepts and Applications in Mechanical Engineering and Mechatronics* (pp. 17–37). Hershey, PA: IGI Global. doi:10.4018/978-1-5225-1639-2.ch002

Hejazi, T., & Hejazi, A. (2017). Monte Carlo Simulation for Reliability-Based Design of Automotive Complex Subsystems. In M. Ram & J. Davim (Eds.), *Mathematical Concepts and Applications in Mechanical Engineering and Mechatronics* (pp. 177–200). Hershey, PA: IGI Global. doi:10.4018/978-1-5225-1639-2.ch009

Hejazi, T., & Poursabbagh, H. (2017). Reliability Analysis of Engineering Systems: An Accelerated Life Testing for Boiler Tubes. In M. Ram & J. Davim (Eds.), *Mathematical Concepts and Applications in Mechanical Engineering and Mechatronics* (pp. 154–176). Hershey, PA: IGI Global. doi:10.4018/978-1-5225-1639-2.ch008

Henao, J., Poblano-Salas, C. A., Vargas, F., Giraldo-Betancur, A. L., Corona-Castuera, J., & Sotelo-Mazón, O. (2021). Principles and Applications of Thermal Spray Coatings. In S. Roy & G. Bose (Eds.), *Advanced Surface Coating Techniques for Modern Industrial Applications* (pp. 31–70). IGI Global. https://doi.org/10.4018/978-1-7998-4870-7.ch002

Henao, J., & Sotelo, O. (2018). Surface Engineering at High Temperature: Thermal Cycling and Corrosion Resistance. In A. Pakseresht (Ed.), *Production, Properties, and Applications of High Temperature Coatings* (pp. 131–159). Hershey, PA: IGI Global. doi:10.4018/978-1-5225-4194-3.ch006

Hrnčič, M. K., Cör, D., & Knez, Ž. (2021). Supercritical Fluids as a Tool for Green Energy and Chemicals. In L. Chen (Ed.), *Handbook of Research on Advancements in Supercritical Fluids Applications for Sustainable Energy Systems* (pp. 761–791). IGI Global. doi:10.4018/978-1-7998-5796-9.ch021

Ibrahim, O., Erdem, S., & Gurbuz, E. (2021). Studying Physical and Chemical Properties of Graphene Oxide and Reduced Graphene Oxide and Their Applications in Sustainable Building Materials. In R. González-Lezcano (Ed.), *Advancements in Sustainable Architecture and Energy Efficiency* (pp. 221–238). IGI Global. https://doi.org/10.4018/978-1-7998-7023-4.ch010

Ihianle, I. K., Islam, S., Naeem, U., & Ebenuwa, S. H. (2021). Exploiting Patterns of Object Use for Human Activity Recognition. In A. Nwajana & I. Ihianle (Eds.), *Handbook of Research on 5G Networks and Advancements in Computing, Electronics, and Electrical Engineering* (pp. 382–401). IGI Global. https://doi.org/10.4018/978-1-7998-6992-4.ch015

Ijemaru, G. K., Ngharamike, E. T., Oleka, E. U., & Nwajana, A. O. (2021). An Energy-Efficient Model for Opportunistic Data Collection in IoV-Enabled SC Waste Management. In A. Nwajana & I. Ihianle (Eds.), *Handbook of Research on 5G Networks and Advancements in Computing, Electronics, and Electrical Engineering* (pp. 1–19). IGI Global. https://doi.org/10.4018/978-1-7998-6992-4.ch001

Ilori, O. O., Adetan, D. A., & Umoru, L. E. (2017). Effect of Cutting Parameters on the Surface Residual Stress of Face-Milled Pearlitic Ductile Iron. *International Journal of Materials Forming and Machining Processes*, *4*(1), 38–52. doi:10.4018/IJMFMP.2017010103

Imam, M. H., Tasadduq, I. A., Ahmad, A., Aldosari, F., & Khan, H. (2017). Automated Generation of Course Improvement Plans Using Expert System. *International Journal of Quality Assurance in Engineering and Technology Education*, *6*(1), 1–12. doi:10.4018/IJQAETE.2017010101

Injeti, S. K., & Kumar, T. V. (2018). A WDO Framework for Optimal Deployment of DGs and DSCs in a Radial Distribution System Under Daily Load Pattern to Improve Techno-Economic Benefits. *International Journal of Energy Optimization and Engineering*, *7*(2), 1–38. doi:10.4018/IJEOE.2018040101

Ishii, N., Anami, K., & Knisely, C. W. (2018). *Dynamic Stability of Hydraulic Gates and Engineering for Flood Prevention*. Hershey, PA: IGI Global. doi:10.4018/978-1-5225-3079-4

Iwamoto, Y., & Yamaguchi, H. (2021). Application of Supercritical Carbon Dioxide for Solar Water Heater. In L. Chen (Ed.), *Handbook of Research on Advancements in Supercritical Fluids Applications for Sustainable Energy Systems* (pp. 370–387). IGI Global. https://doi.org/10.4018/978-1-7998-5796-9.ch010

Jayapalan, S. (2018). A Review of Chemical Treatments on Natural Fibers-Based Hybrid Composites for Engineering Applications. In K. Kumar & J. Davim (Eds.), *Composites and Advanced Materials for Industrial Applications* (pp. 16–37). Hershey, PA: IGI Global. doi:10.4018/978-1-5225-5216-1.ch002

Kapetanakis, T. N., Vardiambasis, I. O., Ioannidou, M. P., & Konstantaras, A. I. (2021). Modeling Antenna Radiation Using Artificial Intelligence Techniques: The Case of a Circular Loop Antenna. In C. Nikolopoulos (Ed.), *Recent Trends on Electromagnetic Environmental Effects for Aeronautics and Space Applications* (pp. 186–225). IGI Global. https://doi.org/10.4018/978-1-7998-4879-0.ch007

Karkalos, N. E., Markopoulos, A. P., & Dossis, M. F. (2017). Optimal Model Parameters of Inverse Kinematics Solution of a 3R Robotic Manipulator Using ANN Models. *International Journal of Manufacturing, Materials, and Mechanical Engineering*, *7*(3), 20–40. doi:10.4018/IJMMME.2017070102

Kelly, M., Costello, M., Nicholson, G., & O'Connor, J. (2021). The Evolving Integration of BIM Into Built Environment Programmes in a Higher Education Institute. In J. Underwood & M. Shelbourn (Eds.), *Handbook of Research on Driving Transformational Change in the Digital Built Environment* (pp. 294–326). IGI Global. https://doi.org/10.4018/978-1-7998-6600-8.ch012

Kesimal, A., Karaman, K., Cihangir, F., & Ercikdi, B. (2018). Excavatability Assessment of Rock Masses for Geotechnical Studies. In N. Ceryan (Ed.), *Handbook of Research on Trends and Digital Advances in Engineering Geology* (pp. 231–256). Hershey, PA: IGI Global. doi:10.4018/978-1-5225-2709-1.ch006

Knoflacher, H. (2017). The Role of Engineers and Their Tools in the Transport Sector after Paradigm Change: From Assumptions and Extrapolations to Science. In H. Knoflacher & E. Ocalir-Akunal (Eds.), *Engineering Tools and Solutions for Sustainable Transportation Planning* (pp. 1–29). Hershey, PA: IGI Global. doi:10.4018/978-1-5225-2116-7.ch001

Kose, U. (2018). Towards an Intelligent Biomedical Engineering With Nature-Inspired Artificial Intelligence Techniques. In U. Kose, G. Guraksin, & O. Deperlioglu (Eds.), *Nature-Inspired Intelligent Techniques for Solving Biomedical Engineering Problems* (pp. 1–26). Hershey, PA: IGI Global. doi:10.4018/978-1-5225-4769-3.ch001

Kostić, S. (2018). A Review on Enhanced Stability Analyses of Soil Slopes Using Statistical Design. In N. Ceryan (Ed.), *Handbook of Research on Trends and Digital Advances in Engineering Geology* (pp. 446–481). Hershey, PA: IGI Global. doi:10.4018/978-1-5225-2709-1.ch013

Kumar, A., Patil, P. P., & Prajapati, Y. K. (2018). *Advanced Numerical Simulations in Mechanical Engineering*. Hershey, PA: IGI Global. doi:10.4018/978-1-5225-3722-9

Kumar, G. R., Rajyalakshmi, G., & Manupati, V. K. (2017). Surface Micro Patterning of Aluminium Reinforced Composite through Laser Peening. *International Journal of Manufacturing, Materials, and Mechanical Engineering*, *7*(4), 15–27. doi:10.4018/IJMMME.2017100102

Kumar, N., Basu, D. N., & Chen, L. (2021). Effect of Flow Acceleration and Buoyancy on Thermalhydraulics of sCO2 in Mini/Micro-Channel. In L. Chen (Ed.), *Handbook of Research on Advancements in Supercritical Fluids Applications for Sustainable Energy Systems* (pp. 161–182). IGI Global. doi:10.4018/978-1-7998-5796-9.ch005

Kumari, N., & Kumar, K. (2018). Fabrication of Orthotic Calipers With Epoxy-Based Green Composite. In K. Kumar & J. Davim (Eds.), *Composites and Advanced Materials for Industrial Applications* (pp. 157–176). Hershey, PA: IGI Global. doi:10.4018/978-1-5225-5216-1.ch008

Kuppusamy, R. R. (2018). Development of Aerospace Composite Structures Through Vacuum-Enhanced Resin Transfer Moulding Technology (VERTMTy): Vacuum-Enhanced Resin Transfer Moulding. In K. Kumar & J. Davim (Eds.), *Composites and Advanced Materials for Industrial Applications* (pp. 99–111). Hershey, PA: IGI Global. doi:10.4018/978-1-5225-5216-1.ch005

Kurganov, V. A., Zeigarnik, Y. A., & Maslakova, I. V. (2021). Normal and Deteriorated Heat Transfer Under Heating Turbulent Supercritical Pressure Coolants Flows in Round Tubes. In L. Chen (Ed.), *Handbook of Research on Advancements in Supercritical Fluids Applications for Sustainable Energy Systems* (pp. 494–532). IGI Global. https://doi.org/10.4018/978-1-7998-5796-9.ch014

Li, H., & Zhang, Y. (2021). Heat Transfer and Fluid Flow Modeling for Supercritical Fluids in Advanced Energy Systems. In L. Chen (Ed.), *Handbook of Research on Advancements in Supercritical Fluids Applications for Sustainable Energy Systems* (pp. 388–422). IGI Global. https://doi.org/10.4018/978-1-7998-5796-9.ch011

Loy, J., Howell, S., & Cooper, R. (2017). Engineering Teams: Supporting Diversity in Engineering Education. In M. Gray & K. Thomas (Eds.), *Strategies for Increasing Diversity in Engineering Majors and Careers* (pp. 106–129). Hershey, PA: IGI Global. doi:10.4018/978-1-5225-2212-6.ch006

Macher, G., Armengaud, E., Kreiner, C., Brenner, E., Schmittner, C., Ma, Z., ... Krammer, M. (2018). Integration of Security in the Development Lifecycle of Dependable Automotive CPS. In N. Druml, A. Genser, A. Krieg, M. Menghin, & A. Hoeller (Eds.), *Solutions for Cyber-Physical Systems Ubiquity* (pp. 383–423). Hershey, PA: IGI Global. doi:10.4018/978-1-5225-2845-6.ch015

Madhu, M. N., Singh, J. G., Mohan, V., & Ongsakul, W. (2021). Transmission Risk Optimization in Interconnected Systems: Risk-Adjusted Available Transfer Capability. In P. Vasant, G. Weber, & W. Punurai (Ed.), *Research Advancements in Smart Technology, Optimization, and Renewable Energy* (pp. 183-199). IGI Global. https://doi.org/10.4018/978-1-7998-3970-5.ch010

Mahendramani, G., & Lakshmana Swamy, N. (2018). Effect of Weld Groove Area on Distortion of Butt Welded Joints in Submerged Arc Welding. *International Journal of Manufacturing, Materials, and Mechanical Engineering*, *8*(2), 33–44. doi:10.4018/IJMMME.2018040103

Makropoulos, G., Koumaras, H., Setaki, F., Filis, K., Lutz, T., Montowtt, P., Tomaszewski, L., Dybiec, P., & Järvet, T. (2021). 5G and Unmanned Aerial Vehicles (UAVs) Use Cases: Analysis of the Ecosystem, Architecture, and Applications. In A. Nwajana & I. Ihianle (Eds.), *Handbook of Research on 5G Networks and Advancements in Computing, Electronics, and Electrical Engineering* (pp. 36–69). IGI Global. https://doi.org/10.4018/978-1-7998-6992-4.ch003

Meric, E. M., Erdem, S., & Gurbuz, E. (2021). Application of Phase Change Materials in Construction Materials for Thermal Energy Storage Systems in Buildings. In R. González-Lezcano (Ed.), *Advancements in Sustainable Architecture and Energy Efficiency* (pp. 1–20). IGI Global. https://doi.org/10.4018/978-1-7998-7023-4.ch001

Mihret, E. T., & Yitayih, K. A. (2021). Operation of VANET Communications: The Convergence of UAV System With LTE/4G and WAVE Technologies. *International Journal of Smart Vehicles and Smart Transportation*, *4*(1), 29–51. https://doi.org/10.4018/IJSVST.2021010103

Mir, M. A., Bhat, B. A., Sheikh, B. A., Rather, G. A., Mehraj, S., & Mir, W. R. (2021). Nanomedicine in Human Health Therapeutics and Drug Delivery: Nanobiotechnology and Nanobiomedicine. In M. Bhat, I. Wani, & S. Ashraf (Eds.), *Applications of Nanomaterials in Agriculture, Food Science, and Medicine* (pp. 229–251). IGI Global. doi:10.4018/978-1-7998-5563-7.ch013

Mohammadzadeh, S., & Kim, Y. (2017). Nonlinear System Identification of Smart Buildings. In P. Samui, S. Chakraborty, & D. Kim (Eds.), *Modeling and Simulation Techniques in Structural Engineering* (pp. 328–347). Hershey, PA: IGI Global. doi:10.4018/978-1-5225-0588-4.ch011

Molina, G. J., Aktaruzzaman, F., Soloiu, V., & Rahman, M. (2017). Design and Testing of a Jet-Impingement Instrument to Study Surface-Modification Effects by Nanofluids. *International Journal of Surface Engineering and Interdisciplinary Materials Science*, *5*(2), 43–61. doi:10.4018/IJSEIMS.2017070104

Moreno-Rangel, A., & Carrillo, G. (2021). Energy-Efficient Homes: A Heaven for Respiratory Illnesses. In R. González-Lezcano (Ed.), *Advancements in Sustainable Architecture and Energy Efficiency* (pp. 49–71). IGI Global. https://doi.org/10.4018/978-1-7998-7023-4.ch003

Msomi, V., & Jantjies, B. T. (2021). Correlative Analysis Between Tensile Properties and Tool Rotational Speeds of Friction Stir Welded Similar Aluminium Alloy Joints. *International Journal of Surface Engineering and Interdisciplinary Materials Science*, *9*(2), 58–78. https://doi.org/10.4018/IJSEIMS.2021070104

Muigai, M. N., Mwema, F. M., Akinlabi, E. T., & Obiko, J. O. (2021). Surface Engineering of Materials Through Weld-Based Technologies: An Overview. In S. Roy & G. Bose (Eds.), *Advanced Surface Coating Techniques for Modern Industrial Applications* (pp. 247–260). IGI Global. doi:10.4018/978-1-7998-4870-7.ch011

Mukherjee, A., Saeed, R. A., Dutta, S., & Naskar, M. K. (2017). Fault Tracking Framework for Software-Defined Networking (SDN). In C. Singhal & S. De (Eds.), *Resource Allocation in Next-Generation Broadband Wireless Access Networks* (pp. 247–272). Hershey, PA: IGI Global. doi:10.4018/978-1-5225-2023-8.ch011

Mukhopadhyay, A., Barman, T. K., & Sahoo, P. (2018). Electroless Nickel Coatings for High Temperature Applications. In K. Kumar & J. Davim (Eds.), *Composites and Advanced Materials for Industrial Applications* (pp. 297–331). Hershey, PA: IGI Global. doi:10.4018/978-1-5225-5216-1.ch013

Mwema, F. M., & Wambua, J. M. (2022). Machining of Poly Methyl Methacrylate (PMMA) and Other Olymeric Materials: A Review. In K. Kumar, B. Babu, & J. Davim (Eds.), *Handbook of Research on Advancements in the Processing, Characterization, and Application of Lightweight Materials* (pp. 363–379). IGI Global. https://doi.org/10.4018/978-1-7998-7864-3.ch016

Mykhailyshyn, R., Savkiv, V., Boyko, I., Prada, E., & Virgala, I. (2021). Substantiation of Parameters of Friction Elements of Bernoulli Grippers With a Cylindrical Nozzle. *International Journal of Manufacturing, Materials, and Mechanical Engineering, 11*(2), 17–39. https://doi.org/10.4018/IJMMME.2021040102

Náprstek, J., & Fischer, C. (2017). Dynamic Stability and Post-Critical Processes of Slender Auto-Parametric Systems. In V. Plevris, G. Kremmyda, & Y. Fahjan (Eds.), *Performance-Based Seismic Design of Concrete Structures and Infrastructures* (pp. 128–171). Hershey, PA: IGI Global. doi:10.4018/978-1-5225-2089-4.ch006

Nautiyal, L., Shivach, P., & Ram, M. (2018). Optimal Designs by Means of Genetic Algorithms. In M. Ram & J. Davim (Eds.), *Soft Computing Techniques and Applications in Mechanical Engineering* (pp. 151–161). Hershey, PA: IGI Global. doi:10.4018/978-1-5225-3035-0.ch007

Nazir, R. (2017). Advanced Nanomaterials for Water Engineering and Treatment: Nano-Metal Oxides and Their Nanocomposites. In T. Saleh (Ed.), *Advanced Nanomaterials for Water Engineering, Treatment, and Hydraulics* (pp. 84–126). Hershey, PA: IGI Global. doi:10.4018/978-1-5225-2136-5.ch005

Nikolopoulos, C. D. (2021). Recent Advances on Measuring and Modeling ELF-Radiated Emissions for Space Applications. In C. Nikolopoulos (Ed.), *Recent Trends on Electromagnetic Environmental Effects for Aeronautics and Space Applications* (pp. 1–38). IGI Global. https://doi.org/10.4018/978-1-7998-4879-0.ch001

Nogueira, A. F., Ribeiro, J. C., Fernández de Vega, F., & Zenha-Rela, M. A. (2018). Evolutionary Approaches to Test Data Generation for Object-Oriented Software: Overview of Techniques and Tools. In M. Khosrow-Pour, D.B.A. (Ed.), Incorporating Nature-Inspired Paradigms in Computational Applications (pp. 162-194). Hershey, PA: IGI Global. https://doi.org/ doi:10.4018/978-1-5225-5020-4.ch006

Nwajana, A. O., Obi, E. R., Ijemaru, G. K., Oleka, E. U., & Anthony, D. C. (2021). Fundamentals of RF/Microwave Bandpass Filter Design. In A. Nwajana & I. Ihianle (Eds.), *Handbook of Research on 5G Networks and Advancements in Computing, Electronics, and Electrical Engineering* (pp. 149–164). IGI Global. https://doi.org/10.4018/978-1-7998-6992-4.ch005

Ogbodo, E. A. (2021). Comparative Study of Transmission Line Junction vs. Asynchronously Coupled Junction Diplexers. In A. Nwajana & I. Ihianle (Eds.), *Handbook of Research on 5G Networks and Advancements in Computing, Electronics, and Electrical Engineering* (pp. 326–336). IGI Global. https://doi.org/10.4018/978-1-7998-6992-4.ch013

Orosa, J. A., Vergara, D., Fraguela, F., & Masdías-Bonome, A. (2021). Statistical Understanding and Optimization of Building Energy Consumption and Climate Change Consequences. In R. González-Lezcano (Ed.), *Advancements in Sustainable Architecture and Energy Efficiency* (pp. 195–220). IGI Global. https://doi.org/10.4018/978-1-7998-7023-4.ch009

Osho, M. B. (2018). Industrial Enzyme Technology: Potential Applications. In S. Bharati & P. Chaurasia (Eds.), *Research Advancements in Pharmaceutical, Nutritional, and Industrial Enzymology* (pp. 375–394). Hershey, PA: IGI Global. doi:10.4018/978-1-5225-5237-6.ch017

Ouadi, A., & Zitouni, A. (2021). Phasor Measurement Improvement Using Digital Filter in a Smart Grid. In A. Recioui & H. Bentarzi (Eds.), *Optimizing and Measuring Smart Grid Operation and Control* (pp. 100–117). IGI Global. https://doi.org/10.4018/978-1-7998-4027-5.ch005

Padmaja, P., & Marutheswar, G. (2017). Certain Investigation on Secured Data Transmission in Wireless Sensor Networks. *International Journal of Mobile Computing and Multimedia Communications*, *8*(1), 48–61. doi:10.4018/IJMCMC.2017010104

Palmer, S., & Hall, W. (2017). An Evaluation of Group Work in First-Year Engineering Design Education. In R. Tucker (Ed.), *Collaboration and Student Engagement in Design Education* (pp. 145–168). Hershey, PA: IGI Global. doi:10.4018/978-1-5225-0726-0.ch007

Panchenko, V. (2021). Prospects for Energy Supply of the Arctic Zone Objects of Russia Using Frost-Resistant Solar Modules. In P. Vasant, G. Weber, & W. Punurai (Eds.), *Research Advancements in Smart Technology, Optimization, and Renewable Energy* (pp. 149-169). IGI Global. https://doi.org/10.4018/978-1-7998-3970-5.ch008

Panchenko, V. (2021). Photovoltaic Thermal Module With Paraboloid Type Solar Concentrators. *International Journal of Energy Optimization and Engineering*, *10*(2), 1–23. https://doi.org/10.4018/IJEOE.2021040101

Pandey, K., & Datta, S. (2021). Dry Machining of Inconel 825 Superalloys: Performance of Tool Inserts (Carbide, Cermet, and SiAlON). *International Journal of Manufacturing, Materials, and Mechanical Engineering*, *11*(4), 26–39. doi:10.4018/IJMMME.2021100102

Panneer, R. (2017). Effect of Composition of Fibers on Properties of Hybrid Composites. *International Journal of Manufacturing, Materials, and Mechanical Engineering*, *7*(4), 28–43. doi:10.4018/IJMMME.2017100103

Pany, C. (2021). Estimation of Correct Long-Seam Mismatch Using FEA to Compare the Measured Strain in a Non-Destructive Testing of a Pressurant Tank: A Reverse Problem. *International Journal of Smart Vehicles and Smart Transportation*, *4*(1), 16–28. doi:10.4018/IJSVST.2021010102

Paul, S., & Roy, P. (2018). Optimal Design of Power System Stabilizer Using a Novel Evolutionary Algorithm. *International Journal of Energy Optimization and Engineering*, *7*(3), 24–46. doi:10.4018/IJEOE.2018070102

Paul, S., & Roy, P. K. (2021). Oppositional Differential Search Algorithm for the Optimal Tuning of Both Single Input and Dual Input Power System Stabilizer. In P. Vasant, G. Weber, & W. Punurai (Eds.), *Research Advancements in Smart Technology, Optimization, and Renewable Energy* (pp. 256-282). IGI Global. https://doi.org/10.4018/978-1-7998-3970-5.ch013

Pavaloiu, A. (2018). Artificial Intelligence Ethics in Biomedical-Engineering-Oriented Problems. In U. Kose, G. Guraksin, & O. Deperlioglu (Eds.), *Nature-Inspired Intelligent Techniques for Solving Biomedical Engineering Problems* (pp. 219–231). Hershey, PA: IGI Global. doi:10.4018/978-1-5225-4769-3.ch010

Pioro, I., Mahdi, M., & Popov, R. (2017). Application of Supercritical Pressures in Power Engineering. In L. Chen & Y. Iwamoto (Eds.), *Advanced Applications of Supercritical Fluids in Energy Systems* (pp. 404–457). Hershey, PA: IGI Global. doi:10.4018/978-1-5225-2047-4.ch013

Plaksina, T., & Gildin, E. (2017). Rigorous Integrated Evolutionary Workflow for Optimal Exploitation of Unconventional Gas Assets. *International Journal of Energy Optimization and Engineering*, *6*(1), 101–122. doi:10.4018/IJEOE.2017010106

Popat, J., Kakadiya, H., Tak, L., Singh, N. K., Majeed, M. A., & Mahajan, V. (2021). Reliability of Smart Grid Including Cyber Impact: A Case Study. In R. Singh, A. Singh, A. Dwivedi, & P. Nagabhushan (Eds.), *Computational Methodologies for Electrical and Electronics Engineers* (pp. 163–174). IGI Global. https://doi.org/10.4018/978-1-7998-3327-7.ch013

Quiza, R., La Fé-Perdomo, I., Rivas, M., & Ramtahalsing, V. (2021). Triple Bottom Line-Focused Optimization of Oblique Turning Processes Based on Hybrid Modeling: A Study Case on AISI 1045 Steel Turning. In L. Burstein (Ed.), *Handbook of Research on Advancements in Manufacturing, Materials, and Mechanical Engineering* (pp. 215–241). IGI Global. https://doi.org/10.4018/978-1-7998-4939-1.ch010

Rahmani, M. K. (2022). Blockchain Technology: Principles and Algorithms. In S. Khan, M. Syed, R. Hammad, & A. Bushager (Eds.), *Blockchain Technology and Computational Excellence for Society 5.0* (pp. 16–27). IGI Global. https://doi.org/10.4018/978-1-7998-8382-1.ch002

Ramdani, N., & Azibi, M. (2018). Polymer Composite Materials for Microelectronics Packaging Applications: Composites for Microelectronics Packaging. In K. Kumar & J. Davim (Eds.), *Composites and Advanced Materials for Industrial Applications* (pp. 177–211). Hershey, PA: IGI Global. doi:10.4018/978-1-5225-5216-1.ch009

Ramesh, M., Garg, R., & Subrahmanyam, G. V. (2017). Investigation of Influence of Quenching and Annealing on the Plane Fracture Toughness and Brittle to Ductile Transition Temperature of the Zinc Coated Structural Steel Materials. *International Journal of Surface Engineering and Interdisciplinary Materials Science*, *5*(2), 33–42. doi:10.4018/IJSEIMS.2017070103

Robinson, J., & Beneroso, D. (2022). Project-Based Learning in Chemical Engineering: Curriculum and Assessment, Culture and Learning Spaces. In A. Alves & N. van Hattum-Janssen (Eds.), *Training Engineering Students for Modern Technological Advancement* (pp. 1–19). IGI Global. https://doi.org/10.4018/978-1-7998-8816-1.ch001

Rondon, B. (2021). Experimental Characterization of Admittance Meter With Crude Oil Emulsions. *International Journal of Electronics, Communications, and Measurement Engineering*, *10*(2), 51–59. https://doi.org/10.4018/IJECME.2021070104

Rudolf, S., Biryuk, V. V., & Volov, V. (2018). Vortex Effect, Vortex Power: Technology of Vortex Power Engineering. In V. Kharchenko & P. Vasant (Eds.), *Handbook of Research on Renewable Energy and Electric Resources for Sustainable Rural Development* (pp. 500–533). Hershey, PA: IGI Global. doi:10.4018/978-1-5225-3867-7.ch021

Sah, A., Bhadula, S. J., Dumka, A., & Rawat, S. (2018). A Software Engineering Perspective for Development of Enterprise Applications. In A. Elçi (Ed.), *Handbook of Research on Contemporary Perspectives on Web-Based Systems* (pp. 1–23). Hershey, PA: IGI Global. doi:10.4018/978-1-5225-5384-7.ch001

Sahoo, P., & Roy, S. (2017). Tribological Behavior of Electroless Ni-P, Ni-P-W and Ni-P-Cu Coatings: A Comparison. *International Journal of Surface Engineering and Interdisciplinary Materials Science*, *5*(1), 1–15. doi:10.4018/IJSEIMS.2017010101

Sahoo, S. (2018). Laminated Composite Hypar Shells as Roofing Units: Static and Dynamic Behavior. In K. Kumar & J. Davim (Eds.), *Composites and Advanced Materials for Industrial Applications* (pp. 249–269). Hershey, PA: IGI Global. doi:10.4018/978-1-5225-5216-1.ch011

Sahu, H., & Hungyo, M. (2018). Introduction to SDN and NFV. In A. Dumka (Ed.), *Innovations in Software-Defined Networking and Network Functions Virtualization* (pp. 1–25). Hershey, PA: IGI Global. doi:10.4018/978-1-5225-3640-6.ch001

Salem, A. M., & Shmelova, T. (2018). Intelligent Expert Decision Support Systems: Methodologies, Applications, and Challenges. In T. Shmelova, Y. Sikirda, N. Rizun, A. Salem, & Y. Kovalyov (Eds.), *Socio-Technical Decision Support in Air Navigation Systems: Emerging Research and Opportunities* (pp. 215–242). Hershey, PA: IGI Global. doi:10.4018/978-1-5225-3108-1.ch007

Samal, M. (2017). FE Analysis and Experimental Investigation of Cracked and Un-Cracked Thin-Walled Tubular Components to Evaluate Mechanical and Fracture Properties. In P. Samui, S. Chakraborty, & D. Kim (Eds.), *Modeling and Simulation Techniques in Structural Engineering* (pp. 266–293). Hershey, PA: IGI Global. doi:10.4018/978-1-5225-0588-4.ch009

Samal, M., & Balakrishnan, K. (2017). Experiments on a Ring Tension Setup and FE Analysis to Evaluate Transverse Mechanical Properties of Tubular Components. In P. Samui, S. Chakraborty, & D. Kim (Eds.), *Modeling and Simulation Techniques in Structural Engineering* (pp. 91–115). Hershey, PA: IGI Global. doi:10.4018/978-1-5225-0588-4.ch004

Samarasinghe, D. A., & Wood, E. (2021). Innovative Digital Technologies. In J. Underwood & M. Shelbourn (Eds.), *Handbook of Research on Driving Transformational Change in the Digital Built Environment* (pp. 142–163). IGI Global. https://doi.org/10.4018/978-1-7998-6600-8.ch006

Sawant, S. (2018). Deep Learning and Biomedical Engineering. In U. Kose, G. Guraksin, & O. Deperlioglu (Eds.), *Nature-Inspired Intelligent Techniques for Solving Biomedical Engineering Problems* (pp. 283–296). Hershey, PA: IGI Global. doi:10.4018/978-1-5225-4769-3.ch014

Schulenberg, T. (2021). Energy Conversion Using the Supercritical Steam Cycle. In L. Chen (Ed.), *Handbook of Research on Advancements in Supercritical Fluids Applications for Sustainable Energy Systems* (pp. 659–681). IGI Global. doi:10.4018/978-1-7998-5796-9.ch018

Sezgin, H., & Berkalp, O. B. (2018). Textile-Reinforced Composites for the Automotive Industry. In K. Kumar & J. Davim (Eds.), *Composites and Advanced Materials for Industrial Applications* (pp. 129–156). Hershey, PA: IGI Global. doi:10.4018/978-1-5225-5216-1.ch007

Shaaban, A. A., & Shehata, O. M. (2021). Combining Response Surface Method and Metaheuristic Algorithms for Optimizing SPIF Process. *International Journal of Manufacturing, Materials, and Mechanical Engineering, 11*(4), 1–25. https://doi.org/10.4018/IJMMME.2021100101

Shafaati Shemami, M., & Sefid, M. (2022). Implementation and Demonstration of Electric Vehicle-to-Home (V2H) Application: A Case Study. In M. Alam, R. Pillai, & N. Murugesan (Eds.), *Developing Charging Infrastructure and Technologies for Electric Vehicles* (pp. 268–293). IGI Global. https://doi.org/10.4018/978-1-7998-6858-3.ch015

Shah, M. Z., Gazder, U., Bhatti, M. S., & Hussain, M. (2018). Comparative Performance Evaluation of Effects of Modifier in Asphaltic Concrete Mix. *International Journal of Strategic Engineering, 1*(2), 13–25. doi:10.4018/IJoSE.2018070102

Sharma, N., & Kumar, K. (2018). Fabrication of Porous NiTi Alloy Using Organic Binders. In K. Kumar & J. Davim (Eds.), *Composites and Advanced Materials for Industrial Applications* (pp. 38–62). Hershey, PA: IGI Global. doi:10.4018/978-1-5225-5216-1.ch003

Shivach, P., Nautiyal, L., & Ram, M. (2018). Applying Multi-Objective Optimization Algorithms to Mechanical Engineering. In M. Ram & J. Davim (Eds.), *Soft Computing Techniques and Applications in Mechanical Engineering* (pp. 287–301). Hershey, PA: IGI Global. doi:10.4018/978-1-5225-3035-0.ch014

Shmelova, T. (2018). Stochastic Methods for Estimation and Problem Solving in Engineering: Stochastic Methods of Decision Making in Aviation. In S. Kadry (Ed.), *Stochastic Methods for Estimation and Problem Solving in Engineering* (pp. 139–160). Hershey, PA: IGI Global. doi:10.4018/978-1-5225-5045-7.ch006

Siero González, L. R., & Romo Vázquez, A. (2017). Didactic Sequences Teaching Mathematics for Engineers With Focus on Differential Equations. In M. Ramírez-Montoya (Ed.), *Handbook of Research on Driving STEM Learning With Educational Technologies* (pp. 129–151). Hershey, PA: IGI Global. doi:10.4018/978-1-5225-2026-9.ch007

Sim, M. S., You, K. Y., Esa, F., & Chan, Y. L. (2021). Nanostructured Electromagnetic Metamaterials for Sensing Applications. In M. Bhat, I. Wani, & S. Ashraf (Eds.), *Applications of Nanomaterials in Agriculture, Food Science, and Medicine* (pp. 141–164). IGI Global. https://doi.org/10.4018/978-1-7998-5563-7.ch009

Singh, R., & Dutta, S. (2018). Visible Light Active Nanocomposites for Photocatalytic Applications. In K. Kumar & J. Davim (Eds.), *Composites and Advanced Materials for Industrial Applications* (pp. 270–296). Hershey, PA: IGI Global. doi:10.4018/978-1-5225-5216-1.ch012

Skripov, P. V., Yampol'skiy, A. D., & Rutin, S. B. (2021). High-Power Heat Transfer in Supercritical Fluids: Microscale Times and Sizes. In L. Chen (Ed.), *Handbook of Research on Advancements in Supercritical Fluids Applications for Sustainable Energy Systems* (pp. 424–450). IGI Global. https://doi.org/10.4018/978-1-7998-5796-9.ch012

Sözbilir, H., Özkaymak, Ç., Uzel, B., & Sümer, Ö. (2018). Criteria for Surface Rupture Microzonation of Active Faults for Earthquake Hazards in Urban Areas. In N. Ceryan (Ed.), *Handbook of Research on Trends and Digital Advances in Engineering Geology* (pp. 187–230). Hershey, PA: IGI Global. doi:10.4018/978-1-5225-2709-1.ch005

Stanciu, I. (2018). Stochastic Methods in Microsystems Engineering. In S. Kadry (Ed.), *Stochastic Methods for Estimation and Problem Solving in Engineering* (pp. 161–176). Hershey, PA: IGI Global. doi:10.4018/978-1-5225-5045-7.ch007

Strebkov, D., Nekrasov, A., Trubnikov, V., & Nekrasov, A. (2018). Single-Wire Resonant Electric Power Systems for Renewable-Based Electric Grid. In V. Kharchenko & P. Vasant (Eds.), *Handbook of Research on Renewable Energy and Electric Resources for Sustainable Rural Development* (pp. 449–474). Hershey, PA: IGI Global. doi:10.4018/978-1-5225-3867-7.ch019

Sukhyy, K., Belyanovskaya, E., & Sukhyy, M. (2021). *Basic Principles for Substantiation of Working Pair Choice*. IGI Global. doi:10.4018/978-1-7998-4432-7.ch002

Suri, M. S., & Kaliyaperumal, D. (2022). Extension of Aspiration Level Model for Optimal Planning of Fast Charging Stations. In A. Fekik & N. Benamrouche (Eds.), *Modeling and Control of Static Converters for Hybrid Storage Systems* (pp. 91–106). IGI Global. https://doi.org/10.4018/978-1-7998-7447-8.ch004

Tallet, E., Gledson, B., Rogage, K., Thompson, A., & Wiggett, D. (2021). Digitally-Enabled Design Management. In J. Underwood & M. Shelbourn (Eds.), *Handbook of Research on Driving Transformational Change in the Digital Built Environment* (pp. 63–89). IGI Global. https://doi.org/10.4018/978-1-7998-6600-8.ch003

Terki, A., & Boubertakh, H. (2021). A New Hybrid Binary-Real Coded Cuckoo Search and Tabu Search Algorithm for Solving the Unit-Commitment Problem. *International Journal of Energy Optimization and Engineering, 10*(2), 104–119. https://doi.org/10.4018/IJEOE.2021040105

Tüdeş, Ş., Kumlu, K. B., & Ceryan, S. (2018). Integration Between Urban Planning and Natural Hazards For Resilient City. In N. Ceryan (Ed.), *Handbook of Research on Trends and Digital Advances in Engineering Geology* (pp. 591–630). Hershey, PA: IGI Global. doi:10.4018/978-1-5225-2709-1.ch017

Ulamis, K. (2018). Soil Liquefaction Assessment by Anisotropic Cyclic Triaxial Test. In N. Ceryan (Ed.), *Handbook of Research on Trends and Digital Advances in Engineering Geology* (pp. 631–664). Hershey, PA: IGI Global. doi:10.4018/978-1-5225-2709-1.ch018

Valente, M., & Milani, G. (2017). Seismic Assessment and Retrofitting of an Under-Designed RC Frame Through a Displacement-Based Approach. In V. Plevris, G. Kremmyda, & Y. Fahjan (Eds.), *Performance-Based Seismic Design of Concrete Structures and Infrastructures* (pp. 36–58). Hershey, PA: IGI Global. doi:10.4018/978-1-5225-2089-4.ch002

Vargas-Bernal, R. (2021). Advances in Electromagnetic Environmental Shielding for Aeronautics and Space Applications. In C. Nikolopoulos (Ed.), *Recent Trends on Electromagnetic Environmental Effects for Aeronautics and Space Applications* (pp. 80–96). IGI Global. https://doi.org/10.4018/978-1-7998-4879-0.ch003

Vasant, P. (2018). A General Medical Diagnosis System Formed by Artificial Neural Networks and Swarm Intelligence Techniques. In U. Kose, G. Guraksin, & O. Deperlioglu (Eds.), *Nature-Inspired Intelligent Techniques for Solving Biomedical Engineering Problems* (pp. 130–145). Hershey, PA: IGI Global. doi:10.4018/978-1-5225-4769-3.ch006

Verner, C. M., & Sarwar, D. (2021). Avoiding Project Failure and Achieving Project Success in NHS IT System Projects in the United Kingdom. *International Journal of Strategic Engineering, 4*(1), 33–54. https://doi.org/10.4018/IJoSE.2021010103

Verrollot, J., Tolonen, A., Harkonen, J., & Haapasalo, H. J. (2018). Challenges and Enablers for Rapid Product Development. *International Journal of Applied Industrial Engineering, 5*(1), 25–49. doi:10.4018/IJAIE.2018010102

Wan, A. C., Zulu, S. L., & Khosrow-Shahi, F. (2021). Industry Views on BIM for Site Safety in Hong Kong. In J. Underwood & M. Shelbourn (Eds.), *Handbook of Research on Driving Transformational Change in the Digital Built Environment* (pp. 120–140). IGI Global. https://doi.org/10.4018/978-1-7998-6600-8.ch005

Yardimci, A. G., & Karpuz, C. (2018). Fuzzy Rock Mass Rating: Soft-Computing-Aided Preliminary Stability Analysis of Weak Rock Slopes. In N. Ceryan (Ed.), *Handbook of Research on Trends and Digital Advances in Engineering Geology* (pp. 97–131). Hershey, PA: IGI Global. doi:10.4018/978-1-5225-2709-1.ch003

You, K. Y. (2021). Development Electronic Design Automation for RF/Microwave Antenna Using MATLAB GUI. In A. Nwajana & I. Ihianle (Eds.), *Handbook of Research on 5G Networks and Advancements in Computing, Electronics, and Electrical Engineering* (pp. 70–148). IGI Global. https://doi.org/10.4018/978-1-7998-6992-4.ch004

Yousefi, Y., Gratton, P., & Sarwar, D. (2021). Investigating the Opportunities to Improve the Thermal Performance of a Case Study Building in London. *International Journal of Strategic Engineering*, *4*(1), 1–18. https://doi.org/10.4018/IJoSE.2021010101

Zindani, D., & Kumar, K. (2018). Industrial Applications of Polymer Composite Materials. In K. Kumar & J. Davim (Eds.), *Composites and Advanced Materials for Industrial Applications* (pp. 1–15). Hershey, PA: IGI Global. doi:10.4018/978-1-5225-5216-1.ch001

Zindani, D., Maity, S. R., & Bhowmik, S. (2018). A Decision-Making Approach for Material Selection of Polymeric Composite Bumper Beam. In K. Kumar & J. Davim (Eds.), *Composites and Advanced Materials for Industrial Applications* (pp. 112–128). Hershey, PA: IGI Global. doi:10.4018/978-1-5225-5216-1.ch006

About the Contributors

Waleed Fouad Abobatta acts as a Professor at Horticulture Research Institute, he's a member of the Scientific Committee for Greenhouses Plantation, Agriculture Research Center, Egypt (2014-2020), and he is a member of the Scientific Team for National Campaign for Navel Orange Improvement (H R I), he's Expert in International Citrus Society, California University, Riverside, also, Abobatta is a Scientific Committee Member of international and national conferences, also, he's the coordinator for various international conferences and workshops Abobatta is a supervisor of M. Sc. and Ph. D. Thesis in various branches of agricultural science. Abobatta published more than 85 books, book chapters, and research papers until now, he has been serving as an editorial board member in several respected journals, and he's a member of the Board of Environment Friends Society (H R I). Abobatta is a consultant of Nano application in the agricultural sector and one of Jojoba experts in the MENA region. He is Deputy General Secretary of the Arab Federation for the Protection of Wild and Marine Life.

* * *

Ashok Chaubey graduated and Doctoral degree (D.Phil) from University of Allahabad, Allahabad and presently working as Professor, Department of Zoology and Coordinator, Department of Environmental Science, Chaudhary Charan Singh University, Meerut. He has been honoured with 30 awards notably few are G.D. Bhale Rao Medal (2018), Prof. GB Shinde Gold Medal (2018), Excellence in Research Award (2019) and recently he has been conferred Life time Achievement Award (2022) by the Environment Welfare Society, Khajuraho. Prof. Chaubey is Fellow of various scientific societies notably FIAZ (2021), FSLSc (2018), FAZRA (2017), FHS (2017); life member of 12 Scientific Academic Bodies; has organized 10 symposia/workshops; attended/participated in more than 75 National and International conferences and delivered lead talk/keynote lectures/chaired and co-chaired the scientific sessions; have completed 6 major research projects; 12 PhD and 52 MPhil have been awarded under his supervision; published 114 research papers in peer reviewed National and International Journals and 05 edited books. Currently working on Entomopathogenic Nematodes (EPN), their mass production and development of EPN based bio-pesticides against insect pests. Since 2015 till date participating in Asian Waterbird Census (AWC) in Hastinapur Wildlife Sanctuary Wetlands, Meerut organized every year by Wetlands International, Netherlands.

Wasan S. Hussain is an Asst. Prof, Ph.D. Degree in Botany at College of Sciences /Mosul University 2017, A faculty member in the Biology Dept., College of Sciences, Mosul university/Iraq, she has 30 research papers published in Scopus, international and local magazines, and three books authored—a member of the editorial board of a number of international journals. Orcid https://orcid.org/0000-0002-1537-9979.7708e4ca-fe3d-4817-b25f-7057aaa75a03

Marian Malak is a Lecturer of Entomology Faculty of Education, Alexandria University, Egypt.

El-Sayed Ewis Omran is a professor of Soil Science (RS & GIS) at the Faculty of Agriculture, Suez Canal University, Egypt. In June 1996 he obtained a MSc in soil science at Suez Canal University. From June 2003 he was studying at Wageningen University, The Netherlands. Meanwhile, he graduated also as MSc in geo-information science at Wageningen University in May 2005. From June 2005 to June 2007, he is doing his PhD research at the Center for Geo-Information and Remote Sensing of Wageningen University, The Netherlands. From 2007 till 2012, he works as a lecturer of soil science, Soil and Water Department, Faculty of Agriculture, Suez Canal University. From 2012 till 2017, he works as Associate Professor of soil pedology (RS& GIS). From 2017 until present, he works as Professor of soil pedology (RS& GIS). From 2018 until 2021, he works as a Director for Surveying, GIS and Remote Sensing Center, Aswan University. From 2018 to 2019, he works as Vice Dean for Community and Environmental Development, Institute of African Research and Studies and Nile Basin Countries, Aswan University. From 2019 until 2021, he works as a Dean for Institute of African Research and Studies and Nile Basin Countries, Aswan University. He supervised more than 25 master's and doctoral theses. He participated in several international projects since 2007 until now. He published more than 35 papers in national and international journals and conferences and contributed in more than 70 book chapters. He is the editor of several volumes at the Springer Handbooks. Three volumes were published and two volumes will be published during the year 2023. He is a member of the scientific committee of several international conferences and participated in more than 50 conferences and workshops as well. He is a reviewer for more than 30 scientific journals that indexed in Scopus and WoS. He performed more than 200 reviews for journals. He is an associate Editor of 6 journals and Editor for 3 journals. He is an arbitrator on the Standing Scientific Committee for Soil Science. His research interests are diverse but his focus includes climate change research, application of Vis/Near- and Mid- Infrared Spectroscopy, Hyperspectral Imaging, Spatial data sharing, Neural network and Network analysis.

Mohamed Raheem is a Professor at Pests & Plant Protection Department, Agricultural and Biological Research Institute, National Research Centre, Cairo, Egypt.

Magdi Ismaiel Saif, Ph.D., is an associate professor of Pomology (Doctor of Pomology), Horticulture department, Faculty of Agriculture, Suez Canal University (SCU), Egypt. He got his Ph.D. from Russian State Agrarian University, Moscow, Russia (2002-2006). Visiting researcher at Division of Natural Products Chemistry, Institute of Natural Medicine, University of Toyama, Japan (April-November 2016). Teaching Horticultural (Pomology) courses at the Faculty of Agriculture in SCU. He attended many scientific meetings, training courses and workshops. Areas of expertise are fruit propagation, fruit cultivation, fruit production, post-harvest and quality of fruits, Viticulture, orchard management, tissue culture, physiological studies of fruit trees and genetic and molecular biology of fruits with published articles and work in progress projects and researches. He has different skills like fruit propagation (cuttings, grafting and tissue culture), fruit production (irrigation, fertilization, pruning and improving fruit quality using orchard management and growth regulators), agricultural feasibility studies, statistical analysis and computer skills.

Chidanandamurthy Thippeswamy Swamy is a Research Officer (Microbiology Dept.), Captain Srinivasa Murthy Central Ayurveda Research Institute, India.

Kajol Yadav has Graduation, Post-Graduation and M.Phil. degree from Chaudhary Charan Singh University Meerut and presently Pursuing PhD from Nematology Laboratory, Chaudhary Charan Singh University. Her M.Phil. research work was on entomopathogenic nematode and their associated entomopathogenic bacteria and presently doing research on Mollusks Parasitic Nematodes and soil nematodes. She has been awarded with 4 Best Presentation awards and 1 Young women scientist award in National and International Conferences respectively. She has been published 2 research paper in peer reviewed international journal and 2 book chapter in national book.

Index

F

G

L

M

N

O

P

R

S

T

Recommended Reference Books

IGI Global's reference books are available in three unique pricing formats:
Print Only, E-Book Only, or Print + E-Book.

Order direct through IGI Global's Online Bookstore at **www.igi-global.com** or through your preferred provider.

ISBN: 9781668445488
EISBN: 9781668445501
© 2022; 334 pp.
List Price: US$ **240**

ISBN: 9781668467503
EISBN: 9781668467527
© 2022; 296 pp.
List Price: US$ **240**

ISBN: 9781799889007
EISBN: 9781799889021
© 2022; 340 pp.
List Price: US$ **215**

ISBN: 9781799898511
EISBN: 9781799898535
© 2022; 276 pp.
List Price: US$ **250**

ISBN: 9781799896913
EISBN: 9781799896937
© 2022; 219 pp.
List Price: US$ **215**

ISBN: 9781668436868
EISBN: 9781668436875
© 2022; 2,064 pp.
List Price: US$ **2,085**

Do you want to stay current on the latest research trends, product announcements, news, and special offers?
Join IGI Global's mailing list to receive customized recommendations, exclusive discounts, and more.
Sign up at: **www.igi-global.com/newsletters.**

Publisher of Timely, Peer-Reviewed Inclusive Research Since 1988

IGI Global
PUBLISHER of TIMELY KNOWLEDGE

www.igi-global.com Sign up at www.igi-global.com/newsletters facebook.com/igiglobal twitter.com/igiglobal

www.ingramcontent.com/pod-product-compliance
Lightning Source LLC
LaVergne TN
LVHW081330300926
849161LV00008B/1936